MULTIPHYSICS MODELING

USING *COMSOL®5* AND *MATLAB®*

SECOND EDITION

PREFACE

The purpose of this second edition is to update the model building instructions to comply with COMSOL® Version 5.6, MATLAB® R2021a, and later. The MATLAB material contained in the Appendix is revised and updated to facilitate LIVELINK® interaction with COMSOL 5.x. This second edition also introduces five new hands-on model building and problem-solving techniques. In this new book, scientists, engineers, biophysicists, and other readers interested in exploring the behavior of potential physical device structures built on a computer (a virtual prototype), can develop exploratory models, before going to the workshop or laboratory and attempting to physically create the whatever-it-is (a real prototype).

The models presented herein are built within the context of the currently well-established laws of the physical world (applied physics) and are explored in light of widely applied, well-known, First Principle Analysis techniques. As with any other method of problem solution (mathematically computed answer), the information obtained (derived) through the use of such modeled solutions, as these computer simulations (virtual prototypes), can ultimately be only as accurate as the materials properties values and the fundamental assumptions employed to build (create) these simulations.

The primary advantage of the combination of computer simulation (virtual prototyping) and First Principles Analysis to explore artifacts (device structures) is that the modeler can try as many different approaches to the solution of the same underlying problem as are needed in order to get it right (or close thereto) before fabrication of the device components and the assembled device (real prototype) in the workshop or laboratory for the first time.

ACKNOWLEDGMENTS

I would like to thank David Pallai of Mercury Learning and Information for his ongoing encouragement. I would also like to thank the many staff members of COMSOL, Inc. for their help and encouragement.

I would especially like to thank Beverly E. Pryor, my wife, for her patience and encouragement during the creation of the manuscript of this book. Any residual errors in this work are mine and mine alone.

Roger W. Pryor, Ph.D.

October 2021

INTRODUCTION

COMSOL Multiphysics software is a powerful, Partial Differential Equation (PDE) solution engine. The basic COMSOL Multiphysics 5.x software has over twenty-five (25) add-on modules that expand the capabilities of the basic software into a broad collection of application areas: AC/DC, acoustics, batteries and fuel cells, CFD, chemical reaction engineering, electro-deposition, geomechanics, heat transfer, MEMS, microfluidics, plasma, RF, structural mechanics and subsurface flow, to name a few. The COMSOL Multiphysics software also has other supporting software, such as the Optimization Module, the Material Library Module, the CAD Import Module and LiveLink™ interfaces for several engineering software programs.

In this book, scientists, engineers, and others interested in exploring the behavior of different physical device structures through computer modeling are introduced to the techniques of hands-on building and solving models through the direct application of the basic COMSOL Multiphysics software, along with some samples using the AC/DC, heat transfer, rf, semiconductor, and structural mechanics modules. The next to the last technical chapter explores the use of Perfectly Matched Layers in the RF Module. The final technical chapter explores the use of the Bioheat Equation in the Heat Transfer and RF Modules.

The models presented herein are built within the context of the physical world (applied physics) and are presented in light of First Principle Analysis techniques. The demonstration models emphasize the fundamental concept that the information derived from the modeling solutions through the use of these computer simulations is only as good as the materials coefficients and the fundamental assumptions employed in building the models.

The combination of computer simulation and First Principle Analysis gives the modeler the opportunity to try a variety of approaches to the solution of the same problem as needed in order to get the design right or nearly right in the workshop or laboratory before the first device components are fabricated and tested. The modeler can also use the physical device test results to modify the model parameters and arrive at an improved solution more rapidly than by simply using the cut and try methodology.

CHAPTER TOPICS

The eleven (11) technical chapters in this book demonstrate to the reader the hands-on technique of model building and solving. The COMSOL concepts and techniques used in these chapters are shown in Figure Int.1. The COMSOL modules employed in the various models in specific chapters are shown in Figure Int.2, and the physics concepts and techniques employed in the various models in specific chapters are shown in Figure Int.3.

Concept/Technique Chapter:	1	2	3	4	5	6	7	8	9	10	11
0D Modeling			•								
1D Modeling	•	•		•							
2D Axisymmetric Coordinates								•			•
2D Axisymmetric Modeling						•		•			•
2D Modeling				•		•	•		•		
3D Modeling									•		
Animation				•	•	•					
Bioheat Equation											•
Boolean Operations – geometry				•	•		•	•	•	•	
Boundary Conditions	•	•		•	•	•	•	•	•	•	•
Conductive Media DC				•	•	•		•	•	•	
Coupled Multiphysics Analysis				•				•	•	•	•

Concept/Technique Chapter:	1	2	3	4	5	6	7	8	9	10	11
Cylindrical Coordinates						•		•		•	
Deformed Mesh – Moving Mesh					•						
Domain Plot Parameter								•	•	•	•
Electromagnetics				•	•			•	•	•	•
Electronic Circuit Modeling			•								
Electrostatic Potentials				•					•		
Fillet corners									•		
Floating Contacts					•						
Free Mesh Parameters					•	•	•	•	•	•	•
Frequency Domain							•	•	•	•	•
Global Equations					•		•	•	•	•	•
Heat Transfer Coefficient	•	•		•		•	•		•		•
Laplacian Operator							•				
Lumped Parameters			•						•		
Magnetostatic Modeling					•			•	•		
Materials Library	•	•					•	•	•	•	
Mathematics – Coefficient Form PDE				•							
Mathematics – General Form PDE				•							
Maximum Element Size				•			•	•		•	•
Mixed Materials Modeling								•			•
Mixed Mode Modeling							•	•			

(continued)

Concept/Technique Chapter:	1	2	3	4	5	6	7	8	9	10	11
Out-of-Plane Thickness				•	•		•	•		•	
Parametric Solutions				•	•	•	•	•	•		
Perfectly Matched Layers										•	
Pointwise constraints					•		•		•		
Quasi-Static Solutions					•	•	•				
Scalar Expressions							•				
Scalar Variables							•		•		
Spherical Coordinates				•							
Static Solutions		•				•	•		•		
Streamline Plot									•		
Terminal Boundary Condition					•				•		
Transient Analysis		•			•		•	•			
Triangular Mesh					•	•	•	•		•	•
Work Plane									•		

Figure Int.1 COMSOL Concepts and Techniques

Module Chapter:	1	2	3	4	5	6	7	8	9	10	11
Basic	•	•	•	•	•	•	•	•	•	•	•
AC/DC			•	•	•		•	•	•		
Heat Transfer	•	•		•		•	•		•		•
RF								•		•	•
Semiconductor				•							
Structural Mechanics									•		

Figure Int.2 COMSOL Modules Employed

Physics Concepts Chapter:	1	2	3	4	5	6	7	8	9	10	11
Anisotropic Conductivity					•						
Antennas											•
Bioheat Equation											•
Boltzmann Thermodynamics						•					
Complex AC Theory							•	•			
Concave Mirror										•	
Distributed Resistance			•				•		•		
Electrochemical Polishing					•						
Electromagnetic induction (Inductance)				•			•		•		
Electrostatic Potentials in Different Geometric Configurations				•					•		
Energy Concentrator										•	
Faraday's Law			•		•				•		
Fick's Laws						•	•				
First Estimate Review	•			•	•	•	•	•	•	•	•
Fourier Analysis						•	•				
Fourier's Law						•	•				
Free-Space Permittivity								•	•		
Good First Approximation	•		•	•	•	•	•	•	•	•	•

(continued)

Physics Concepts Chapter:	1	2	3	4	5	6	7	8	9	10	11
Hall Effect					•						
Heat Conduction	•	•	•	•		•	•		•	•	•
Information Transmission				•				•			
Insulated Containers						•					
Joule Heating						•	•		•		•
Kirchoff's Laws (Current, Voltage)			•								
Lorentz Force					•						
Magnetic Field					•				•		
Magnetic Permeability					•				•		
Magnetic Vector Potential					•			•	•		
Magnetostatics									•		
Maxwell-Faraday Equation			•								
Maxwell's Equations			•			•		•	•	•	
Microwave Irradiation											•
Newton's Law of Cooling						•					
Ohm's Law			•				•	•	•		•
Optical (Laser) Irradiation											•
Pennes Equation											•
Perfectly Matched Layers: 2D Planar, 3D Cartesian Cylindrical and Spherical										•	
Perfusion											•
Planck's Constant						•					

Physics Concepts Chapter:	1	2	3	4	5	6	7	8	9	10	11
Semiconductor Dual Carrier Types				•	•						
Semiconductor, Density-Gradient (DG) Theory				•							
Semiconductor, Schrodinger-Poisson (SP) Theory				•							
Soliton Waves				•							
Telegraph Equation				•							
Vector Dot Product Current				•	•			•	•		

Figure Int.3 Physics Concepts

The information in these three figures link the overall presentation of this book to the underlying modeling, mathematical and physical concepts. In this book, in contrast to other books with which the reader may be familiar, key ancillary information is, in most cases, contained in the notes.

NOTE *Please be sure to read, carefully consider, and apply, as needed, each note.*

CHAPTER 1 MODELING METHODOLOGY

Chapter 1 provides an overview of the modeling process by discussing the fundamental considerations involved: the hardware (computer platform), the coordinate systems (physics), the implicit assumptions (lower dimensionality considerations), and First Principles Analysis (physics). Three relatively simple 1D models are presented that build and solve, for comparison, single-, double- and triple-pane thermal insulation window structures. Comments are also included on common sources of modeling errors.

CHAPTER 2 MATERIALS PROPERTIES

Chapter 2 discusses various sources of materials properties data, including the COMSOL Material Library, basic and expanded module, as well as print and Internet sources. A multi-pane thermal insulation window structure model demonstrates three techniques for entering material properties: user-defined direct entry, user-defined parameters, and material definitions. Also included are instructions for building a user-defined material library for storage within COMSOL 5.x.

CHAPTER 3 0D ELECTRICAL CIRCUIT INTERFACE

COMSOL 5.x uses zero-dimensional models to provide for the modeling of electrical circuitry. The models in this chapter illustrate techniques for modeling various basic circuits: a resistor-capacitor series circuit, an inductor-resistor series circuit, and a series resistor, parallel inductor-capacitor circuit. Considerations for the proper setup of the circuits are discussed along with the basics of problem formulation and the implicit assumptions built into COMSOL 5.x relative to electrical circuit modeling.

CHAPTER 4 1D MODELING

The first part of Chapter 4 models the 1D KdV Equation. The KdV Equation is a powerful tool that is used to model soliton wave propagation in diverse media (e.g., physical waves in liquids, electromagnetic waves in transparent media, etc.). It is easily and simply modeled with a 1D PDE mode model.

The second part of Chapter 4 models the 1D Telegraph Equation. The Telegraph Equation is a powerful tool that is used to model wave propagation in diverse transmission lines. The Telegraph Equation can be used to thoroughly characterize the propagation conditions of coaxial lines, twin pair lines, microstrip lines, etc. The Telegraph Equation is easily and simply modeled with a 1D PDE mode model.

The third part of Chapter 4 is a 1D Spherically Symmetric Transport model that illustrates the technique of simplifying models with spherical components from 3D to 1D by assuming that they are essentially symmetrical.

The fourth part of Chapter 4 is an Advanced 1D Silicon Inversion Layer Model using DG and SP Theory Methodologies. It is the purpose of this model to demonstrate the modeling techniques needed to reproduce the calculated inversion layer electron density below the gate oxide, as a function of depth curves.

CHAPTER 5 2D MODELING

The first half of Chapter 5 models the surface smoothing process by using a 2D Electrochemical Polishing Model. This model is a powerful tool that can be used for diverse surface smoothing projects (e.g., microscope samples, precision metal parts, medical equipment and tools, large and small metal drums, thin analytical samples, vacuum chambers, etc.).

The second half of Chapter 5 models Hall Effect magnetic sensors. The 2D Hall Effect Model is a powerful tool that can be used to model such sensors when used for sensing fluid flow, rotating and/or linear motion, proximity, current, pressure, orientation, etc.

CHAPTER 6 2D AXISYMMETRIC MODELING

Modeling a 3D device that is symmetrical on one axis by treating it as a 2D Axisymmetric object simplifies the model for quicker first approximation results.

The first half of Chapter 6 discusses a 2D Axisymmetric Heat Conduction in a Cylinder Model and demonstrates the use of contour plotting of the solver results to show non-linear temperature distribution in the cylinder.

The second half of Chapter 6 models transient heat transfer in a niobium sphere immersed in a medium of constant temperature by using a 2D Axisymmetric model.

CHAPTER 7 2D SIMPLE AND ADVANCED MIXED-MODE MODELING

In this chapter, simple and advanced mixed-mode 2D models are presented. Such 2D models are typically more conceptually complex than the models that were presented in earlier chapters of this text. 2D simple and advanced mixed-mode models have proven to be very valuable to the science and engineering communities, both in the past and currently, as first-cut evaluations of potential systemic physical behavior under the influence of mixed external stimuli. The 2D mixed-mode model responses and other such ancillary information can be gathered and screened early in a project for a first-cut evaluation. That initial information can potentially be used later as guidance in building higher-dimensionality (3D) field-based (electrical, magnetic, etc.) models.

The first part of Chapter 7 uses a 2D Electrical Impedance Sensor Model to demonstrate this technique. The concept of electrical impedance, as used in alternating current (AC) theory, is an expansion on the basic concept of resistance as illustrated by Ohm's Law, in direct current (DC) theory.

The second part of Chapter 7 uses a 2D Axisymmetric Metal Layer on a Dielectric Block Model to demonstrate more aspects of the technique. The modeler was introduced to Fick's laws for the diffusion (mass transport) of a first item (e.g., a gas, a liquid, etc.) through a second item (e.g., another gas, liquid, etc.). In the case of this model, the diffusing item is heat.

In the first part of Chapter 7 implemented above, a copper layer was approximated by the Thin Conductive Layer function in the Heat Transfer in Solids Interface. In this, Advanced Model, the third study implemented in this chapter, the copper layer will be a geometrical copper layer. The modeler will now be able to compare the results of the earlier models, to the results of this Advanced model calculation, on nominally the same modeling problem, employing diamond as the substrate. The modeler can now determine the relative trade-offs required when he chooses different materials to model one methodology or another.

These models are examples of the Good First Approximation type of models because they demonstrate the significant power of relatively simple physical principles, such as Ohm's Law, Joule's Laws, and Fick's Laws, when applied in the COMSOL Multiphysics Modeling environment. The equations can, of course, be modified by the addition of new terms, insulating materials, heat loss through convection, etc.

CHAPTER 8 2D COMPLEX MIXED MODE MODELING

In this chapter, three new primary analysis concepts are introduced to the modeler: 2D electromagnetic impedance calculation for a planar, two-wire geometry (side by side), for a concentric two-wire geometry (coaxial cable), and a 2D Axisymmetric model of the transient behavior of a Concentric 2 Wire Geometry (Coaxial Cable).

CHAPTER 9 3D MODELING

In this chapter, the modeler is introduced to three new modeling concepts: the Terminal boundary condition lumped parameters and coupled thermal, electrical and structural multiphysics analysis. The Terminal boundary condition and the lumped parameter concepts are employed in the solution of the 3D Spiral Coil Microinductor Model. The fully coupled multiphysics solution is employed in the 3D Linear Microresistor Beam Model.

The lumped parameter (lumped element) modeling approach approximates a spatially distributed collection of diverse physical elements by a collection of topologically (series and/or parallel) connected discrete elements.

This technique is commonly employed for first approximation models in electrical, electronic, mechanical, heat transfer, acoustic and other physical systems.

CHAPTER 10 PERFECTLY MATCHED LAYER MODELS

One of the fundamental difficulties underlying electromagnetic wave equation calculations is dealing with a propagating wave after the wave interacts with a boundary (reflection). If the boundary of a model domain is terminated in an abrupt fashion, unwanted reflections will typically be incorporated into the solution, potentially creating undesired and possibly erroneous model solution values. Fortunately, for the modeler of today, there is a methodology that works sufficiently well that it essentially eliminates reflection problems at the domain boundary. That methodology is the Perfectly Matched Layer.

Chapter 10 includes two models, the 2D Concave Metallic Mirror PML Model and the 2D Energy Concentrator PML Model, to demonstrate the use of the Perfectly Matched Layer methodology.

CHAPTER 11 BIOHEAT MODELS

The Bioheat Equation plays an important role in the development and analysis of new therapeutic medical techniques (e.g., killing of tumors). If the postulated method raises the local temperature of the tumor cells, without excessively raising the temperature of the normal cells, then the proposed method will probably be successful. The results (estimated time values) from the model calculations will significantly reduce the effort needed to determine an accurate experimental value. The guiding principle needs to be that tumor cells die at elevated temperatures. The literature cites temperatures that range from 42 °C (315.15 K) to 60 °C (333.15 K).

The first half of Chapter 11 models the Bioheat Equation as applied with a photonic heat source (laser).

The second half of Chapter 11 models the Bioheat Equation as applied with a microwave heat source.

CHAPTER 1

MODELING METHODOLOGY USING COMSOL MULTIPHYSICS 5.x

In This Chapter

GUIDELINES FOR NEW COMSOL MULTIPHYSICS 5.x MODELERS

NOTE *First, for purposes of clarity in this book, when the term "COMSOL Multiphysics 5.x software" or "5.x" occurs, that term means: COMSOL Multiphysics software version 5.6 or later.*

Second, the user of this text should be sure to READ ALL NOTES in this book, as the Notes contain information that is not presented elsewhere and should facilitate comprehension and, hopefully, minimize modeling errors for modelers unfamiliar with 5.x.

When building models, new or otherwise, it is VERY IMPORTANT to SAVE EARLY and OFTEN.

Hardware Considerations

There are two basic rules for selecting hardware that will support successful modeling with 5.x. The first rule is that the modeler should be sure to determine the minimum system requirements their version of 5.x requires before borrowing or buying a computer to run his new modeling software. This book introduces the use of 5.x for modeling within the Microsoft Windows® and the Macintosh OS X® operating systems environments.

COMSOL Multiphysics 5.x software supports shared memory parallelism under both the Microsoft Windows and the Macintosh OS X operating systems. Neither distributed memory nor cluster computing will be covered in this book.

The number of platform cores is equal to the number of coprocessors designed into the computer (e.g., 1, 2, 4, 8, 12, etc.).

NOTE — *Shared memory parallelism means that multiple cores in the same computer share the same memory array. Distributed memory parallelism means multiple computers with multiple cores are connected in a cluster with the problem divided into multiple parallel computational sub-segments {1.1} distributed over the computational units of the cluster.*

The second rule of successful modeling is the modeler should run 5.x on the best platform with the highest processor speed and the most memory obtainable; the bigger, the faster, the better. It is the general rule that the speed of model processing increases in proportion to instruction size (32 bit, 64 bit), the core speed, the number of platform cores, and to the amount of usable, available memory.

The platform this author uses is an Apple Mac Pro®, running Mac OS X version 10.15.x. That platform has 12-2.7 GHz cores and 64 GB of RAM and is configured for 64-bit processing and runs at the 64-bit rate when using 5.x. If the new modeler desires a different 64-bit operating system than Macintosh OS X, then they will need to choose a LINUX® platform, using UNIX® or a PC with a 64-bit Microsoft Windows operating system, as specified in the COMSOL Operating System Requirements (*https://www.comsol.com/system-requirements*).

NOTE

The 2.7 GHz specification is the operating speed of each of the cores and the 64 GB is the total shared Random Access Memory (RAM). 64-bit refers to the width of a processor instruction.

Currently available Macintosh Pro hardware can now be obtained with up to 28 cores and 768 GB of shared ECC {1.2}.

Once the best available processor is obtained, within the constraints of your budget, install your copy of 5.x, following the installer instructions.

After installation, as a matter of caution, the modeler should restart the processor and, once stable, start the COMSOL Multiphysics 5.x software. 5.x will then present the modeler with a configurable Graphical User Interface (GUI). For computer users not familiar with the GUI concept, information is presented primarily in the form of pictures with supplemental text, not exclusively as text. See Figure 1.1a (Mac) and Figure 1.1b (PC) to observe the details of the 5.x GUI interface.

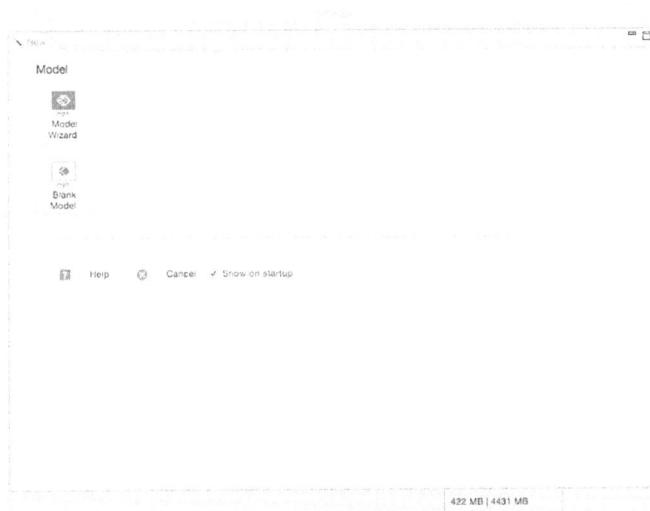

FIGURE 1.1(a) Default COMSOL desktop display on a Mac immediately after startup.

FIGURE 1.1(b) Default COMSOL desktop display on a PC immediately after startup.

Simple Model Setup Overview

Figures 1.1a&b show the default COMSOL Desktop Display immediately after startup, for both the Mac and the PC. The startup screen for the Mac and the PC are the same in 5.x on the two platforms. COMSOL Multiphysics Version 5.x displays two (2) buttons. The two buttons on the Mac and the PC are the same. Those buttons are the "Model Wizard" button and the "Blank Model" button.

In 5.x, after the Model Wizard button has been selected, a row of buttons is displayed that allows the modeler to select the desired geometry (3D, 2D Axisymmetric, 2D, 1D Axisymmetric, 1D, and 0D) appropriate to the model under development. In the case of the 3D coordinate system, the coordinates that are pictorially shown in the Graphics window are based on the Right-Hand-Rule.

NOTE

The Right-Hand-Rule is derived, as it states, from your right hand. Look at your right hand, point the thumb up, point your index (first) finger away from your body, at a right angle (90 degrees) to your thumb and point your middle (second) finger, at a right angle to the thumb, tangent to your body. Your thumb represents the Z-axis, your index (first) finger represents the X-axis, and your middle (second) finger represents the Y-axis.

In a 3D coordinate system that obeys the Right-Hand-Rule, X rotates into Y and generates Z. Z is the direction in which a Right-Handed Screw would advance (move into the material into which it is being screwed).

The modeler should note that the default COMSOL Desktop for 5.x, built by using the Model Wizard button, displays four active windows: Model Builder, Settings, Graphics, and Messages. The Desktop Display can, of course, be modified as needed to show more or fewer (see COMSOL Desktop Environment {1.3}). In this book, the Desktop Display will be used in the default configuration, except when modifications are required.

NOTE

The most singularly important currently active window for the modeler, at this point, is the Model Wizard window. Before anything else can happen relative to creating a new model, 5.x requires that the modeler choose a new coordinate system.

The Model Wizard window of the Desktop Display shows six coordinate system (Space Dimension) choices available for selection: 3D, 2D Axisymmetric, 2D, 1D Axisymmetric, 1D, and 0D. See Figures 1.2a&b to observe the interface details.

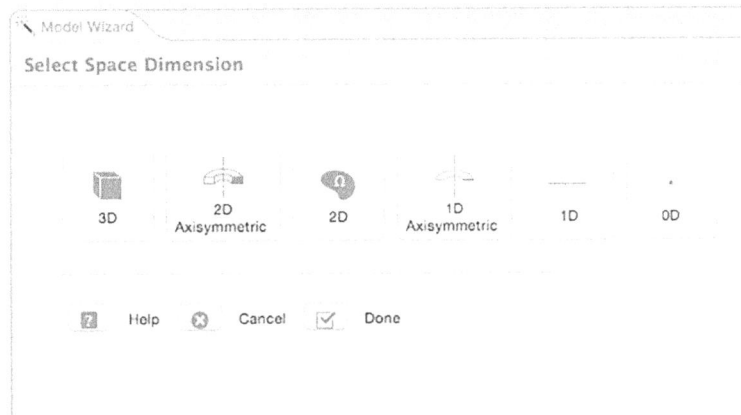

FIGURE 1.2(a) Default model wizard window for model coordinate system (space dimension) selection for a Mac.

Figure 1.2a shows the Model Wizard window for model coordinate system (Space Dimension) selection for a Mac.

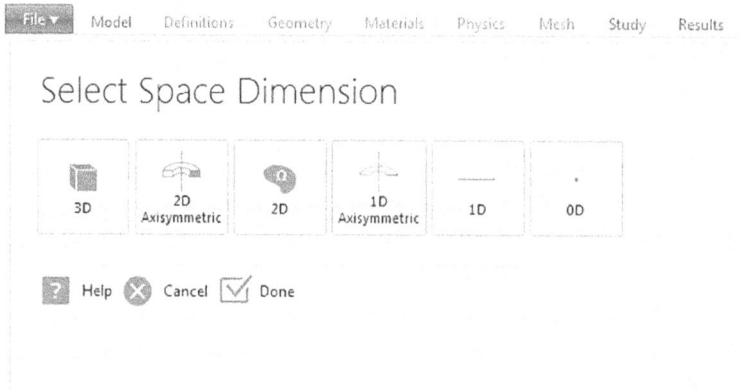

FIGURE 1.2(b) Default model wizard window for model coordinate system (space dimension) selection for a PC.

Figure 1.2b shows the Model Wizard window for model coordinate system (Space Dimension) selection for a PC.

Geometric coordinate systems in COMSOL range from the extremely complex to the nominally very simple. The default geometries are the standard Cartesian geometries 3D, 2D, and 1D, as shown in Figures 1.3, 1.4, and 1.5.

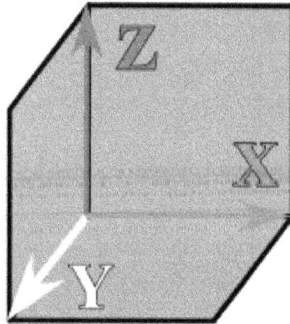

FIGURE 1.3 3D Cartesian coordinate geometry.

Figure 1.3 shows a Right-Hand 3D Cartesian Coordinate Geometry Axes configuration.

FIGURE 1.4 2D Cartesian coordinate geometry.

Figure 1.4 shows a 2D Cartesian Coordinate Geometry Axes configuration.

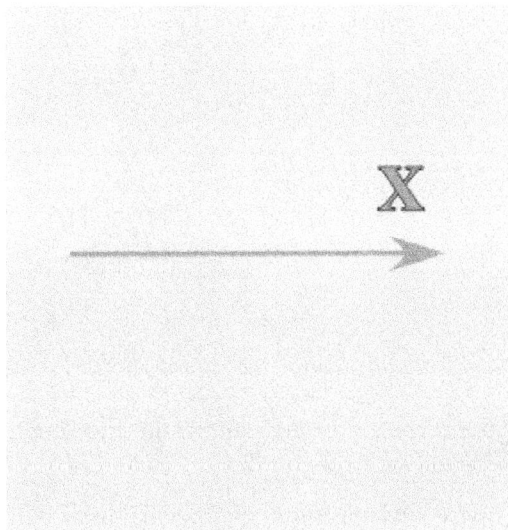

FIGURE 1.5 1D Cartesian coordinate geometry.

Figure 1.5 shows a 1D Cartesian Coordinate Geometry Axis configuration.

Comprehension of the 3D, 2D, 1D Cartesian Coordinate Systems is relatively easy for the modeler, assuming that the modeler has some previous engineering or science training. These coordinate systems are simply an orthogonal combination of rectilinearly incremented axes comprising 1, 2, or 3 measurement dimensions.

NOTE

The 2D and 1D Axisymmetric Geometries are somewhat more complex. Both the 2D and 1D Axisymmetric Geometries comprise cylindrical coordinate systems. The primary benefit derived through the use of the axisymmetric coordinate systems is that they allow the modeler to reduce the effective model dimensionality and the calculational complexity of a problem, for example, 3D → 2D, through the assumption of axial symmetry and rotationally uniform boundary conditions.

Due to the complex nature of the underlying assumptions in the 1D Axisymmetric Coordinate System and its applications, the modeler is referred to the COMSOL literature {1.4} for further details.

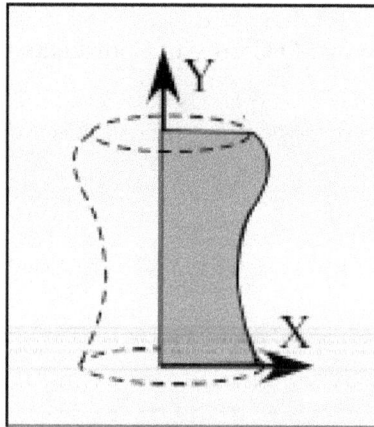

FIGURE 1.6 2D axisymmetric coordinate geometry (space dimension).

Figure 1.6 shows a 2D Axisymmetric coordinate geometry axis (Space Dimension) configuration.

The last Space Dimension choice available is 0D. In the case of 0D, the underlying equations in the multiphysics model have either no relational dependence on geometrical factors or react in a homogeneous and isotropic manner (effectively geometrically relationally independent).

The 0D Space Dimension will be used in this book only to explore electrical and/or electronic circuit model behavior through the use of SPICE calculations.

Select A Space Dimension:

For this first demonstration of the Desktop Display, Select (Click) the 1D selection button. See Figure 1.7.

NOTE

The term Select (Click) means for the modeler to place the display screen computer cursor in proximity to or superimposed on the item to be selected and then tap (momentarily depress) the physical mouse activation key (button).

The selection of the 1D button will set the dimensionality to 1D and cause the Model Wizard display to switch to the Select Physics page. See Figure 1.8.

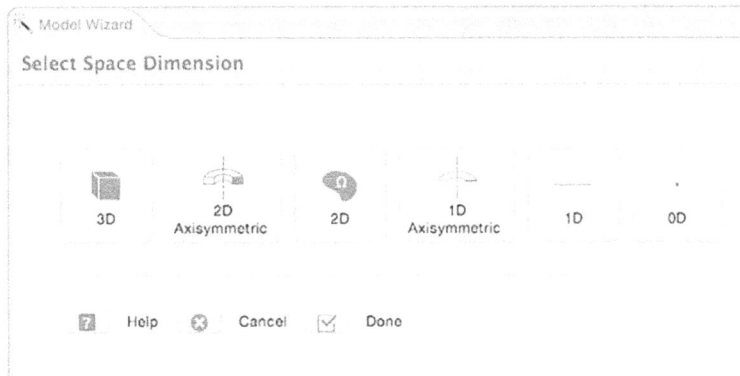

FIGURE 1.7 1D Button.

Figure 1.7 shows the dimensionality buttons (1D button)

Figure 1.8 shows the Select Physics page. See Figure 1.8.

Select Physics

Search

▼ Recently Used
    Heat Transfer in Solids (ht)
    Heat Transfer in Fluids (ht)
▶ AC/DC
▶ Acoustics
▶ Chemical Species Transport
▶ Heat Transfer
▶ Radio Frequency
▶ Semiconductor
▶ Mathematics

Added physics interfaces:

Space dimension   Study

Help   Cancel   Done

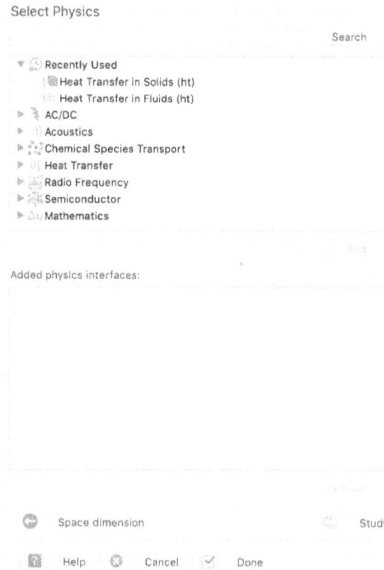

FIGURE 1.8 Select physics page.

Figure 1.8 shows the Select Physics page.

Once a Physics Interface is clicked, the Desktop Display changes to show a description of the Physics Interface and the Add Physics window, as shown in Figure 1.9a. In the case of Figure 1.9a, all of the licensed COMSOL Multiphysics Physics Interfaces are displayed,. If the modeler has not licensed all of the available modules, only the licensed modules are displayed.

FIGURE 1.9(a) Desktop display with the description of the physics interface and the add physics window.

Click the Add button below the Select Physics window.

Figure 1.9b shows the Desktop Display with the Added Physics in the Add Physics window.

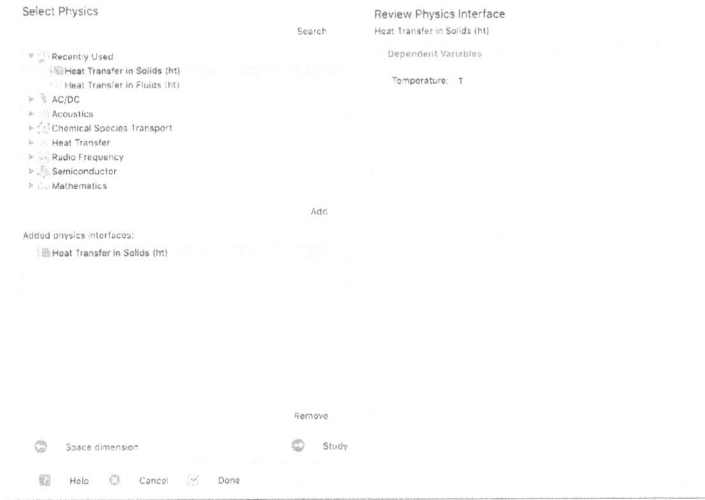

FIGURE 1.9(b) Desktop display with the selected physics interface in the add physics window.

A Physics Interface is the set of relevant equations, typical initial conditions, independent variable(s), dependent variable(s), default settings, boundary conditions, etc., that are appropriate for the solution of a particular problem type for a given branch or sub-branch of physics (e.g., acoustics, electromagnetics, heat transfer, etc.).

NOTE *As is indicated by the name Multiphysics, several different branches or sub-branches of physics can be applied, as needed, simultaneously or serially to achieve the solution of a given model. Diverse models that demonstrate the application of the Multiphysics concept through the incorporation of multiple different Physics Interfaces in a given model will be explored in detail as this book progresses.*

The modeler has now been shown fundamentally how to select a set of Space Coordinates and how to access Physics Interfaces.

If the modeler is using a Macintosh, Select > COMSOL Multiphysics menu > Quit COMSOL Multiphysics.

If the modeler is using a PC, Select > File > Exit.

Basic Problem Formulation and Implicit Assumptions

A first-cut problem solution is the equivalent of a back-of-the-envelope or on-a-napkin solution. Problem solutions of that type are more easily formulated, more quickly built, and typically provide a first estimate of whether the final solution to the full problem will be deemed to be (should possibly be) within reasonable time constraints or budgetary bounds. Creating first-cut solutions will often allow the 5.x modeler to easily decide whether or not it is worth the additional effort and cost needed to build a fully implemented higher dimensionality model.

A modeler can generate a first-cut problem solution as a reasonable first estimate, by choosing initially to use a lower dimensionality coordinate space than 3D (e.g., 1D, 2D Axisymmetric, etc.). By choosing a low-dimensionality Space Dimension, a modeler can significantly reduce the ultimate time needed to achieve a detailed final solution for the chosen prototype model. However, both new modelers and experienced modelers must be especially careful to fully understand the underlying (implicit) assumptions, unspecified conditions, and default values that are automatically incorporated into the model by simply selecting a lower dimensionality Space Dimension.

Reality, as we currently understand it, comes in four basic dimensions, three space dimensions (X-Y-Z), and one time dimension (t) (e.g., X-Y-Z-t, r-φ-θ-t, etc.). Relativistic effects can typically be neglected in most cases, except where high velocity or ultra-high accuracy is involved. Models with relativistic effects will not be covered in this book.

NOTE *Relativistic effects typically only become a concern for bodies in motion with a velocity approaching that of the speed of light (~3.0 × 10^8 m/s) or for ultra-high resolution time calculations at somewhat lower velocities.*

The types of calculations presented within this book are typically for steady-state models, quasi-static models, and for relatively low-velocity transient model solutions.

In a steady-state model, the controlling parameters are defined as numerical constants and the model is allowed to converge to reach the final equilibrium state defined by the specified constants.

NOTE *In the quasi-static methodology, a model solution to a problem is found by finding an initial steady-state solution, not the final steady-state solution. Then, the modeling constants are incrementally modified. That incremental change in the constants moves the modeling solution toward the desired final steady-state solution.*

In a transient solution model, all of the appropriate variables in the model are a function of time. The model solution builds from a set of suitable initial conditions, through a set of incremental intermediate solutions, to a final solution.

1D WINDOW HEAT FLOW MODELS

Consider, for example, a brief comparison between a relatively simple 1D heat flow model and the identical problem as a 3D model. The following models are those of a single pane, a dual pane, or a triple pane window mounted in the wall of a building on a typical winter's day. The basic question to be answered is: Why use a dual pane window rather than a single pane window or a triple pane window?

1D 1 Pane Window Heat Flow Model

Run 5.x > Click the Model Wizard button > Select 1D in the Model Wizard.

Click the Heat Transfer twistie.

Select > Heat Transfer in Solids *(ht)* in the Add Physics window.

Click the Add button (see Figure 1.10).

FIGURE 1.10 Model wizard – add physics window.

Figure 1.10 shows the Model Wizard – Add Physics window.

NOTE *The modeler should note that Clicking on the twistie for any Physics Interface expands the menu associated with that Physics Interface. That expansion allows the modeler to then Select the appropriate Physics Interface necessary to perform the needed model calculations. The Selected Physics Interface is added to the Selected Physics window by Clicking the Add button. If necessary, the added Selected Physics Interface can be removed by Clicking the Remove button.*

Click > Study (right pointing arrow).

Select > General Studies > Stationary.

NOTE *The modeler should note that Stationary and Steady-State are equivalent terms.*

Click > Done button (see Figure 1.11).

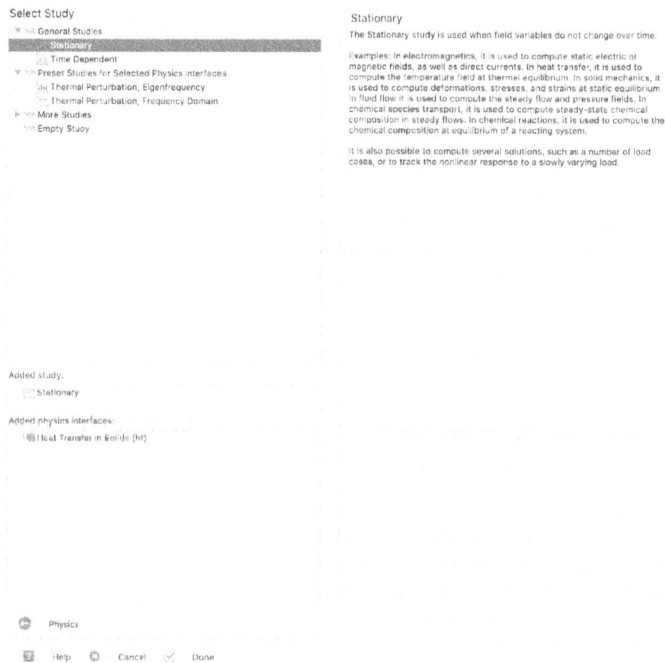

FIGURE 1.11 Done button in the select study type window.

Figure 1.11 shows the Done button in the Select Study Type window for the completion of Model Wizard process, resulting in the Initial Model Build. See Figure 1.12.

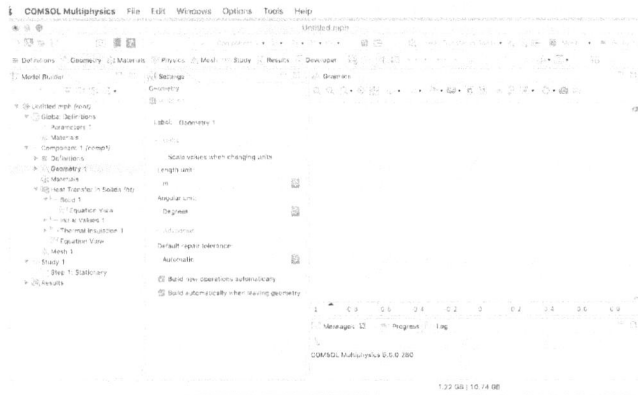

FIGURE 1.12 Initial model build for the 1D window heat transfer.

Figure 1.12 shows Initial Model Build for the 1D Window Heat Transfer Desktop Display.

Save the Model as > MM2E5X_1D_W1Pane.mph. See Figure 1.13.

FIGURE 1.13 Saved initial model build MM2E5X_1D_W1Pane.mph.

Figure 1.13 shows the saved Initial Model Build MM2E5X_1D_W1Pane.mph.

NOTE *The modeler should note that both the Desktop Display and the Model Builder windows show the saved file name.*

Now that the Initial Model Build has been saved, the modeler needs to enter the constant values needed for this model.

In the Model Builder window of the Desktop Display, Click > Global Definitions > Parameters 1. See Figure 1.14.

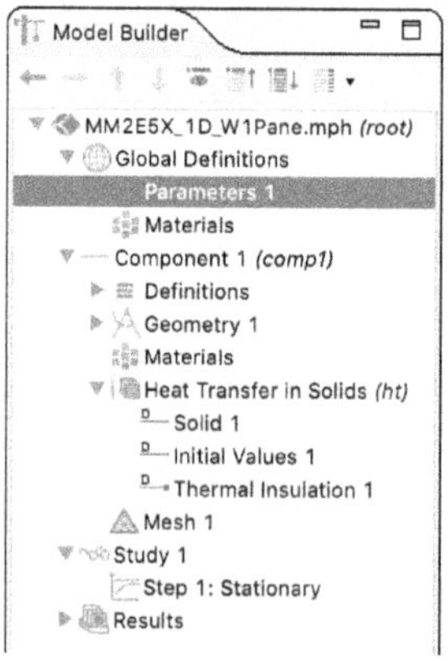

FIGURE 1.14 Global Definitions Parameters 1 selection menu.

Figure 1.14 shows the Global Definitions Parameters 1 Selection menu.

Once the Parameters 1 menu item has been selected, the Desktop Display will be modified to incorporate a Settings window. The Settings window contains a Parameters Entry Table, listing the following attributes for each parameter: Name, Expression, Value, and Description (see Figure 1.15).

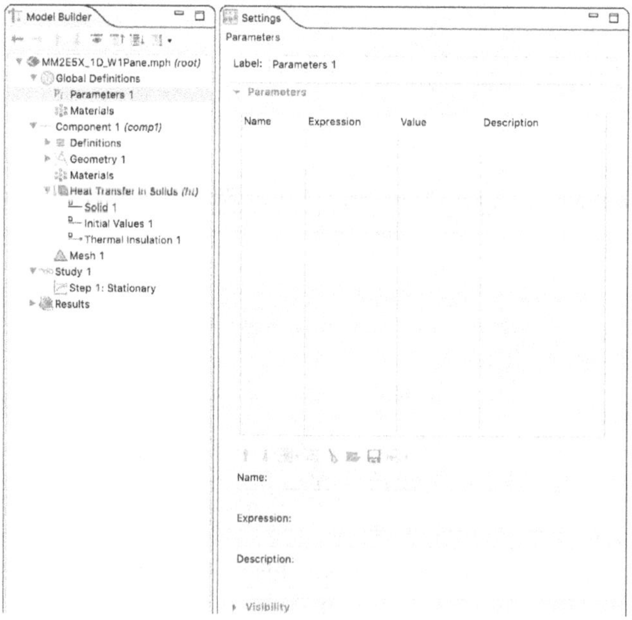

FIGURE 1.15 The desktop display – settings window with Parameters 1 entry table.

Figure 1.15 shows the Desktop Display – Settings window with a Parameters 1 Entry Table.

Parameter information can be entered directly into the fields in the Parameters 1 Entry Table or into the parameter entry fields below the Parameters Entry Table. Control buttons are located between the Parameters Entry Table window and the parameter entry fields. See Figure 1.16.

FIGURE 1.16 Control button array for the Parameters 1 entry table.

Figure 1.16 shows the control button array for the Parameters 1 Entry Table.

These are the control button array functions, from left to right: Up Arrow (Move Up), Down Arrow (Move Down), Blue Arrow (Move to), Blue X (Delete), Broom (Clear Table), Open Folder Icon (Load from File), Disk Icon (Save to File), and Bar Cursor (Insert Expression Here).

NOTE

When entering information into a Parameters field, first Click on the table entry field desired. Then type the desired piece of information into Name field below the Control Button Array. Follow this procedure for each new line of parametric text.

To enter the remaining parameter information for the model, Click on the subsequent entry window in the next column and enter (type) each additional piece of the information, as shown in Table 1.1.

The modeler can transition to the next information entry point in the tables by clicking on the next field.

TABLE 1.1 1D 1 Pane Window Parameters.

Name	Expression	Description
T_in	70 [degF]	Interior Temperature
T_out	0 [degF]	Exterior Temperature
p	1 [atm]	Air Pressure

These entries in the Parameters Entry Table define the interior temperature, the exterior temperature, and the air pressure for use in this model. See Figure 1.17.

FIGURE 1.17 The settings window with a filled parameters entry table.

Figure 1.17 shows a close-up of the Settings window with a filled Parameters Entry Table.

Click on the Save to File icon (disk) symbol in the control button array and enter MM2E5X_1D_W1Pane_Param.txt to save these parameters for future use.

The next step in building a model of a single pane window is to define the geometry through which the heat will flow from the heated interior to the less heated exterior.

NOTE

The modeler should note at this point, that in the next few steps a 1D path for heat flow will be created that is represented by a line progressing from left to right. Shortly, later, boundary conditions will be added. The modeler should also note that what was previously designated as Model 1 (mod1) is now designated as Component 1 (comp1).

Right-Click in the Model Builder window on Geometry 1 under Component 1 *(comp1)* to display the selection list. See Figures 1.18 and 1.19.

The term Right-Click means for the modeler to place the display screen computer cursor in proximity to or superimposed on the item to be selected and then tap (momentarily depress) the right physical mouse activation key (button).

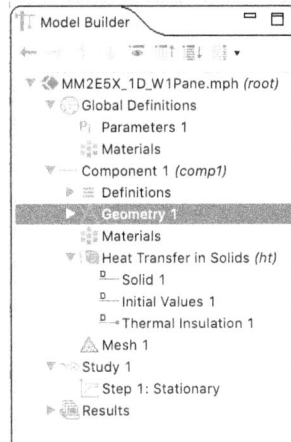

FIGURE 1.18 Model builder showing Geometry 1.

Figure 1.18 shows the Geometry 1 entry in the Model Builder window.

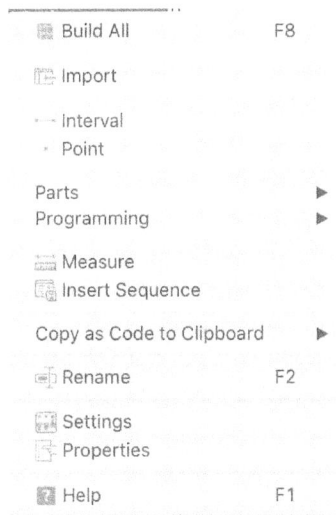

FIGURE 1.19 Geometry 1 selection list.

Figure 1.19 shows the Geometry 1 selection list for selecting the Interval option.

Select > Interval. See Figure 1.20.

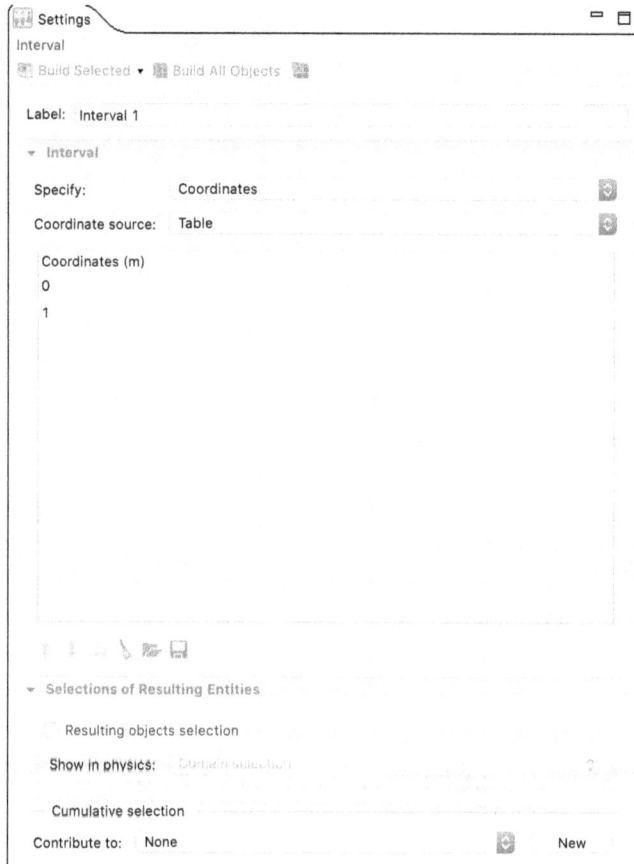

FIGURE 1.20 Settings – interval window.

Figure 1.20 shows the Settings – Interval window.

In the Settings Window > (make sure the Interval Window is open) Click on the Specify pull-down and select Interval lengths.

Enter 0[m] as the Left endpoint.

In the Settings Window > (make sure the Interval Window is open) Click on the Length source: pull-down and select Vector.

Enter 5e-3[m] in the Lengths text field, as shown in Figure 1.21.

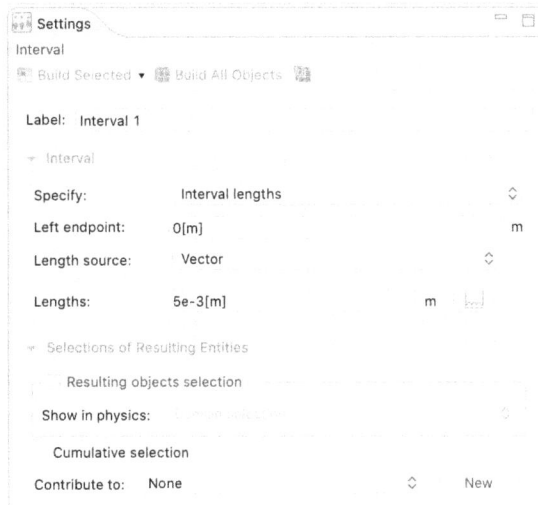

FIGURE 1.21 Settings – interval window, filled.

Figure 1.21 shows the Settings – Interval window with the values filled in.

Click > Build All Objects.

The geometry for the interval will be built and displayed in the Graphics window.

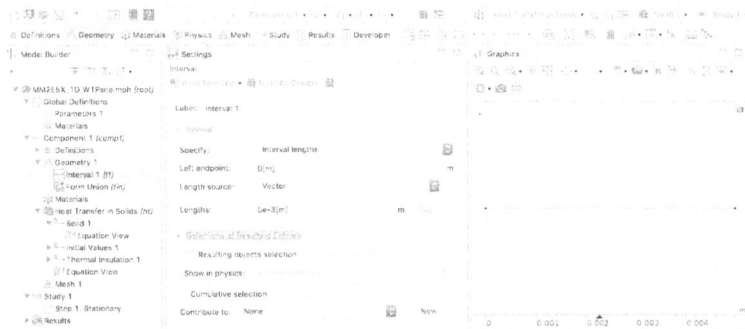

FIGURE 1.22 Desktop display window with built geometry.

Figure 1.22 shows the Desktop Display window with the Built Geometry.

Now that the Geometry has been built, the next step is to define the material(s) which comprise the model. In this case, it is relatively easy because the model only needs to use one (1) material. Also, the material properties of that material are resident in the Built-In Materials Library.

Right-Click in the Model Builder window on Materials under Component 1 *(comp1)* to display the Materials Selection List. See Figures 1.23 and 1.24.

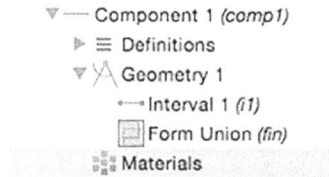

FIGURE 1.23 Model builder showing materials selection button.

Figure 1.23 shows the Materials Selection button entry in the Model Builder window.

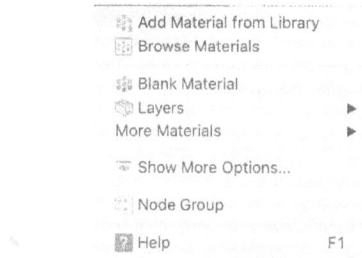

FIGURE 1.24 Materials selection function list.

Figure 1.24 shows the Materials Selection Function List for selecting the Material Browser.

Select > Add Material from Library. See Figure 1.25.

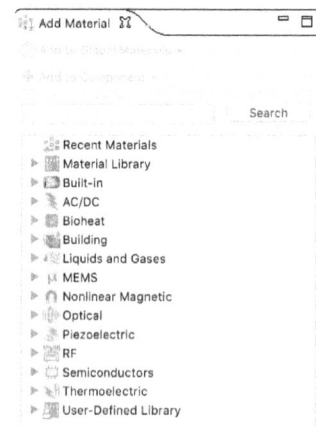

FIGURE 1.25 Add material library.

Figure 1.25 shows the Add Material Library.

Click the twistie {1.5} pointing at Built-In in the Add Material Library window. See Figures 1.26 and 1.27.

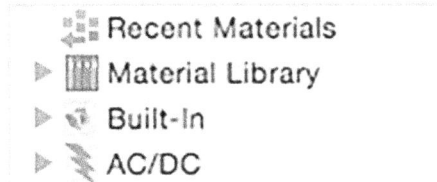

FIGURE 1.26 Twistie before built-in.

Figure 1.26 shows the twistie before the entry for the Built-In material library.

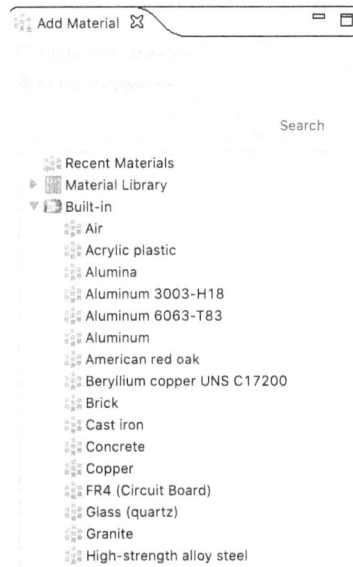

FIGURE 1.27 Expanded built-in material list.

Figure 1.27 shows the beginning of the expanded list of materials in the Built-In material library.

Scroll the material list until Silica glass appears. See Figure 1.28.

High-strength alloy steel
Iron
Magnesium AZ31B
Mica
Molybdenum
Nimonic alloy 90
Nylon
Polysilicon
Lead Zirconate Titanate (PZT-5H)
Silica glass
Silicon
Solder, 60Sn-40Pb
Steel AISI 4340
Structural steel
Titanium beta-21S
Tungsten
Water, liquid

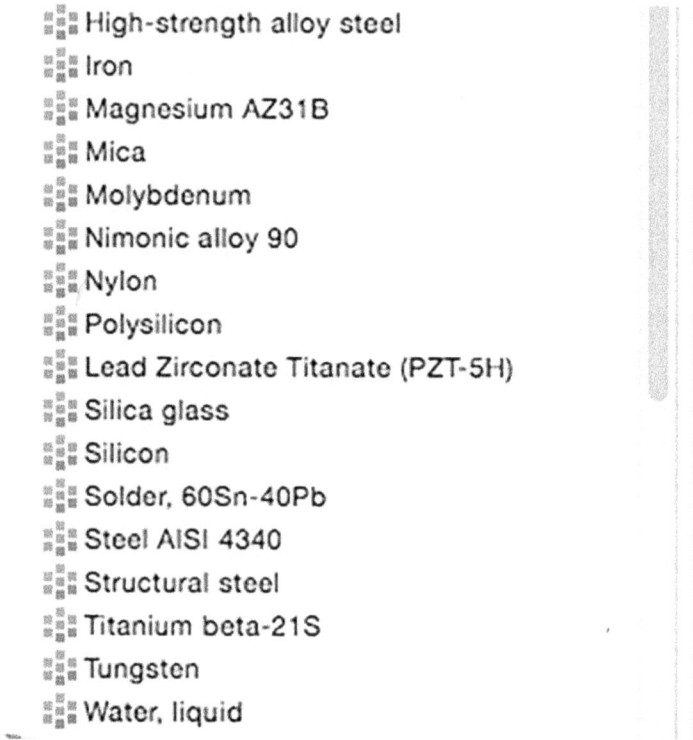

FIGURE 1.28 Scrolled list with silica glass visible.

Figure 1.28 shows the Scrolled List with the Silica glass entry visible.

Right-Click > Silica glass. See Figure 1.29.

Add to Global Materials
— Add to Component 1 (comp1)

FIGURE 1.29 Add material to Component 1.

Figure 1.29 shows the Add Material to Component 1 selection in the pop-up menu.

Select > Add to Component 1. See Figure 1.30.

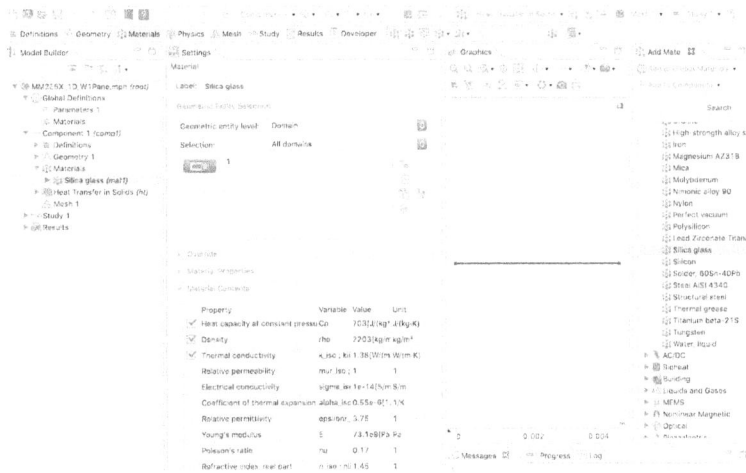

FIGURE 1.30 Silica glass material added to Component 1.

Figure 1.30 shows the Silica glass material added to Component 1.

Components in 5.x may comprise multiple geometric entities. Each geometric entity comprises connected manifolds. The geometric entities of the maximum dimension are called Domains. The order of precedence for 3D is: Domain, Boundary, Edge, and Point. As the dimensionality is reduced (e.g. 3D → 2D), the list is shortened from the right {1.6}.

NOTE

Since this model is 1D and there are only Domain and Boundary in this model, 5.x automatically assigns the Silica Glass material's properties to the available domain. If there were more than 1 domain, 5.x would assign this first material selection to all of the available domains.

As the modeler builds more complex and difficult models, he will need to pay close attention to the assignment of materials properties and to verify that the correct materials are added to the model and that they are assigned to the proper domains.

Also, the modeler should verify that the material selected has the necessary physical properties to allow for the solution of the problem in question. In this case, 5.x has determined which physical properties are needed for the current model and displays a green check mark next to the needed properties in the Material Contents window of the Settings - Material window. See Figure 1.31.

NOTE *The modeler should scroll-down in the Settings – Material window to see the green check marks in the Material Contents window.*

Property	Variable	Value	Unit	Property group
☑ Heat capacity at constant pres...	Cp	703[J/(k...	J/(kg·K)	Basic
☑ Density	rho	2203[kg...	kg/m³	Basic
☑ Thermal conductivity	k_iso ;...	1.38[W/...	W/(m·K)	Basic
Relative permeability	mur_is...	1	1	Basic
Electrical conductivity	sigma...	1e-14[S...	S/m	Basic
Coefficient of thermal expansion	alpha_i...	0.55e-6[...	1/K	Basic
Relative permittivity	epsilo...	3.75	1	Basic
Young's modulus	E	73.1e9[...	Pa	Young's modulus ar
Poisson's ratio	nu	0.17	1	Young's modulus ar
Refractive index, real part	n_iso ;...	1.45	1	Refractive index

FIGURE 1.31 Checked silica glass material properties in material contents window.

Figure 1.31 shows the checked Silica glass material properties in the Material Contents window.

Click > Add Material X (Close the Add Material Window)

Click > In the Model Builder window on Materials under Component 1 *(comp1)* to display the Settings Materials window and verify the Silica glass domain assignment.

Since the model geometry has been built and the appropriate material properties have been assigned to Domain 1, the next step in the modeling process is to specify both the domain and the boundary conditions for this model.

Click on the twistie (as needed) next to the Heat Transfer in Solids *(ht)* Physics Interface to expand that section of the Model Builder. See Figure 1.32.

The modeler should notice that when the twistie is Clicked, the menu expands, and the implicit assumptions are made available for the Heat Transfer in Solids (ht) Module. The modeler will need to explore this region for each model created, and each Physics Interface used to verify that the implicit assumptions made by 5.x either satisfy his needs or should be modified.

NOTE *For the inquisitive experienced modeler, further information about the underlying equations, the implicit assumptions, and the initial values can be obtained by Clicking each associated twistie and perusing the available information. However, the new modeler should save some of his inquisitiveness for later explorations.*

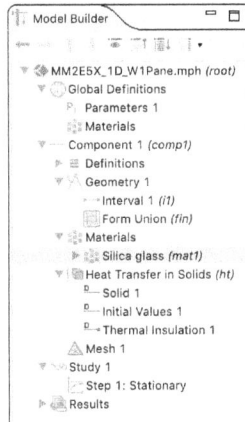

FIGURE 1.32 Expanded heat transfer in solids *(ht)* in model builder window.

Figure 1.32 shows the expanded Heat Transfer in Solids *(ht)* entry in the Model Builder window.

At this point in the development of this Component, the modeler needs to verify the assignment of values to the domain. In the Model Builder window under Heat Transfer in Solids *(ht)*,

Click > Solid 1. See Figure 1.33.

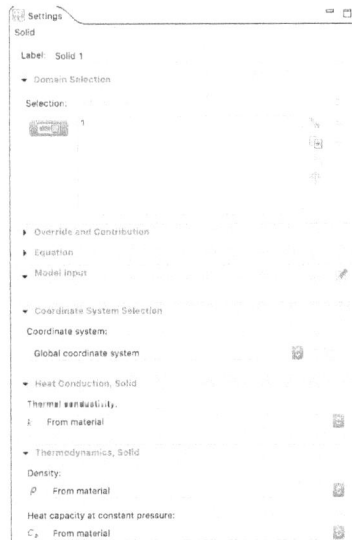

FIGURE 1.33 Settings for the Solid 1 window.

Figure 1.33 shows the Settings for the Solid 1 window.

The modeler can now verify that the selected material has been assigned to Domain 1 and that the material's properties are available for use in the equations to calculate the model solution.

Also, if the modeler were to Click on Initial Values 1, he would find that the Temperature is set to Room (20 °C = 293.15[K]).

If the modeler were now to Click on Thermal Insulation 1, he would find that both endpoints (1, 2) are insulated.

Those values are the result of the implicit assumptions remarked on earlier.

At this point, the modeler needs to set the desired modeling conditions for each endpoint (Boundary) and utilize the previously entered parameters that are needed to calculate the transfer of heat through this single pane window.

Right-Click > Heat Transfer in Solids *(ht)* in the Model Builder. See Figure 1.34.

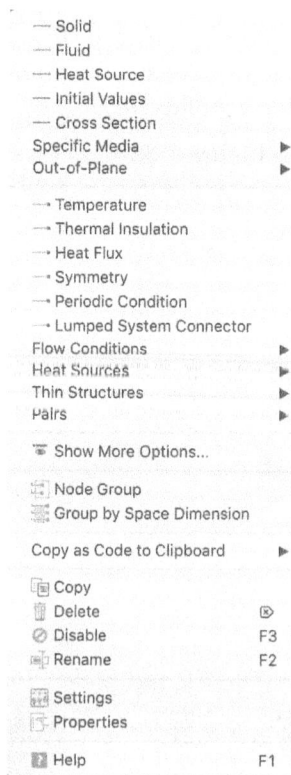

— Solid
— Fluid
— Heat Source
— Initial Values
— Cross Section
Specific Media ▶
Out-of-Plane ▶

— Temperature
— Thermal Insulation
— Heat Flux
— Symmetry
— Periodic Condition
— Lumped System Connector
Flow Conditions ▶
Heat Sources ▶
Thin Structures ▶
Pairs ▶

Show More Options...

Node Group
Group by Space Dimension

Copy as Code to Clipboard ▶

Copy
Delete ⊗
Disable F3
Rename F2

Settings
Properties

Help F1

FIGURE 1.34 Heat transfer in solids selection window.

Figure 1.34 shows the Heat Transfer in Solids Selection window, which lists the conditions that may be defined for this Physics Interface.

Select > Heat Flux. See Figure 1.35a.

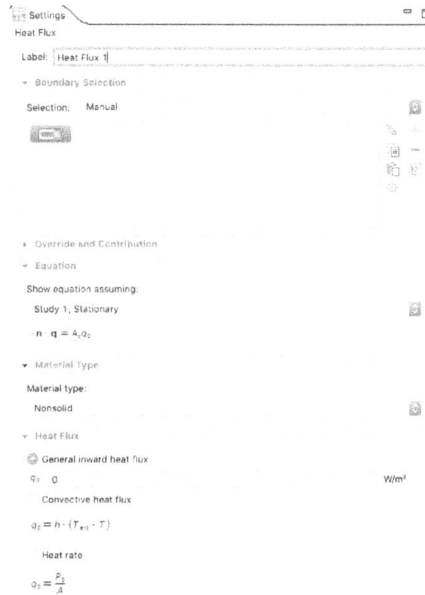

FIGURE 1.35(a) Initial boundaries heat flux window.

Figure 1.35a shows the Initial Boundaries Heat Flux window.

Verify that the Material type: is set to From material (otherwise, Click the pull-down bar and change the setting to From material).

Click Zoom Extents on the Graphics window Toolbar.

See Figure 1.35b.

FIGURE 1.35(b) Zoom extents button.

Figure 1.35a shows the Initial Boundaries Heat Flux window.

In the Graphics window, Click > Point 1 (leftmost boundary).

In the Settings – Heat Flux window, Click > Convective heat flux radio button.

NOTE *A typical value for the convection heat transfer coefficient (h) in the presence of a gas ranges from 2 to 25 W/(m² · K){1.7}. In this case, a typical value of 15 W/(m² · K) is chosen.*

Type 15 in the Heat transfer coefficient entry window.

Type T_in in the External temperature entry window.

Figure 1.36 shows the Settings – Heat Flux window as finally configured for Boundary Point 1.

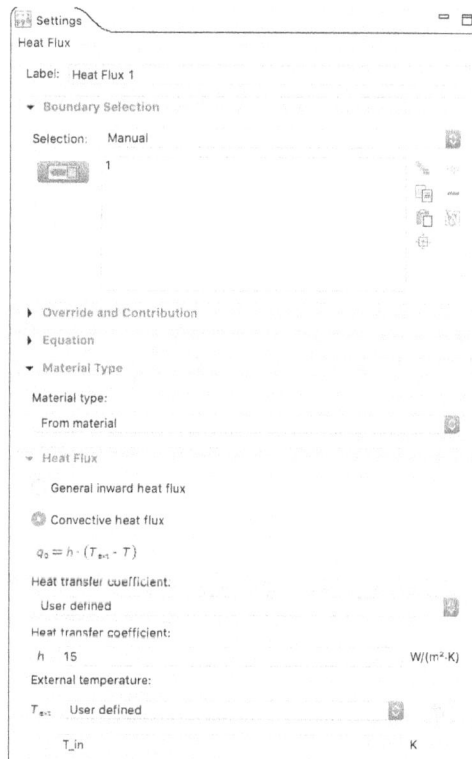

FIGURE 1.36 Filled Boundary Point 1 heat flux settings.

Right-Click > Heat Transfer in Solids *(ht).*

Select > Heat Flux.

Verify that the Material type: is set to From material (otherwise, Click the pull-down bar and change the setting to From material).

In the Graphics window, Click > Point 2 (rightmost boundary).

In the Settings – Heat Flux window, Click > Convective heat flux.

Type 15 in the Heat transfer coefficient entry window.

Type T_out in the External temperature entry window.

Figure 1.37 shows the Settings – Heat Flux window as finally configured for Boundary Point 2.

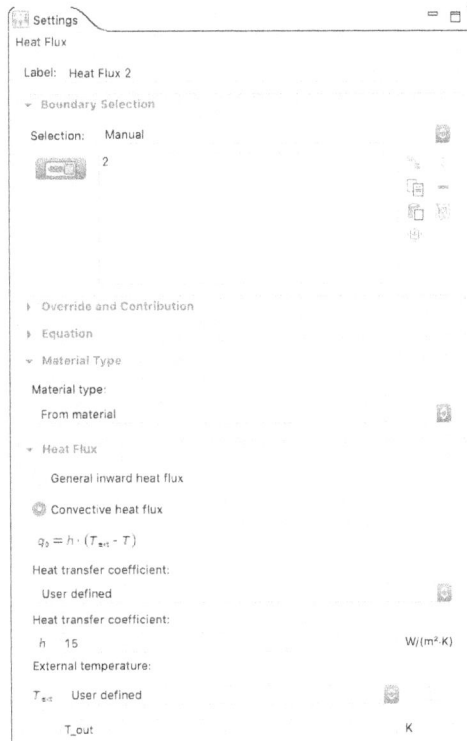

FIGURE 1.37 Filled Boundary Point 2 heat flux settings.

Since this first model is relatively simple, the easiest path is to allow 5.x to automatically mesh the model.

Right-Click > Mesh 1.

Select > Build All.

Right-Click > Graphics Window Background.

The Domain-Boundary Selection Window is shown in Figure 1.38a.

Select: Domains.

✓ — Select Domains
⋯→ Select Boundaries
✗ Disable Mouse Selection

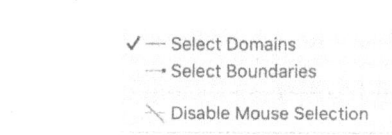

FIGURE 1.38(a) Domain-boundary selection window.

The meshing results are shown in Figure 1.38b.

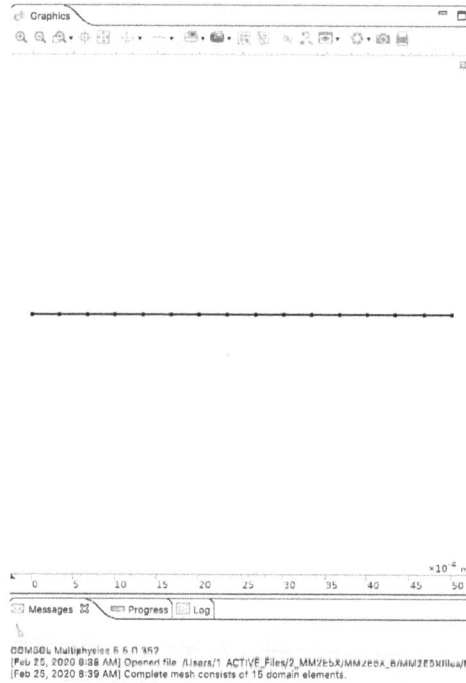

FIGURE 1.38(b) Meshed 1D domain.

Figure 1.38b shows the meshed 1D domain.

Meshing is the process by which the active geometrical space (domain) of a model is subdivided into a collection of sufficiently smaller spaces so that linear or higher-order polynomial approximations can be used as a reasonable

NOTE *analog of the functional physical behavior being modeled {1.8, 1.9}.*

In this case, based on a First Principles Analysis, the heat flow is linear within the domain.

Now that the model has been meshed, the modeler can proceed to compute the solution.

Right-Click > Study 1 in the Model Builder window.

Select > Compute.

NOTE *5.x automatically selects the appropriate Solver for this problem, computes the solution, and displays the results in the Graphics window.*

The results of this modeling computation are shown in Figure 1.39.

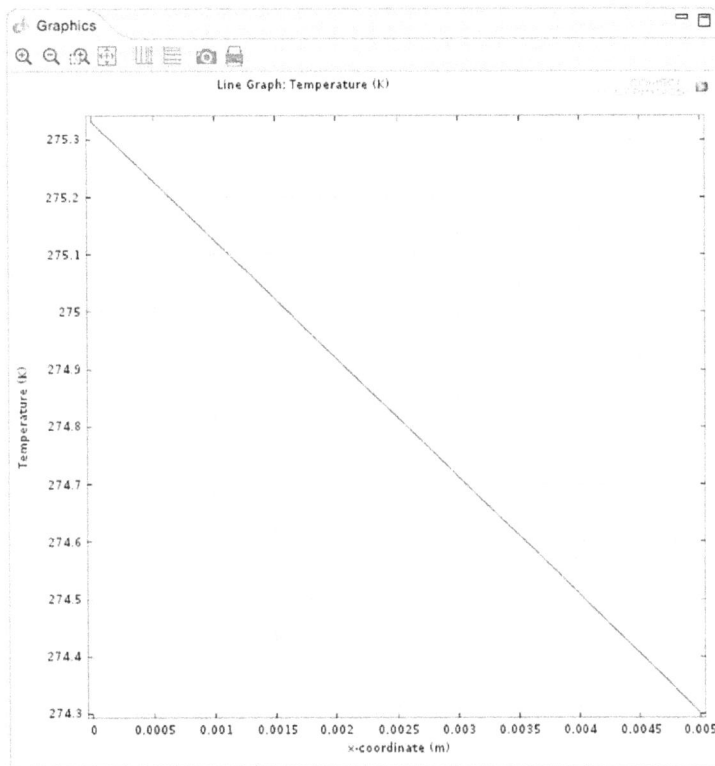

FIGURE 1.39 Initial calculated results for a 1 Pane window.

Figure 1.39 shows the Graphics plot window of the calculated results for a 1 Pane Window.

The 1D Plot displayed in the Graphics window shows the temperature in degrees Kelvin as a function of the distance from the inside surface of the Silica glass of the 1 Pane Window to the outside surface.

Since the modeler originally specified the initial input parameters in degrees Fahrenheit, the modeler now needs to adjust the instructions of 5.x so that the Graphics window will display the results of this modeling calculation in the desired units.

Under Results, Click > Temperature (ht) twistie.

Click > Line Graph.

In the Settings – Line Graph window,

Under Y-Axis Data, Click > Unit pull-down menu > Select > degF > Click > Plot. See Figure 1.40.

NOTE *The Plot button is one of the symbols located at the upper left corner of the Settings window.*

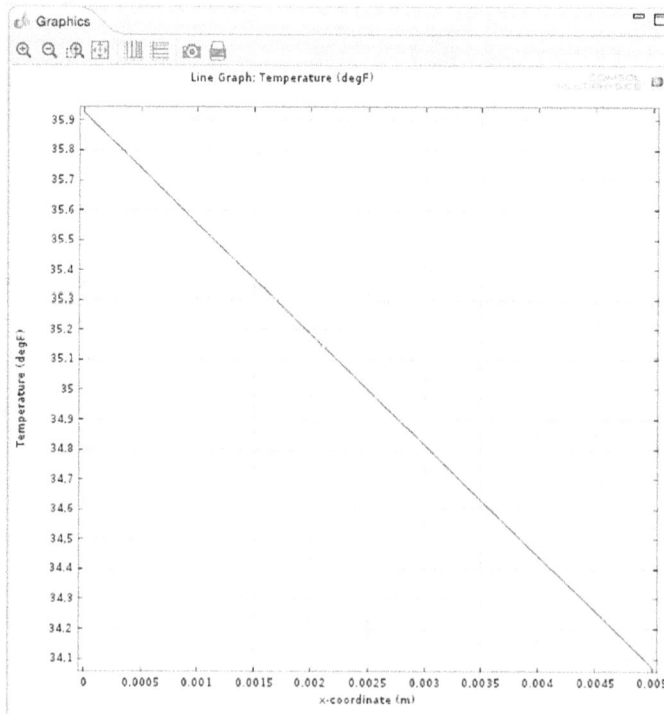

FIGURE 1.40 Calculated results for a 1 pane window in degrees Fahrenheit.

Figure 1.40 shows the Desktop Display Graphics window with the calculated results for a 1 Pane Window in degrees Fahrenheit.

The Modeler should, at this point, be sure to Click > Save and thus retain all the previous modeling work.

1D 1 Pane Window Model Analysis and Conclusions

TABLE 1.2 1 Pane Window Calculation Results.

Name	Coordinate Location (x)	Temperature
Interior Surface	0.0e–3[m]	≈35.926 °F
Midpoint	2.5e–3[m]	≈35.000 °F
Exterior Surface	5.0e–3[m]	≈34.074 °F
Interior ΔT	0.0e–3[m]	≈34.074 °F
Exterior ΔT	5.0e–3[m]	≈34.074 °F
Window ΔT	0.0–5.0e–3[m]	≈1.852 °F
Room ΔT	0.0 to –∞ [m]	≈34.074 °F
Exterior ΔT	5.0e–3 to ∞ [m]	≈34.074 °F

The modeler should notice that this calculation shows that for a 1 Pane Window, the temperature drop across the Pane is approximately 2 degrees Fahrenheit. It also shows that the Interior Surface will be perceptibly cold to the touch. The 1 Pane Window will, due to the temperature difference between the Interior Window Pane surface and the room air, cause condensation of the water vapor in the room on the surface.

1D 2 Pane Window Heat Flow Model

Now that the modeler has had an initial experience in model development, it is time to consider, for example, how the addition of a second Pane in series with the first Pane will alter the heat flow through the window, given the same environmental conditions. That comparison is relatively simple to implement in a 1D heat flow model. The following model is similar to the first model, but sufficiently different that the modeler should build it with care.

Run 5.x > Click the Model Wizard button > Select 1D in the Model Wizard.

Click the Heat Transfer twistie.

Select > Heat Transfer in Solids *(ht)* in the Add Physics window.

Click > Add button.

Click > Study (right pointing arrow).

Select General Studies > Stationary.

Click > Done button.

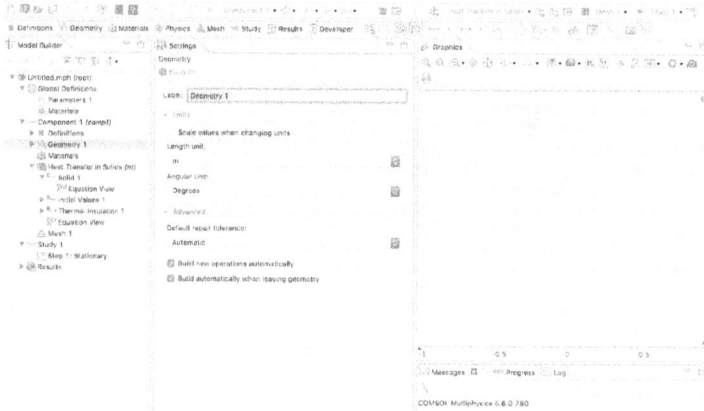

FIGURE 1.41 Initial 5.x model build for the 1D 2 pane window heat transfer.

Figure 1.41 shows the initial 5.x model build for the 1D 2 Pane Window Heat Transfer Desktop Display.

Save the Model as > MM2E5X_1D_W2Pane.mph. See Figure 1.42.

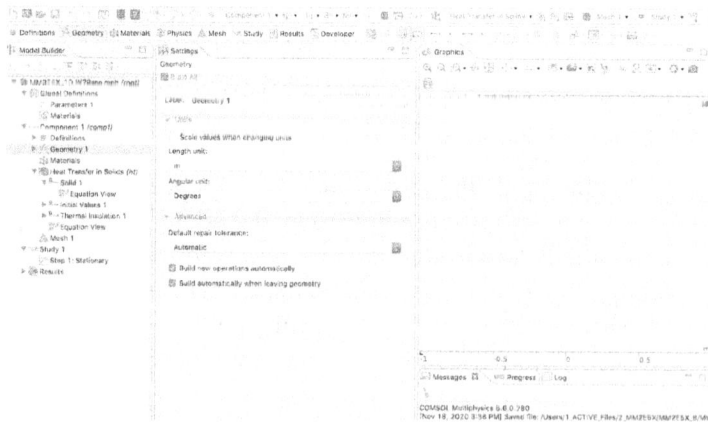

FIGURE 1.42 Saved initial model build MM2E5X_1D_W2Pane.mph.

Figure 1.42 shows the saved initial model build for MM2E5X_1D_W2Pane.mph.

Now that the model has been saved, the modeler needs to enter the constant values needed for this model.

In the Model Builder window of the Desktop Display,

Click > Global Definitions > Parameters 1.

Once the Parameters 1 menu item has been selected, the Desktop Display will be modified to incorporate a Settings window. The Settings window contains a Parameters Entry Table where the following attributes can be entered for each Parameter: Name, Expression, Value, and Description.

Parameter information can be entered directly into the fields in the Parameters Entry Table or into the parameter entry fields below the table. Control buttons are located between the Parameters Entry Table window and the parameter entry fields.

To enter the parameters, Click on the first entry window in the Name column and enter (type) each piece of the information, as shown in Table 1.3.

TABLE 1.3 1D 2 Pane Window Parameters.

Name	Expression	Description
T_in	70[degF]	Interior Temperature
T_out	0[degF]	Exterior Temperature
h_sg	15[W/(m^2*K)]	Heat Transfer Coefficient
p	1[atm]	Air Pressure

These entries in the Parameters Table define the interior temperature, the exterior temperature, the heat transfer coefficient, and the air pressure for use in this model.

Click on the Save to File icon (Disk), a symbol in the control button array, and enter MM2E5X_1D_W2Pane_Param.txt to save these parameters for future use.

The next step in building a model of a 2 Pane Window is to define the geometry through which the heat will flow from the heated interior to the less heated exterior.

Right-Click in the Model Builder window on Geometry 1 under Component 1 *(comp1)*.

Select > Interval.

In the Interval 1 Settings Window > Click on the Specify pull-down and select Interval lengths.

Enter 0.0 as the Left endpoint.

In the Interval 1 Settings Window > Click on the Length source pull-down and select Vector.

Enter 5e-3[m] in the Lengths field.

Click > Build Selected.

Right-Click in the Model Builder window on Geometry 1 under Component 1 *(comp1)*.

Select > Interval.

In the Interval 2 Settings Window > Click on the Specify pull-down and select Interval lengths.

Enter 5e-3[m] as the Left endpoint.

In the Interval 2 Settings Window > Click on the Length source pull-down and select Vector.

Enter 15e-3[m] in the Lengths field.

Click > Build All Objects.

Click > Zoom Extents (the square icon with the blue arrows in the Graphics Window Control Bar).

Right-Click in the Model Builder window on Geometry 1 under Component 1 *(comp1)*.

Select > Interval.

In the Interval 3 Settings Window > Click on the Specify pull-down and select Interval lengths.

Enter 20e-3[m] as the Left endpoint.

In the Interval 3 Settings Window > Click on the Length source pull-down and select Vector.

Enter 5e-3[m] in the Lengths field.

Click > Build All Objects.

Click > Zoom Extents.

See Figure 1.43.

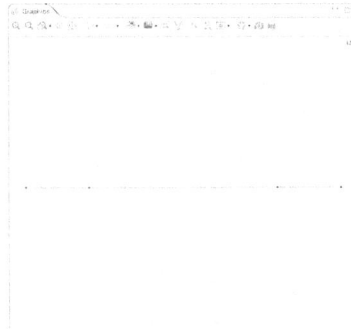

FIGURE 1.43 Desktop display window with the 2 Pane built geometry.

Figure 1.43 shows the Desktop Display window with the 2 Pane Built Geometry.

Now that the 2 Pane Geometry has been built, the next step is to define the materials comprising the model. In this case, it is relatively easy because the model only needs to use two (2) materials. Also, the material properties of those materials are resident in the Built-In Materials Library.

NOTE

The 2 Pane Window Model physically comprises 2 Panes of Silica glass separated by a "Dead Air Space." The dimensions of the "Dead Air Space" are such that convection currents are suppressed. Thus, thermal conduction plays the major role in the conduction of heat from the interior surface of the Window to the exterior surface of the Window.

Right-Click in the Model Builder window on Materials under Component 1 *(comp1).*

Select > Add Material from Library.

Click the twistie pointing at Built-In in the Add Material window.

Scroll the Materials List until Silica glass appears.

Right-Click > Silica glass.

Select > Add to Component 1. See Figure 1.44.

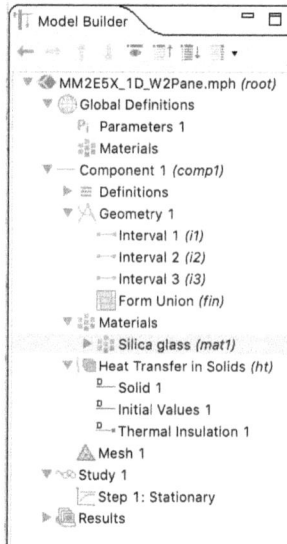

FIGURE 1.44 Silica glass material added to Component 1.

Figure 1.44 shows the Silica glass material has been added to the model.

NOTE

Because this model is 1D and there are only Domain and Boundary in this model, 5.x automatically assigns the Silica glass material's properties to all the available domains (1, 2, 3).

In this case, the materials assignment of Silica glass to Domain 2 is not correct and must be adjusted.

Click > 2 in the Settings – Material – Geometric Entity Selection window.

Click > Minus (-) button in the rightmost vertical control button array. See Figure 1.45.

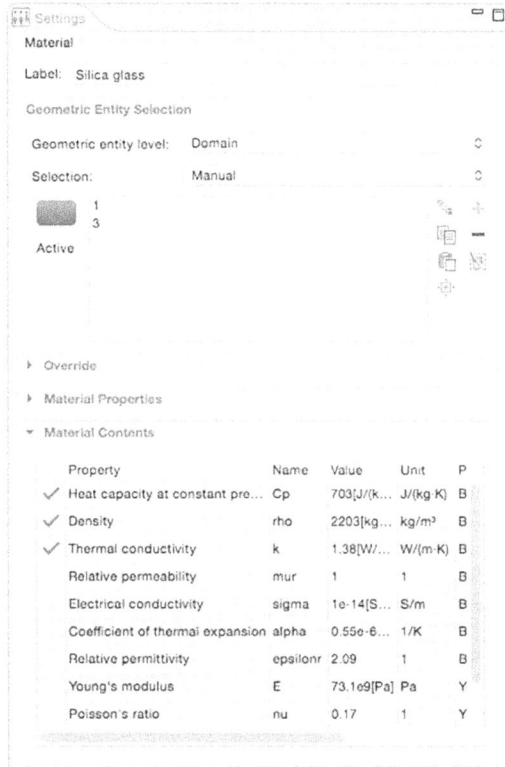

FIGURE 1.45 Silica glass material removed from Domain 2.

Figure 1.45 shows that the Silica glass material has been removed from Domain 2.

The next step is to add the material Air to the model.

Right-Click in the Model Builder window on Materials under Component 1 *(comp1)*.

Select > Add Material from Library.

Click the twistie (as needed) pointing at Built-In in the Add Material window.

Scroll the materials list until Air appears.

Right-Click > Air.

Select > Add Material to Component 1. See Figure 1.46.

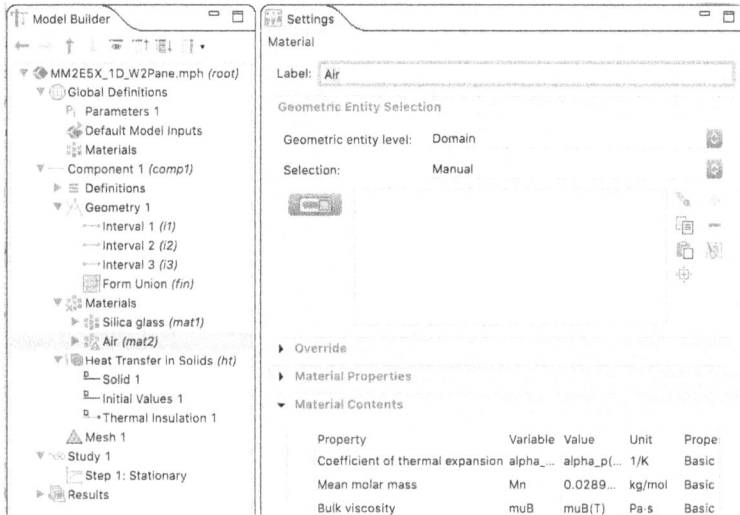

FIGURE 1.46 Material air added to Component 1.

Figure 1.46 shows the material Air has been added to the model.

NOTE *In this case, since there are multiple Domains, 5.x does not know where to assign the material Air. Thus the material Air is left unassigned.*

The materials assignment of Air to Domain 2 must be performed manually.

The Paste Selection operation is made available as follows:

Click > Clipboard Icon (text window appears).

NOTE *Enter:Required text (in this case, the number 2)*

Close text window.

Number appears in the Selection window.

In the Settings Material window, Paste > 2.

Click > Add Material X (close the add material window).

See Figure 1.47.

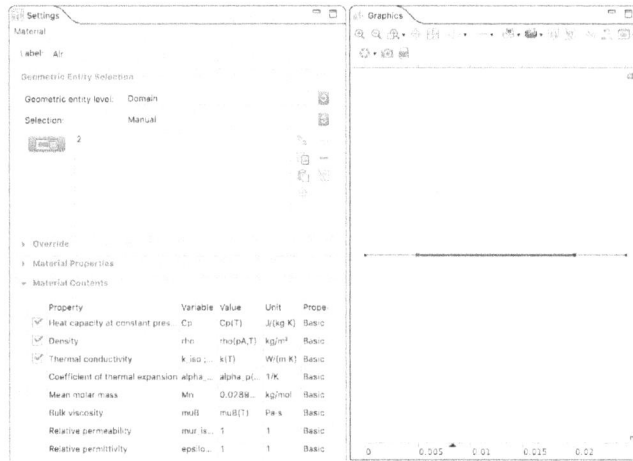

FIGURE 1.47 Material air added to Domain 2.

Figure 1.47 shows the material Air has been added to Domain 2.

NOTE *The modeler should verify that the material selected has the necessary physical properties to allow for the solution of the problem in question. In this case, 5.x has determined which physical properties are needed for the current model and displays a green checkmark next to the needed properties in the Material Contents window of the Settings - Material window. See Figure 1.48.*

Material Contents

Property	Variable	Value	Unit	Propei
✓ Heat capacity at constant pres...	Cp	Cp(T)	J/(kg·K)	Basic
✓ Density	rho	rho(pA,T)	kg/m³	Basic
✓ Thermal conductivity	k_iso ;...	k(T)	W/(m·K)	Basic
Coefficient of thermal expansion	alpha_...	alpha_p(...	1/K	Basic
Mean molar mass	Mn	0.0289...	kg/mol	Basic
Bulk viscosity	muB	muB(T)	Pa·s	Basic
Relative permeability	mur_is...	1	1	Basic
Relative permittivity	epsilo...	1	1	Basic
Dynamic viscosity	mu	eta(T)	Pa·s	Basic
Ratio of specific heats	gamma	1.4	1	Basic

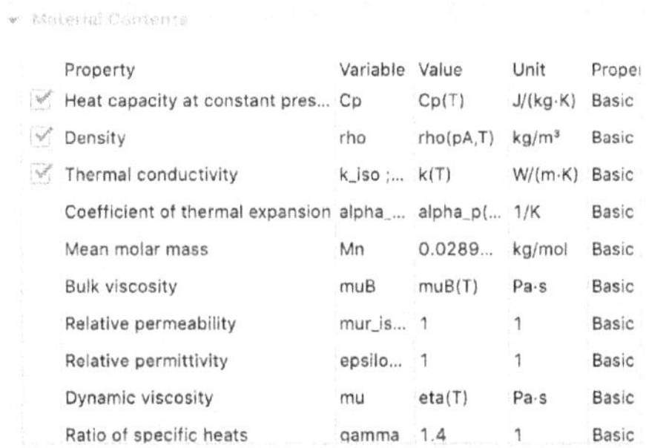

FIGURE 1.48 Material properties of air in Domain 2 verified.

Figure 1.48 shows the material properties of Air in Domain 2 have been verified.

Since the model geometry has been built and the appropriate material properties have been assigned to Domains 1, 2, and 3, the next step in the modeling process is to specify both the Domain and the Boundary conditions for this model.

Thus, the modeler needs to utilize the previously entered parameters that are needed to calculate the transfer of heat through this 2 Pane Window.

Right-Click > Heat Transfer in Solids *(ht)*.

Select > Heat Flux.

Click the Zoom Extents button on the Graphics window Toolbar.

In the Graphics window, Click > Point 1 (leftmost boundary).

In the Heat Flux Settings window in the Material Type,

Click > Material type pull-down

Select: From material

In the Settings – Heat Flux window,

Click > Convective heat flux radio button.

Type h_sg in the Heat transfer coefficient entry window.

Type T_in in the External temperature entry window.

Figure 1.49 shows the Settings - Heat Flux window as finally configured for Boundary Point 1.

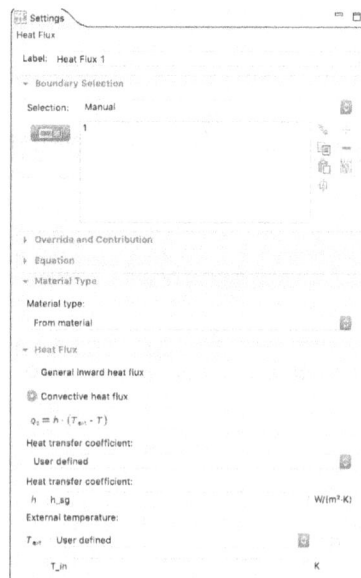

FIGURE 1.49 Filled Boundary Point 1 heat flux window.

Right-Click > Heat Transfer in Solids *(ht)*.

Select > Heat Flux.

Click Zoom Extents on the Graphics window Toolbar.

In the Graphics window, Click > Point 4 (rightmost boundary).

In the Heat Flux Settings window in the Material Type,

 Click > Material type pull-down,

Select: From material

In the Settings – Heat Flux window,

Click > Convective heat flux radio button.

Type h_sg in the Heat transfer coefficient entry window.

Type T_out in the External temperature entry window.

Figure 1.50 shows the Settings – Heat Flux window as finally configured for Boundary Point 4.

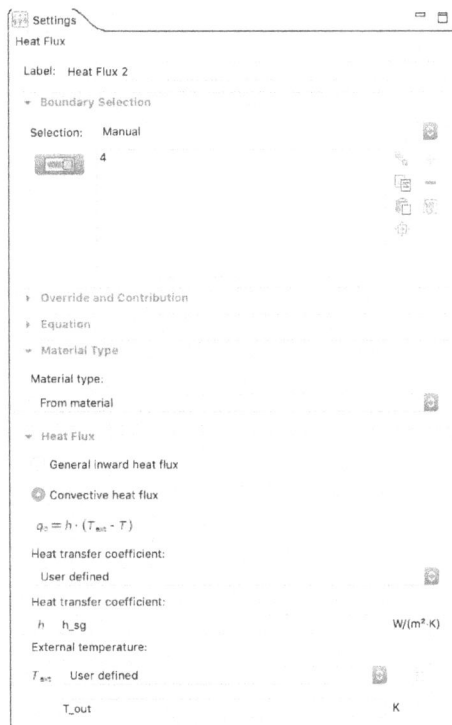

FIGURE 1.50 Filled Boundary Point 4 heat flux window.

Since this model is relatively simple, the easiest path is to allow 5.x to automatically mesh the model.

Click > Mesh 1.

Right-Click on the Blank Background of the Graphics window, Select Domains.

Select > All Domains in the Graphics window

Click > the Build All button in the Mesh Settings window Toolbar.

The meshing results are shown in Figure 1.51.

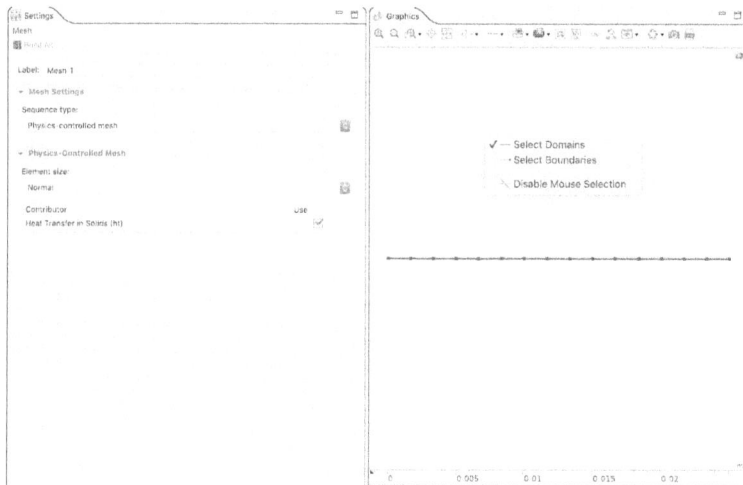

FIGURE 1.51 Meshed 1D domain.

Figure 1.51 shows the meshed 1D domain.

Now that the model has been meshed, the modeler can proceed to compute the solution.

Right-Click > Study 1.

Select > Compute.

NOTE   *5.x automatically selects the appropriate Solver for this problem, computes the solution and displays the results in the Graphics window.*

The results of this modeling computation are shown in Figure 1.52.

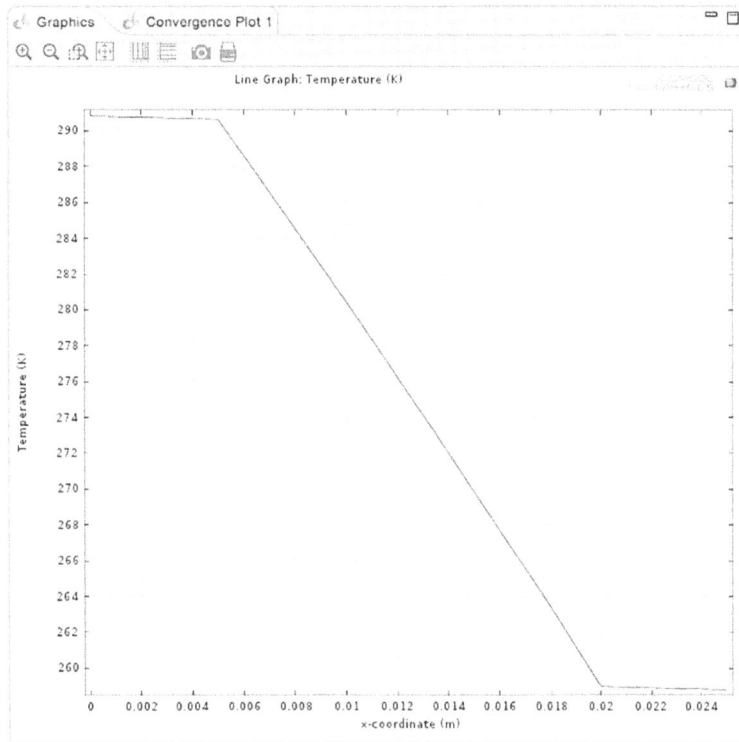

FIGURE 1.52 Initial calculated results for a 2 Pane window.

Figure 1.52 shows the initial Desktop Display window of the calculated results for a 2 Pane Window.

The 1D Plot displayed in the Desktop Display Graphics window shows the temperature in degrees Kelvin as a function of the distance from the inside surface of Silica glass of the First Pane of the 2 Pane Window to the outside surface of Silica glass of the second Pane of the 2 Pane Window.

Since the modeler originally specified the initial input parameters in degrees Fahrenheit, the modeler now needs to adjust the instructions of 5.x so that the Graphics window will display the results of this modeling calculation in the desired units.

Under Results, Click > Temperature (ht) twistie (as needed).

Click > Line Graph.

In the Settings – Line Graph window,

Under Y-Axis Data, Click > Unit pull-down menu > Select > degF.

Click > Plot.

Figure 1.53 shows the Desktop Display window of the calculated results for a 2 Pane Window in degrees Fahrenheit.

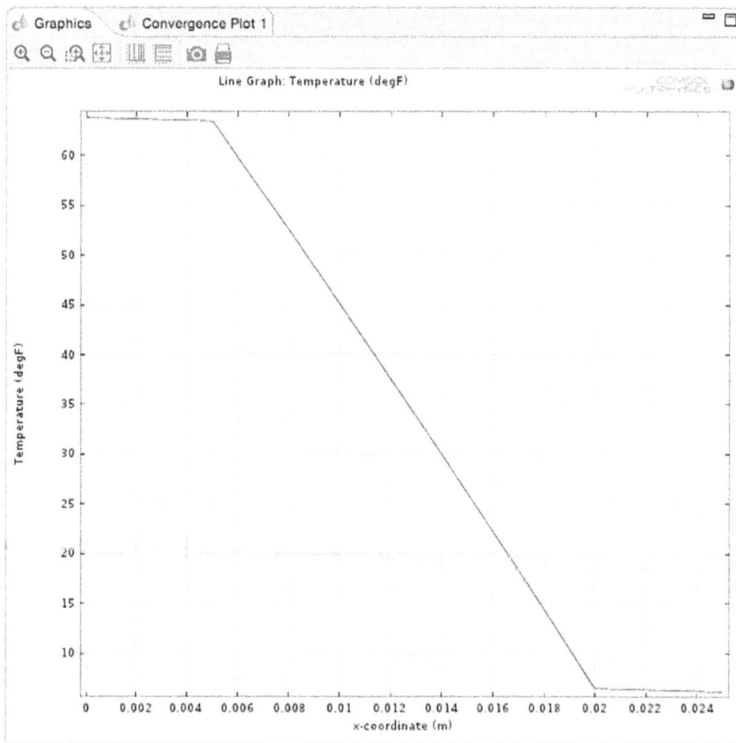

FIGURE 1.53 Calculated results for a 2 Pane window in degrees Fahrenheit.

1D 2 Pane Window Model Analysis and Conclusions

TABLE 1.4 Pane Window Calculation Results.

Name	Coordinate Location (x)	Temperature
Interior Surface	0.0e–3[m]	≈63.84 °F
Midsurface 1	5.0e–3[m]	≈63.50 °F
Midsurface 2	20.0e–3[m]	≈6.50 °F
Exterior Surface	25.0e–3[m]	≈6.16 °F
Midpoint	12.5e–3[m]	≈35.000 °F
Interior Pane ΔT	0.0e–3[m]	≈0.34 °F
Exterior Pane ΔT	5.0e–3[m]	≈0.34 °F
Window ΔT	0.0–5.0e–3[m]	≈57.68 °F
Room ΔT	0.0 to –∞ [m]	≈6.16 °F
Exterior ΔT	5.0e–3 to ∞ [m]	≈6.16 °F

The modeler should notice this calculation shows that for a 2 Pane Window, the temperature drop across the 2 Panes in this configuration is approximately 58 degrees Fahrenheit. It also shows that the interior surface will be perceptibly cool but not cold to the touch. The 2 Pane Window will, due to the small temperature difference between the interior Window Pane surface and the room air, not cause condensation of the water vapor in the room on the surface, except under highly saturated (high relative humidity) conditions.

1D 3 Pane Window Heat Flow Model

Now that the modeler has had some experience in model development, it is time to consider, for example, how the addition of a third Pane in series with the first two Panes will alter the heat flow through the window, given the same environmental conditions. That comparison is relatively simple to implement in a 1D heat flow model. The following model is similar to the first and second models, but sufficiently different that the modeler should build it with care.

Run 5.x > Click the Model Wizard button,

Select: 1D in the Model Wizard.

Click the Heat Transfer twistie.

Select > Heat Transfer in Solids (*ht*) in the Add Physics window.

Click > Add button.

Click > Study (right pointing arrow).

Select > General Studies > Stationary.

Click > Done button.

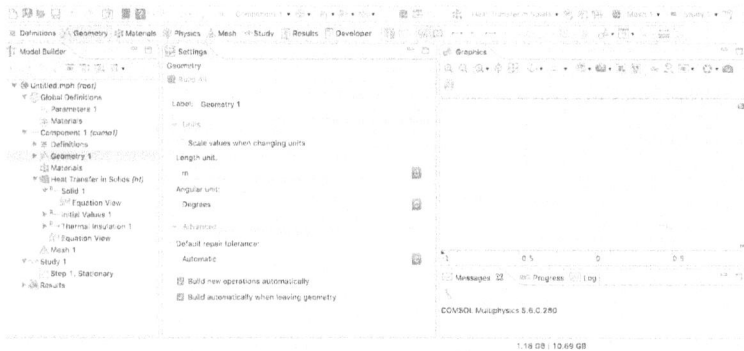

FIGURE 1.54 Initial 5.x model build for the 1D 3 Pane window heat transfer.

Figure 1.54 shows the initial 5.x model build for the 1D 3 Pane Window Heat Transfer Desktop Display.

Save the Model as > MM2E5X_1D_W3Pane.mph. See Figure 1.55.

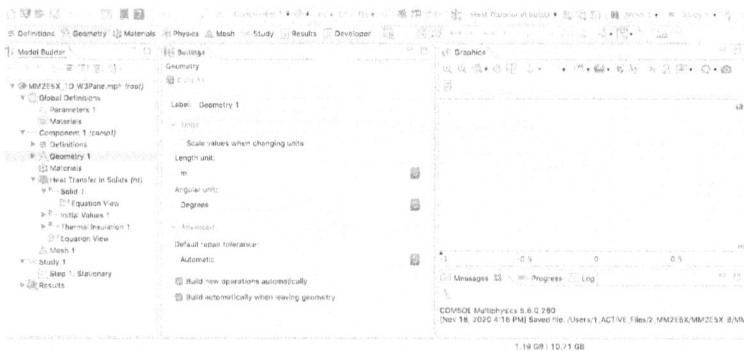

FIGURE 1.55 Saved initial model build MM2E5X_1D_W3Pane.mph.

Figure 1.55 shows the Saved Initial Model Build MM2E5X_1D_W3Pane. mph.

Now that the model has been saved, the modeler needs to enter the constant values needed for this model.

In the Model Builder window of the Desktop Display,

Click > Global Definitions > Parameters 1.

To enter the parameters, Click on the first entry window in the Name column and enter (type) each piece of the information, as shown in Table 1.5.

TABLE 1.5 1D 3 Pane Window Parameters.

Name	Expression	Description
T_in	70[degF]	Interior Temperature
T_out	0[degF]	Exterior Temperature
h_sg	15[W/(m^2*K)]	Heat Transfer Coefficient
p	1[atm]	Air Pressure

These entries in the Parameters Entry Table define the interior temperature, the exterior temperature, the heat transfer coefficient, and the air pressure for use in this model.

Click on the Save to File (Disk) icon, in the control button array, and enter MM2E5X_1D_W3Pane_Param.txt to save these parameters for future use.

The next step in building a model of this 3 Pane Window is to define the geometry through which the heat will flow from the heated interior to the less heated exterior.

Right-Click in the Model Builder window on Geometry 1 under Component 1 *(comp1)*.

Select > Interval.

In the Interval 1 Settings Window > Click on the Specify pull-down and select Interval lengths.

Enter 0.0 as the Left endpoint.

In the Interval 1 Settings Window > Click on the Length source pull-down and select Vector.

Enter 5e-3[m] in the Lengths field.

Click > Build Selected.

Right-Click in the Model Builder window on Geometry 1 under Component 1 *(comp1)*.

Select > Interval.

In the Interval 2 Settings Window > Click on the Specify pull-down and select Interval lengths.

Enter 5e-3[m] as the Left endpoint.

In the Interval 2 Settings Window > Click on the Length source pull-down and select Vector.

Enter 15e-3[m] in the Lengths field.

Click > Build Selected.

Click > Zoom Extents. See Figure 1.56.

FIGURE 1.56 Zoom extents button.

Figure 1.56 shows the Zoom Extents button, which changes the dimensions of the axes in the Graphics window to reflect the value range of the geometry.

Right-Click in the Model Builder window on Geometry 1 under Component 1 *(comp1)*.

Select > Interval.

In the Interval 3 Settings Window > Click on the Specify pull-down and select Interval lengths.

Enter 20e-3[m] as the Left endpoint.

In the Interval 3 Settings Window > Click on the Length source pull-down and select Vector.

Enter 5e-3[m] in the Lengths field.

Click > Build Selected.

Click > Zoom Extents.

Right-Click in the Model Builder window on Geometry 1 under Component 1 *(comp1)*.

Select > Interval.

In the Interval 4 Settings Window > Click on the Specify pull-down and select Interval lengths.

Enter 25e-3[m] as the Left endpoint.

In the Interval 4 Settings Window > Click on the Length source pull-down and select Vector.

Enter 15e-3[m] in the Lengths field.

Click > Build Selected.

Click > Zoom Extents.

Right-Click in the Model Builder window on Geometry 1 under Component 1 *(comp1)*.

Select > Interval.

In the Interval 5 Settings Window > Click on the Specify pull-down and select Interval lengths.

Enter 40e-3[m] as the Left endpoint.

In the Interval 5 Settings Window > Click on the Length source pull-down and select Vector.

Enter 5e-3[m] in the Lengths field.

Click > Build All Objects.

Click > Zoom Extents.

Now that the 3 Pane (5 Interval) Geometry has been built, the next step is to define the materials comprising the Model. In this case, it is relatively easy because the model only needs to use two (2) materials. Also, the material properties of those materials are resident in the Built-In Materials Library.

NOTE

The 3 Pane Window Model physically comprises 3 Panes of Silica Glass separated by 2 "Dead Air Spaces." The dimensions of the "Dead Air Spaces" are such that convection currents are suppressed. Thus, thermal conduction plays the major role in the conduction of heat from the interior surface of the Window to the exterior surface of the Window.

Right-Click in the Model Builder window on Materials under Component 1 *(comp1)*.

Select > Add Material from Library.

Click the twistie pointing at Built-In in the Add Material window.

Scroll the materials list until Silica glass appears.

Right-Click > Silica glass.

Select > Add to Component 1. See Figure 1.57a.

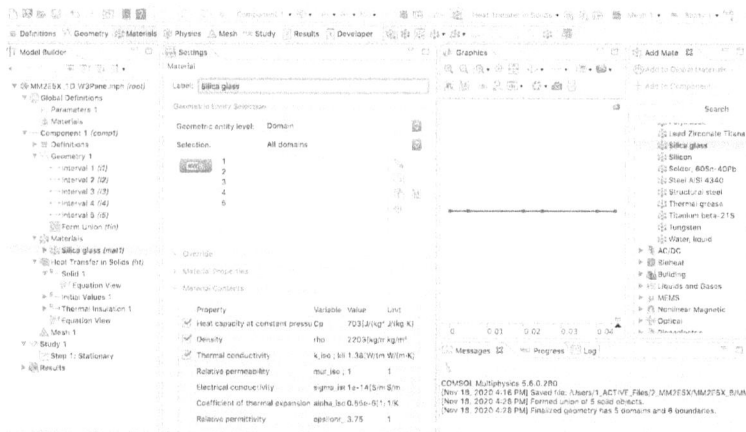

FIGURE 1.57(a) Silica glass material added to 3 Pane model.

Figure 1.57a shows the Silica glass material has been added to the 3 Pane Model.

NOTE

Since this model is 1D, there are 5 Domains and this is the first material added, 5.x automatically assigns the Silica glass material's properties to all the available domains (1, 2, 3, 4, 5).

In this case, the materials assignment of Silica glass to Domains 2 & 4 is not correct and must be adjusted.

Click > 2 in the Settings – Material – Geometric Entity Selection window.

Click > Minus (-) button in the rightmost vertical control button array.

Click > 4 in the Settings – Material – Geometric Entity Selection window.

Click > Minus (-) button in the rightmost vertical control button array.

The next step is to add the material Air to the model.

Scroll the materials list until Air appears.

Right-Click > Air.

Select > Add to Component 1.

Click > Add Material X (Close the Add Material window)

In this case, since there are multiple Domains, 5.x does not know where to assign the material Air. Thus, the material Air is left unassigned.

In this case, the materials assignment of Air to Domains 2 & 4 must be performed manually.

Right-Click > On the Blank Background of the Graphics window > Choose > New Domain Selection

See Figure 1.57b.

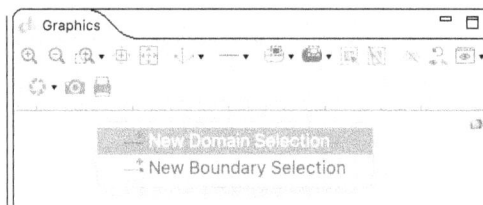

FIGURE 1.57(b) Domain selection pop-up window.

Figure 1.57b shows the Domain Selection pop-up window.

In the Graphics window, Click > Domain 2.

Next in the Graphics window, Click > Domain 4.

In the Settings Material > Geometric Entity Selection > Click > the Plus (+) button.

See Figure 1.57c.

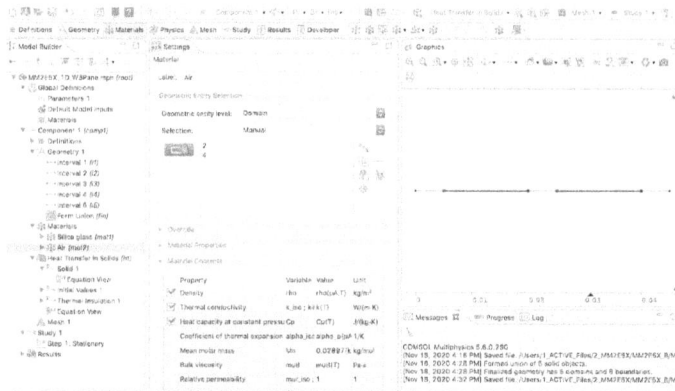

FIGURE 1.57(c) Domains 2 and 4 with air material.

Figure 1.57c shows the Graphics window with Air Material.

Since the model geometry has been built and the appropriate material properties have been assigned to Domains 1, 2, 3, 4, & 5, the next step in the modeling process is to specify both the domain and the boundary conditions for this model.

Thus, the modeler needs to utilize the previously entered parameters that are needed to calculate the transfer of heat through this 3 Pane Window.

Right-Click > Heat Transfer in Solids *(ht)*.

Select > Heat Flux.

Click the Zoom Extents button on the Graphics window Toolbar.

In the Graphics window, Click > Point 1 (leftmost boundary).

In the Heat Flux Settings window in the Material Type, Click > Material type pull-down > Select > From material

In the Settings – Heat Flux window, Click > Convective heat flux radio button.

Type h_sg in the Heat transfer coefficient entry window.

Type T_in in the External temperature entry window.

Right-Click > Heat Transfer in Solids *(ht)*.

Select > Heat Flux.

In the Graphics window, Click > Point 6 (rightmost boundary).

In the Heat Flux Settings window in the Material Type, Click > Material type pull-down > Select > From material

In the Settings – Heat Flux window, Click > Convective heat flux radio button.

Type h_sg in the Heat transfer coefficient entry window.

Type T_out in the External temperature entry window.

Since this model is relatively simple, the easiest path is to allow 5.x to automatically mesh the model.

Right-Click > Graphics Window Background.

Select > New Domain Selection.

Hold down the shift key > Select all Domains (1-5) in the Graphics window > Release Shift Key

Right-Click > Mesh 1.

Select > Build All.

The meshing results are shown in Figure 1.58.

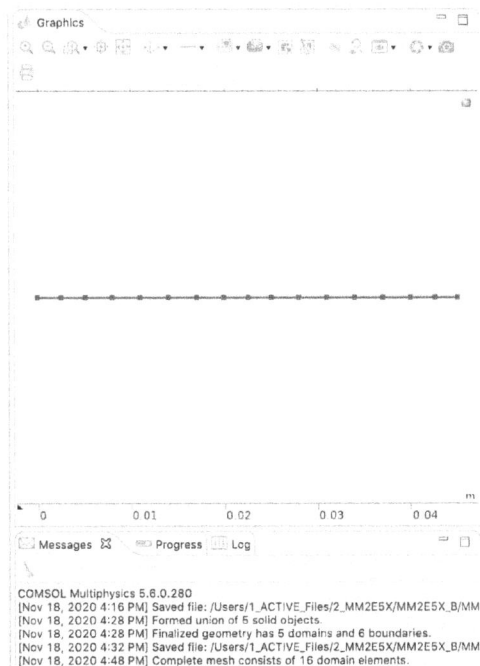

FIGURE 1.58 Meshed 1D 3 Pane, 5 Domain window model.

Figure 1.58 shows the meshed 1D 3 Pane, 5 Domain Window model.

Now that the model has been meshed, the modeler can proceed to compute the solution.

Right-Click > Study 1.

Select > Compute.

NOTE
5.x automatically selects the appropriate Solver for this problem, computes the solution, and displays the results in the Graphics window.

The results of this modeling computation are shown in Figure 1.59.

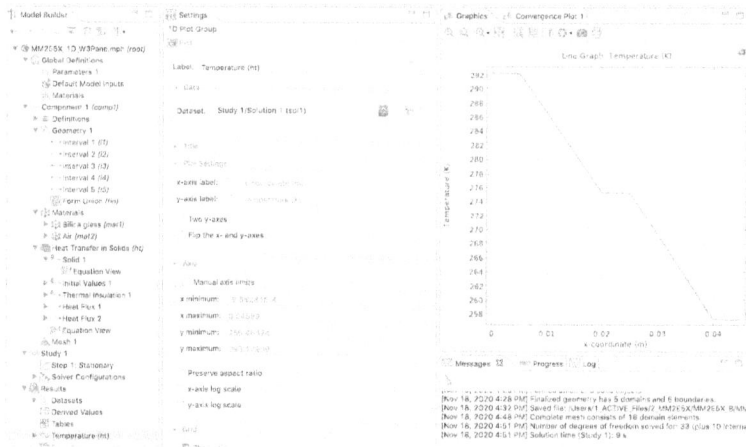

FIGURE 1.59 Initial calculated results for a 3 Pane window.

Figure 1.59 shows the Desktop Display window of the initial calculated results for a 3 Pane Window.

The 1D Plot displayed in the Desktop Display Graphics window shows the temperature in degrees Kelvin as a function of the distance from the inside surface of Silica glass of the first Pane of the 3 Pane Window to the outside surface of Silica glass of the third Pane of the 3 Pane Window.

Since the modeler originally specified the initial input parameters in degrees Fahrenheit, the modeler now needs to adjust the instructions of 5.x so that the Graphics window will display the results of this modeling calculation in the desired units.

Under Results, Click > Temperature (ht) twistie (as needed).

Click > Line Graph.

In the Settings – Line Graph window.

Under Y-Axis Data, Click > Unit pull-down menu > Select > degF.

Click > Plot.

Figure 1.60 shows the Desktop Display window of the Calculated Results for a 3 Pane Window in degrees Fahrenheit.

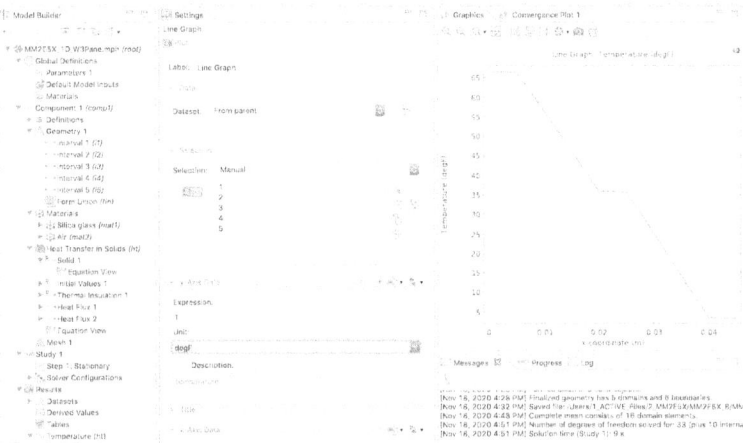

FIGURE 1.60 Calculated results for a 3 Pane window in degrees Fahrenheit.

1D 3 Pane Window Model Analysis and Conclusions

TABLE 1.6 Pane Window Calculation Results.

Name	Coordinate Location (x)	Temperature
Interior Surface	0.0e-3 [m]	≈66.62 °F
Midsurface 1	5.0e-3 [m]	≈66.43 °F
Midsurface 2	20.0e-3 [m]	≈35.96 °F
Midsurface 3	25.0e-3 [m]	≈35.77 °F
Midsurface 4	40.0e-3 [m]	≈3.57 °F
Exterior Surface	45.0e-3 [m]	≈3.38 °F
Midpoint	22.5e-3 [m]	≈35.87 °F

Name	Coordinate Location (x)	Temperature
Interior Pane ΔT	0.0e-3 [m]	≈0.19 °F
Middle Pane ΔT	20.0e-3 [m]	≈0.19 °F
Exterior Pane ΔT	40.0e-3 [m]	≈0.19 °F
Window ΔT	0.0–45.0e-3 [m]	≈63.24 °F
Room ΔT	0.0 to -∞ [m]	≈3.38 °F
Exterior ΔT	45.0e-3 to ∞ [m]	≈3.38 °F

The modeler should notice that this calculation shows that for a 3 Pane Window, the temperature drop across the 3 Panes in this configuration is approximately 63 °F. It also shows that the interior surface will be perceptibly cool but not cold to the touch. The 3 Pane Window will, due to the small temperature difference between the interior Window Pane surface and the room air, not cause condensation of the water vapor in the room on the surface, except under highly saturated (high relative humidity) conditions.

1D Window Model 1 Pane, 2 Pane and 3 Pane Analysis and Conclusions

TABLE 1.7 All Window Calculation Results.

Name	Window Δ Temperature
1 Pane	1.85 °F
2 Pane	57.68 °F
3 Pane	63.24 °F

It can be readily seen that the 2 Pane Window adds significant advantage in heat loss reduction and reduces condensation (fogging) on the window. The incremental advantage compared to the added cost of adding the third Pane is probably not a good relative investment. Other factors would need to be considered to derive significant perceived benefit.

FIRST PRINCIPLES AS APPLIED TO MODEL DEFINITION

First Principles Analysis derives from the fundamental laws of nature. In the case of models in this book or from any other source, the modeler should be able to demonstrate that the calculated results derived from the models are consistent with the laws of physics and the basic observed properties of materials. In the case of this Classical Physics Analysis, the laws of conservation in physics require that what goes in (as mass, energy, charge, etc.) must come out (as mass, energy, charge, etc.) or must accumulate within the boundaries of the model. To do otherwise violates fundamental principles.

NOTE *In the COMSOL Multiphysics software, the default interior boundary conditions are set to apply the conditions of continuity in the absence of sources (e.g., heat generation, charge generation, molecule generation, etc.) or sinks (e.g., heat loss, charge recombination, molecule loss, etc.).*

The careful modeler must be able to determine by inspection of the model that the appropriate factors have been considered in the development of the specifications for the various geometries, for the material properties of each domain and for the boundary conditions. He must also be knowledgeable of the implicit assumptions and default specifications that are normally incorporated into the COMSOL Multiphysics software model, when a model is built using the default settings.

Consider, for example, the three Window models developed earlier in this chapter. By choosing to develop those models in the simplest 1D geometrical mode, the implicit assumption is the heat flow occurs in only one direction. That direction is basically normal to the surface of the window and from the high temperature (interior temperature) to the low temperature (exterior temperature). That assumption essentially eliminates the consideration of heat flow along other paths. It also assumes the materials are homogeneous and isotropic and there are no thin thermal barriers at the surfaces of the panes. None of these assumptions are typically true in the general case. However, by making such assumptions, it is possible to easily build a 1D First Approximation Model.

NOTE *A First Approximation Model is one that captures all the essential features of the problem that needs to be solved, without dwelling excessively on small details. A good First Approximation Model will yield an answer that enables the modeler to determine if he needs to invest the time and the resources required to build a more highly detailed model.*

SOME COMMON SOURCES OF MODELING ERRORS

There are four primary sources of modeling errors: insufficient preparation, insufficient attention to detail, insufficient understanding of the basic principles required for the creation of an adequate model, and the lack of a comprehensive understanding of what is required to define an adequate model to fulfill the needs.

Primarily, the most frequent modeling errors are those that result from the modeler exercising insufficient attention to either the development of the details of the model or the incorporation into the model of conceptual errors and/or the generation of keying errors during data/parameter/formula entry.

One major source of errors occurs during the process of naming variables. The modeler should be careful to NEVER GIVE THE SAME NAME TO HIS VARIABLES AS COMSOL GIVES TO THE DEFAULT VARIABLES. COMSOL Multiphysics software seeks a value for the designated variable everywhere within its operating domain.

NOTE

If two or more variables have the same name, an error is created. Also, it is best to avoid human errors by using uniquely distinguishable characters in variable names. For example, avoid using the lower case L, the number 1, and the upper case I, which in some fonts are relatively indistinguishable. Similarly, avoid the upper case O and the number 0. Give your variables meaningful names (T_in, T_out, T_hot, etc.). Also, variable names are case-sensitive, i.e. T_in is not the same as T_IN.

The first rule in model development is to clearly understand the exact problem to be solved and to specify in detail what aspects of the problem the model will address. The specification of the nature of the problem should include a list of the magnitude of the relative contributions from the physical properties vital to the functioning of the anticipated model and their relative degree of interaction. Always build the model to solve for the minimum desired information. It will be faster and the model can always be expanded later.

NOTE

Examples of typical physical properties that are potentially/probably coupled in any developed model are heat and geometrical expansion/contraction (liquid, gas, solid), current flow and heat generation/reduction, phase change and geometrical expansion/contraction (liquid, gas, solid) and/or heat generation/reduction, chemical reactions, etc. Be sure to investigate your problem and understand what you are building first, then build your model carefully.

Having done the primary analysis and written a hierarchical list, the modeler should then estimate the best physical, least coupled, lowest dimensionality modeling approach to achieve the most meaningful First Approximation Model.

REFERENCES

1. Blaise Barney, Lawrence Livermore National Laboratory, Introduction to Parallel Computing, *https://computing.llnl.gov/tutorials/parallel_comp/*

2. *https://www.apple.com/mac-pro/*

3. COMSOL Multiphysics Reference Manual, p. 59

4. COMSOL Multiphysics Reference Manual, p. 478

5. *http://www-03.ibm.com/support/techdocs/atsmastr.nsf/Web/TwHelp*

6. COMSOL Multiphysics Reference Manual, p. 156

7. F. P. Incropera and D. P. Dewitt, *Fundamentals of Heat and Mass Transfer*, Fifth Edition, John Wiley & Sons, Hoboken, NJ, 2002, ISBN 0-471-38650-2, p. 8

8. D. W. Pepper and J. C. Heinrich, *The Finite Element Method*, Second Edition, Taylor & Francis, New York, 2006, ISBN 1-59169-027-7, p. 1

9. COMSOL Multiphysics Reference Manual, p. 464

SUGGESTED MODELING EXERCISES

1. Build, mesh and solve the 1D 1 Pane Window problem presented earlier in this chapter.

2. Build, mesh and solve the 1D 2 Pane Window problem presented earlier in this chapter.

3. Build, mesh and solve the 1D 3 Pane Window problem presented earlier in this chapter.

4. Change the material comprising the Panes, then build, mesh and solve the problems. Analyze, compare and contrast the results with the new material to the results of Problems 1, 2, and 3.

5. Change the material (Gas) between the Panes, then build, mesh and solve the problems. Analyze, compare and contrast the results with the new material to the results of Problems 1, 2, and 3.

CHAPTER 2

MATERIALS PROPERTIES USING COMSOL MULTIPHYSICS 5.x

In This Chapter

- Materials Properties Guidelines and Considerations
- COMSOL Materials Properties Sources
- Other Materials Properties Sources
- Material Property Entry Techniques
    - Multi-Pane Window Model
- References

MATERIALS PROPERTIES GUIDELINES AND CONSIDERATIONS

The selection of materials for a given device or process is crucial to the success of the model for that device or process. When that model reflects the content and behavior of the actual device or process, then the conclusions drawn from the model will be most reliable. Once the first approximation of the device or process has been defined, begin choosing the materials.

NOTE *The modeler does not need to describe all of the properties of the material. Once the essential physical functions have been determined, only those properties that are used by the model's physics interfaces need to be supplied.*

A large number of materials and their properties are available, as general searches of the Internet will demonstrate. The modeler will need to exercise some selection criteria based on practical considerations to narrow the material choices to a manageable number.

The search for reliable and accurate material property values may encounter one or more of the following issues: 1) a source may not have all of the properties needed; 2) a source may supply the value in an unconventional, or at least not the desired, unit of measure, requiring conversion; 3) once in the desired units, the property value may prove to be incorrect based on a knowledge of values for other almost identical materials; and 4) errors may occur during transcription from the source into the model. Care must be taken to verify all property values before running the model.

This chapter will discuss the two COMSOL material property libraries available in 5.x. This chapter also covers other sources of property values, how to introduce those values into a model, and how to create your own User Defined Library of materials properties with 5.x.

COMSOL MATERIALS PROPERTIES SOURCES

COMSOL provides two sources of materials properties: a basic library that is included in the COMSOL Multiphysics software, and a Material Library module that provides many more materials and referenced property functions, related to some variable, such as temperature {2.1}. Access to both of these sources is available through the Material Browser and/or the Add Material window as was used in the models in Chapter 1. Use these tools to explore whether the materials needed for your model are already available in the COMSOL Multiphysics Materials Library.

Run 5.x > from the COMSOL Multiphysics Desktop Toolbar, Select > Windows > Material Browser, to display the contents of the Materials Library.

In the Materials search window, Enter > steel in the Material Browser.

The search term may be the name of a material, or in the case of the Material Library module, a UNS number {2.2} or a DIN number {2.3}.

Click > Search.

The results of a successful search will be displayed in the Materials section of the Material Browser window. See Figure 2.1.

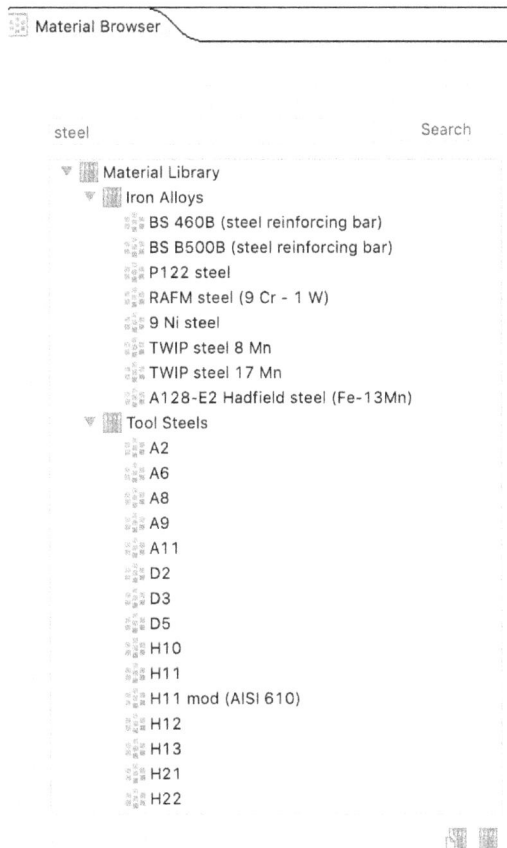

FIGURE 2.1 Material browser search results.

Figure 2.1 shows the results of a Material Browser search.

To view the properties available for one of the selected materials,

Scroll > Built-in (Built-in may be at the bottom of the Scroll range).

Click > Built-In twistie.

Scroll > Structural Steel (if needed)

Select > Structural steel.

In the Material Browser Display window, Material: Structural steel Properties are displayed.

A list of properties and their values are displayed. See Figure 2.2.

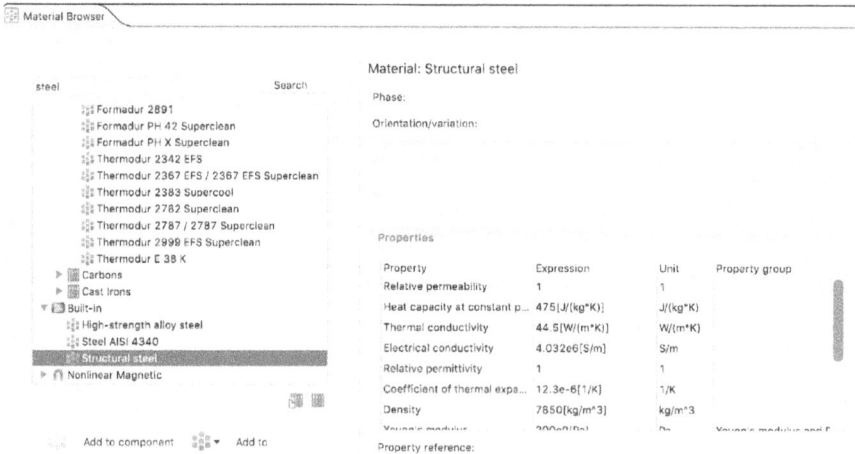

FIGURE 2.2 Material properties list.

Figure 2.2 shows the list of available properties for the selected material, Structural steel.

Quit COMSOL Multiphysics.

OTHER MATERIALS PROPERTIES SOURCES

Materials and their properties are available from many sources, either in printed form or on the Internet.

Manufacturers provide such information for their products in their catalogs.

Some professional societies, such as ASM International {2.4}, provide subscriptions to handbooks of materials information, including material properties.

Online encyclopedias, such as Wikipedia {2.5}, may contain data for some of the more common materials.

Specialized sites, such as MatWeb {2.6}, provide searchable online databases of materials properties, which may be downloadable in a format compatible with particular versions of modeling software.

Some search engines, both general purpose Internet tools and specialized tools designed to be used with a particular website, will allow the modeler to search by a property value range or material type, thus assisting the modeler in finding materials that fit the design criteria.

MATERIAL PROPERTY ENTRY TECHNIQUES

The modeler enters the materials properties data into the model using one of the three techniques: building a material property entry, setting user-defined parameters, or entering user-defined property values directly.

This chapter illustrates the three techniques by means of a model similar to the 1D 3 Pane model in Chapter 1, using the properties of materials not found in the basic COMSOL Multiphysics materials library.

Multipane Window Model

NOTE *The Multipane Window Model physically comprises 3 panes of Vycor silica glass separated by 2 "Dead Air Spaces," one containing argon and the other krypton. The dimensions of the "Dead Air Spaces" are such that convection currents are suppressed. Thus, thermal conduction plays the major role in the conduction of heat from the interior surface of the Window to the exterior surface of the Window.*

Run 5.x > Click the Model Wizard button > Select 1D in the Model Wizard – Space Dimension window.

To display the list of available physics interfaces,

In the Add Physics window, Click > Heat Transfer twistie.

Select > Heat Transfer in Solids (ht).

Click > Add button below the physics interface list.

The Heat Transfer (ht) selection will be displayed in the Selected physics section of the Add Physics window (see Figure 2.3).

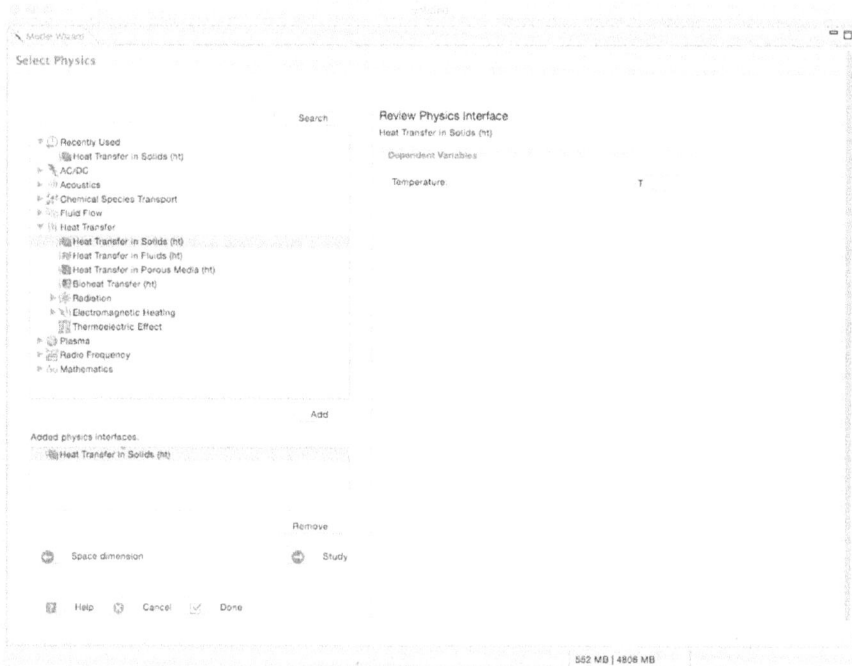

FIGURE 2.3 Model Wizard – Add Physics window.

Figure 2.3 shows the Model Wizard – Add Physics window.

In the Add Physics window, Click > Study (right-pointing arrow).

Select > General Studies > Stationary (see Figure 2.4).

NOTE *The modeler should note that "Stationary" and "Steady State" are equivalent terms.*

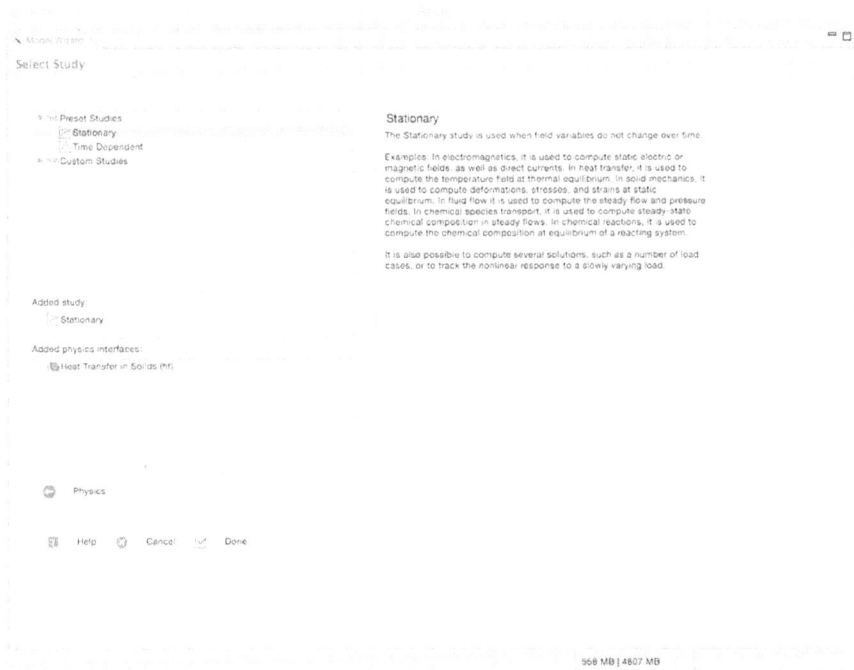

FIGURE 2.4 The select study type window.

Figure 2.4 shows the selection of the Stationary study type and the Done button in the Select Study Type window for the completion of Model Wizard process, resulting in the Initial Model Build.

At the bottom of the Select Study Type window, Click > Done button.

Click > File menu > Save as… to save the Model as > MM2E5X_1D_Multipane.mph. See Figure 2.5.

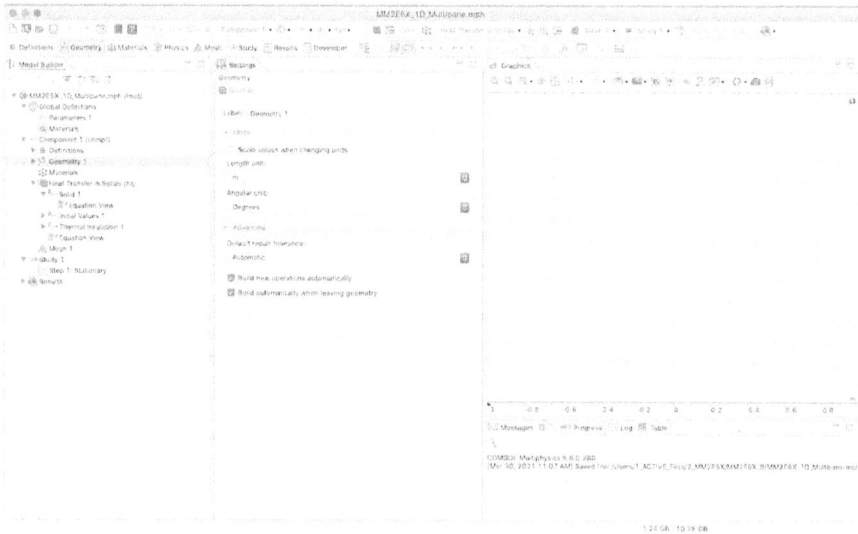

FIGURE 2.5 Saved initial model build of the 1D_Multipane model.

Figure 2.5 shows the saved Initial Model Build of the 1D_Multipane model.

Parameter Setting

One way to enter a material property into a model is to provide it as a parameter, along with the other constant values needed for a model. For example, in this model the properties for the Vycor silica glass to be used in the panes will be entered using this technique. This approach allows the modeler to save the property along with any other constants needed for the model in one file, but it does mean that the property parameter names will need to be entered manually wherever they will be used in the model.

NOTE

Care must be taken to ensure that the name assigned to the material property parameter is unique to the model and is not one of the parameter names already in use by 5.x. An error message will warn the modeler if the name is already defined when entry is attempted.

Now that the Initial Model Build has been saved, the modeler needs to enter the constant values needed for this model.

In the Model Builder window of the Desktop Display,

Click > Global Definitions > Parameter 1.

This action will display the Settings – Parameters 1 window.

To enter the parameters for the model,

Click on the first entry window in the Name column and enter (type) each piece of the information, as shown in Table 2.1.

NOTE

When entering information into a Parameters field, first Click on the field desired and then observe the presence of a blinking cursor. The blinking cursor verifies that 5.x is ready to accept the information. Then type the desired piece of information into the activated field.

The modeler can transition to the next information entry point in the tables by clicking on the next field.

TABLE 2.1 1D Multipane Window Parameters.

Name	Expression	Description
T_in	70[degF]	Interior Temperature
T_out	0[degF]	Exterior Temperature
h_sg	15[W/(m^2*K)]	Heat Transfer Coefficient
p	1[atm]	Air Pressure
kVycor	1.38[W/(m*K)]	Thermal Conductivity of Vycor
CVycor	753.624[J/(kg*K)]	Heat Capacity of Vycor
rhoVycor	2180[kg/m^3]	Density of Vycor

These entries in the Parameters Entry Table define the interior temperature, the exterior temperature, the heat transfer coefficient, the air pressure, and the thermal conductivity, heat capacity, and density of the Vycor glass for use in this model. See Figure 2.6.

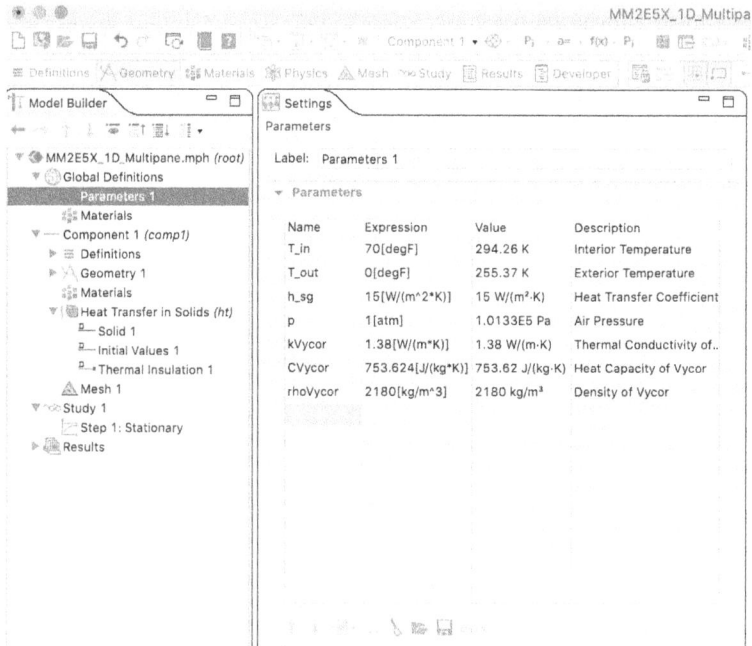

FIGURE 2.6 The settings window with a filled parameters entry table.

Figure 2.6 shows a close-up of the Settings window with a filled Parameters Entry Table.

Click on the Save to File icon (disk), in the control button array below the parameter list, and enter MM2E5X_1D_Multipane_Param.txt to save these parameters for future use in another model.

Geometry Building

The next step in building a model of a Multipane Window is to define the geometry through which the heat will flow from the heated interior to the less heated exterior.

Right-Click in the Model Builder window on Geometry 1 under Component 1 (*comp1*).

Select > Interval.

In the Interval 1 Settings Window > Click on the Specify pull-down and select Interval lengths.

Enter 0.0 as the Left endpoint.

In the Interval 1 Settings Window > Click on the Length source pull-down and select Vector.

Enter 5e-3[m] in the Lengths field.

Click > Build Selected.

Right-Click in the Model Builder window on Geometry 1 under Component 1 (*comp1*).

Select > Interval.

In the Interval 2 Settings Window > Click on the Specify pull-down and select Interval lengths.

Enter 5e-3[m] as the Left endpoint.

In the Interval 2 Settings Window > Click on the Length source pull-down and select Vector.

Enter 15e-3[m] in the Lengths field.

Click > Build Selected.

Click > Zoom Extents.

Right-Click in the Model Builder window on Geometry 1 under Component 1 (*comp1*).

Select > Interval.

In the Interval 3 Settings Window > Click on the Specify pull-down and select Interval lengths.

Enter 20e-3[m] as the Left endpoint.

In the Interval 3 Settings Window > Click on the Length source pull-down and select Vector.

Enter 5e-3[m] in the Lengths field.

Click > Build Selected.

Click > Zoom Extents to see the first three intervals in the Graphics window.

See Figure 2.7.

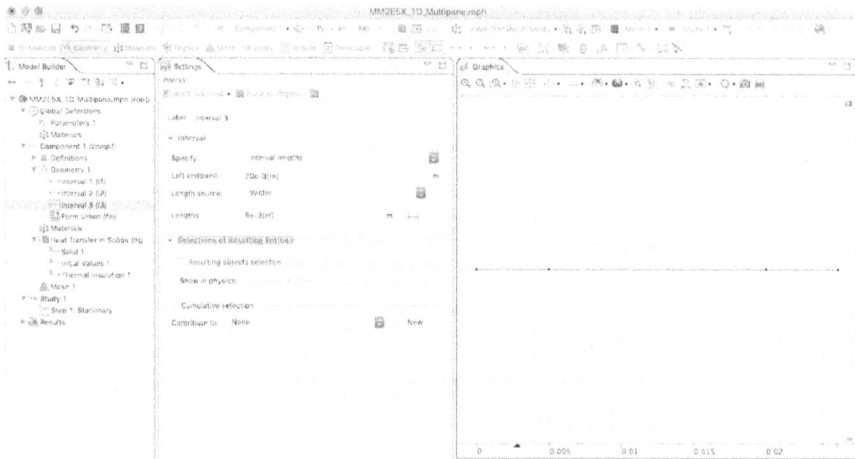

FIGURE 2.7 Zoom extents results.

Figure 2.7 shows the results of clicking the Zoom Extents button, which changes the dimensions of the axes in the Graphics window to reflect the value range of the geometry.

Right-Click in the Model Builder window on Geometry 1 under Component 1 (*comp1*).

Select > Interval.

In the Interval 4 Settings Window > Click on the Specify pull-down and select Interval lengths.

Enter 25e-3[m] as the Left endpoint.

In the Interval 4 Settings Window > Click on the Length source pull-down and select Vector.

Enter 15e-3[m] in the Lengths field.

Click > Build Selected.

Click > Zoom Extents.

Right-Click in the Model Builder window on Geometry 1 under Component 1 (*comp1*).

Select > Interval.

In the Interval 5 Settings Window > Click on the Specify pull-down and select Interval lengths.

Enter 40e-3[m] as the Left endpoint.

In the Interval 5 Settings Window > Click on the Length source pull-down and select Vector.

Enter 5e-3[m] in the Lengths field.

Click > Build All Objects.

Click > Zoom Extents.

Material Definition

Now that the Multipane Geometry has been built, the next step is to define the materials comprising the Model. Because we are assuming the three materials being used are not already stored in the COMSOL Material Library, each material will be defined by a different approach to illustrate the three techniques: material building, user-defined parameters, and user-defined direct entry.

Materials Properties from a Newly Built Material

The required material properties for the Argon gas will be defined using the Material Settings Window.

Right-Click in the Model Builder window on Materials under Component 1 (*comp1*).

Select > Blank Material.

A Material Settings Window will appear on the desktop.

Right-Click > Material 1.

Select > Rename.

Enter Argon in the Rename Material window. See Figure 2.8.

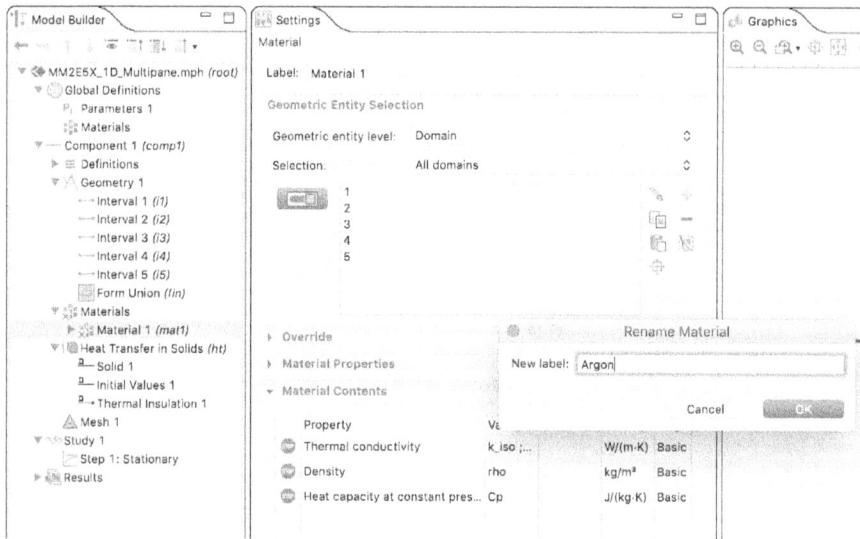

FIGURE 2.8 Material settings – rename material window.

Figure 2.8 shows the Rename Material window related to the Material Settings for Material 1.

Click > OK.

The Material Contents section of the Material Settings window shows the properties required for the Heat Transfer physics interface used in this model. All entries are marked with red stop sign symbols to indicate that the properties have not yet been assigned a value, as shown in Figure 2.9.

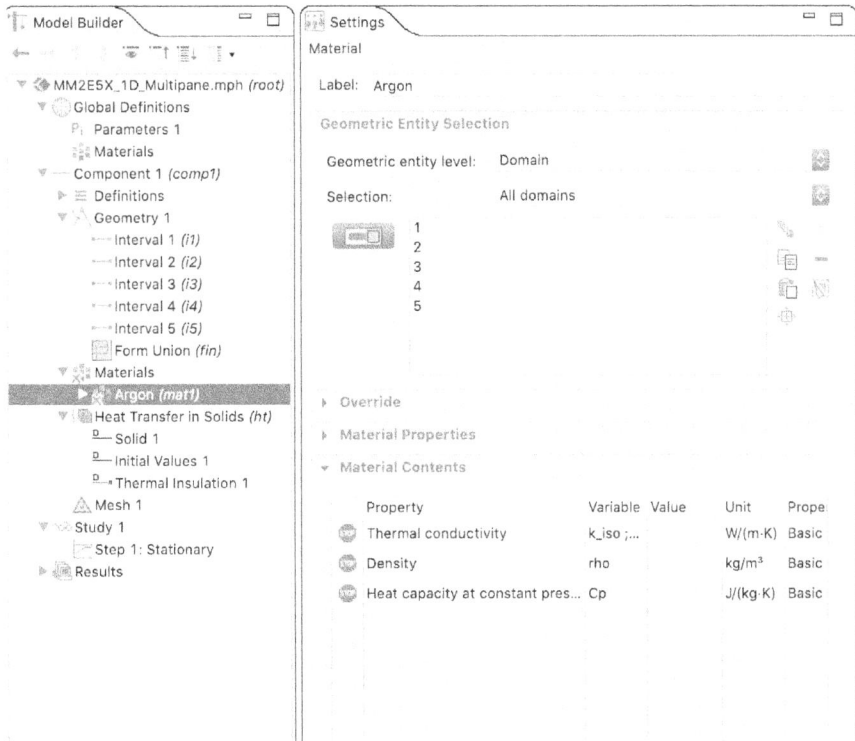

FIGURE 2.9 Material contents section for argon.

Figure 2.9 shows the Material Contents section of the Material Settings for Argon.

Enter the property values for Argon into the entry windows in the Value column of the Material Contents section: thermal conductivity is 1.7e-2[W/(m*K)], density is 1.784[kg/m^3], and heat capacity is 523.0[J/(kg*K)].

The red stop signs change to green check marks when a value has been entered for the property, as shown in Figure 2.10.

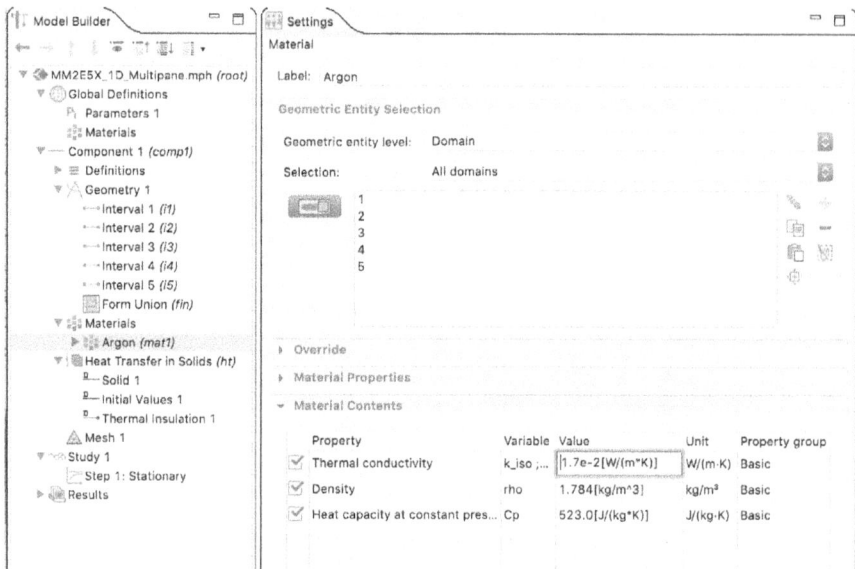

FIGURE 2.10 Material settings – material contents section completed.

Figure 2.10 shows the property values entered in the Material Contents section of the Material Settings for Argon.

A material definition can be saved for use in future models in the User Defined Library in the Materials Browser.

To save a definition,

Right-Click > Component 1 – Materials – Argon.

Select > Add to User Defined Library from the popup menu.

To verify the entry,

Click > COMSOL Multiphysics > Windows > Material Browser Tab.

In the Materials window (scroll as needed), Click > User Defined Library twistie to display the material list (see Figure 2.11).

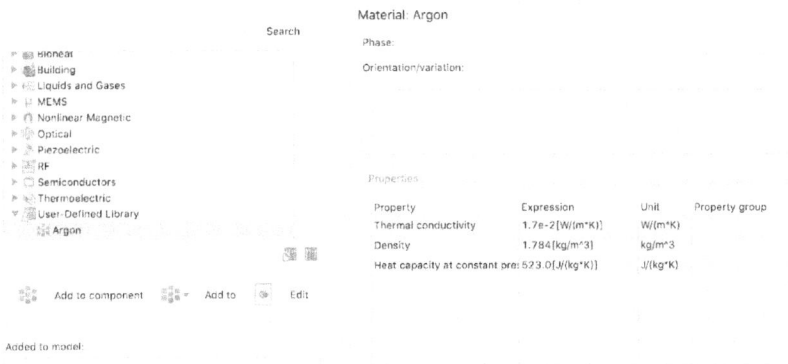

FIGURE 2.11 Material browser – user defined library.

Figure 2.11 shows the material list in the User Defined Library.

Click > Done button to return to the Model Builder

NOTE

Since this model is 1D, there are 5 Domains and this is the first material added, 5.x automatically assigns the Argon material's properties to all the available domains (1, 2, 3, 4, 5).

In this case, the materials assignment of Argon to Domains 1, 3, 4, & 5 is not correct and must be adjusted.

Click > 1 in the Settings – Material – Geometric Entity Selection window.

Click > Minus (-) button in the rightmost vertical control button array.

Click > 3 in the Settings – Material – Geometric Entity Selection window.

Click > Minus (-) button in the rightmost vertical control button array.

Click > 4 in the Settings – Material – Geometric Entity Selection window.

Click > Minus (-) button in the rightmost vertical control button array.

Click > 5 in the Settings – Material – Geometric Entity Selection window.

Click > Minus (-) button in the rightmost vertical control button array.

The Argon material has now been defined, saved, and assigned to the appropriate domain.

Materials Properties from User Defined Parameters

The default source for the properties needed by a model is defined as coming from the material.

To verify the setting,

Click > Component 1 – Heat Transfer in Solids (*ht*) twistie (as needed) in the Model Builder window to display the Heat Transfer entries.

Click > Solid 1.

Note that all five domains are included by default in the Domains – Selections list at the top of the Settings – Heat Transfer in Solids window. Also note that the properties listed at the bottom of the Settings – Heat Transfer in Solids window are all set to the "From material" option from the popup menu next to each property. See Figure 2.12.

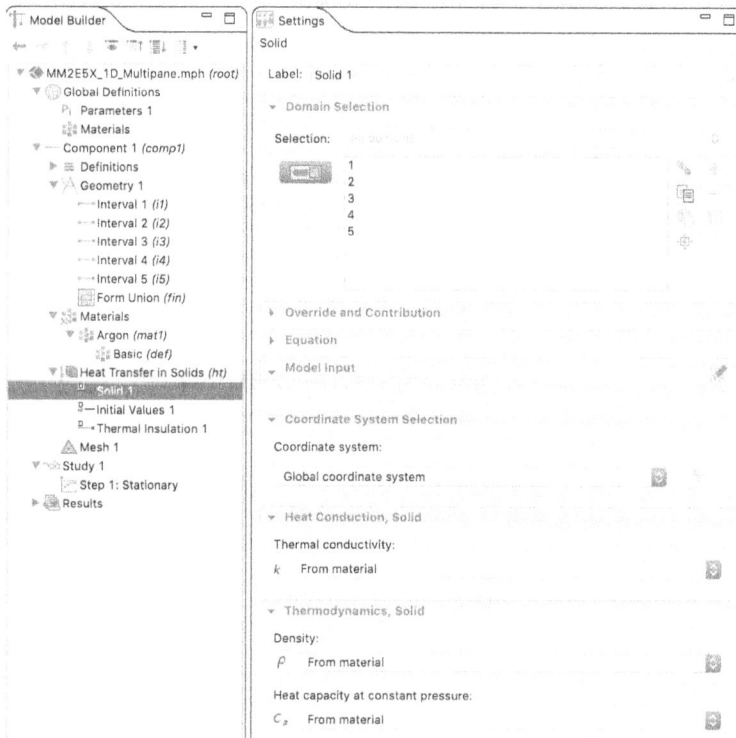

FIGURE 2.12 Settings window – heat transfer in solids.

Figure 2.12 shows the default property settings for the Heat Transfer in Solids physics interface.

That default setting is appropriate for the Argon gas since its material property values do come from the material definition. However, since the properties for the Vycor glass were entered as parameters in a previous section and therefore will not be coming from a material definition, the model needs another set of Heat Transfer in Solids settings to override the "From materials" default.

Add the new settings, as follows:

Right-Click > Component 1 > Heat Transfer in Solids *(ht)* in the Model Builder window.

Select > Solid from the popup menu.

Click > Solid 2.

Enter the material property values.

In the Settings Solid window, Click > the popup menu in the Heat Conduction, Solid > Thermal conductivity:.

Select > User defined.

Type kVycor in the Thermal conductivity entry window.

Click > the popup menu in the Thermodynamics, Solid > Density.

Select > User defined.

Type rhoVycor in the Density entry window.

Click > the popup menu in the Thermodynamics, Solid > Heat capacity at constant pressure:.

Select > User defined.

Type CVycor in the Heat Capacity entry window.

To assign the domains that will use these property values, return to the top of the Settings Solid (Solid 2) window,

Click > Select Domains tool in the Graphics window Toolbar.

In the Graphics window, Click > Interval 1.

Click > Interval 3.

Click > Interval 5.

See Figure 2.13.

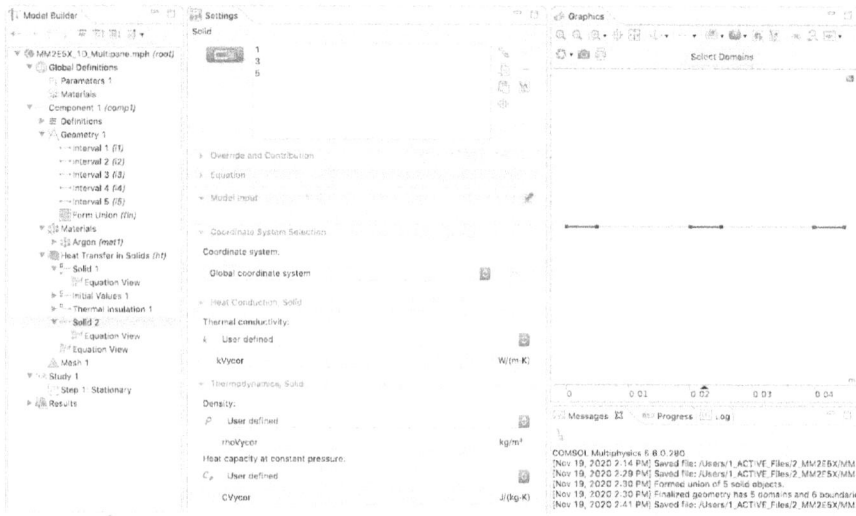

FIGURE 2.13 Property value sources for Vycor glass.

Figure 2.13 shows the property value sources for the Vycor glass domains.

Materials Properties by User Defined Direct Entry

The properties for the Krypton gas were neither entered as parameters nor as a material definition. Therefore, the values will be entered directly into another set of Heat Transfer in Solids settings.

Right-Click > Component 1 > Heat Transfer in Solids *(ht)* in the Model Builder window.

Select > Solid from the popup menu.

Click > Solid 3.

Click > the popup menu in the Heat Conduction, Solid > Thermal conductivity section.

Select > User defined.

Type 8.8e-3[W/(m*K)] in the Thermal conductivity entry window.

Click > the popup menu in the Thermodynamics, Solid > Density section.

Select > User defined.

Type 3.743[kg/m^3] in the Density entry window.

Click > the popup menu in the Thermodynamics, Solid – Heat capacity at constant pressure section.

Select > User defined.

Type 248[J/(kg*K)] in the Heat Capacity entry window.

To assign the domain that will use these property values,

In the Graphics window, Click > Interval 4.

See Figure 2.14.

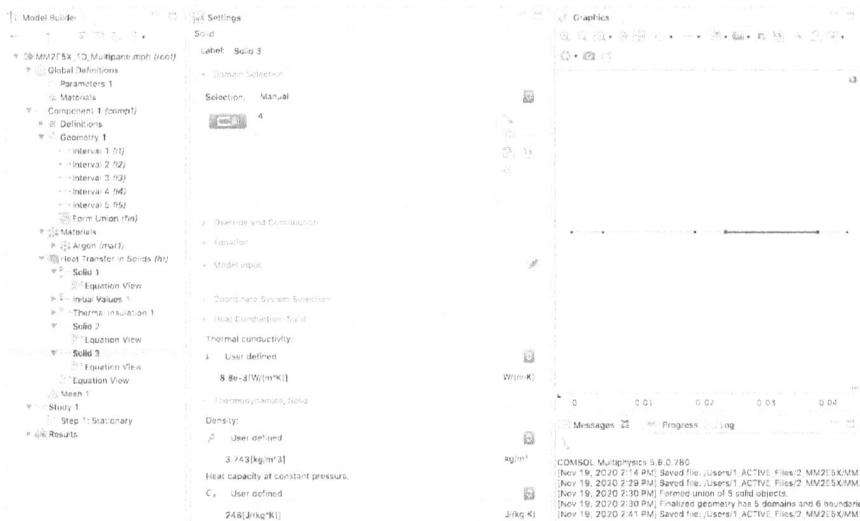

FIGURE 2.14 Property values for krypton gas.

Figure 2.14 shows the material property values for the Krypton gas domain.

In the Model Builder window, Click > Heat Transfer in Solids *(ht)* > Solid 1.

Now the settings in the Domains window show that the material property value sources for all of the domains except Domain 2, the Argon gas, have been overridden by the subsequent entries for the Vycor glass and Krypton gas, as shown in Figure 2.15.

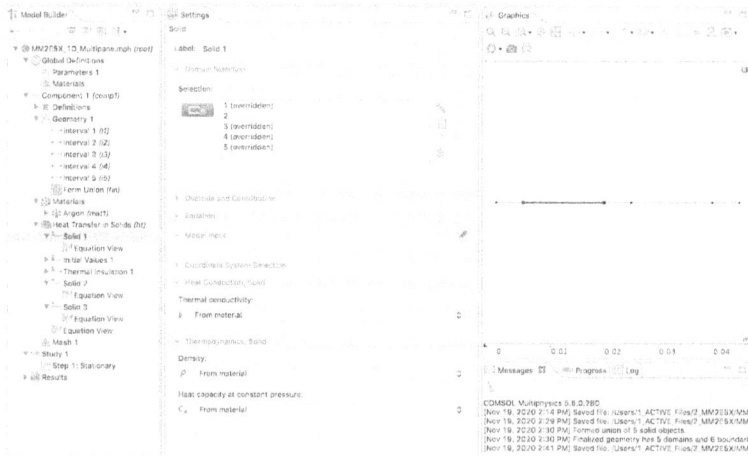

FIGURE 2.15 Default property value source settings – overridden.

Set Boundary Conditions

Since the model geometry has been built and the appropriate material properties have been assigned to the Domains 1 through 5, the next step in the modeling process is to specify the boundary conditions for the Heat Transfer physics in this model.

Right-Click > Heat Transfer in Solids *(ht)* in the Model Builder window.

Select > Heat Flux.

Click > Heat Flux 1 (if needed).

Click > Point 1 (leftmost boundary) in the Graphics window.

Click > Convective heat flux in the Heat Flux window.

Type h_sg in the Heat transfer coefficient entry window.

Type T_in in the External temperature entry window.

Right-Click > Heat Transfer in Solids *(ht)*.

Select > Heat Flux.

Click > Heat Transfer in Solids *(ht)* – Heat Flux 2 in the Model Builder window (if needed).

Click > Point 6 (rightmost boundary) in the Graphics window.

Click > Convective heat flux in the Heat Flux window.

Type h_sg in the Heat transfer coefficient entry window.

Type T_out in the External temperature entry window.

Meshing and Solution Computations

Since this model is relatively simple, allow 5.x to automatically mesh the model.

Right-Click on the blank background in the Graphics window, Select > New Domain Selection

Select > All Domains in the Graphics window.

Right-Click > Mesh 1 in the Model Builder window.

Select > Build All.

The meshing computation results in a complete mesh of 16 elements.

Now that the model has been meshed, the modeler can proceed to compute the solution.

Right-Click > Study 1 in the Model Builder window.

Select > Compute.

NOTE

5.x automatically selects the appropriate Solver for this problem, computes the solution, and displays the results in the Graphics window.

The 1D Plot displayed in the Graphics window shows the temperature in degrees Kelvin as a function of the distance from the inside surface of Vycor glass of the first pane of the Multipane window to the outside surface of Vycor glass of the third pane of the Multipane window.

Since the modeler originally specified the input parameters in degrees Fahrenheit, the modeler now needs to adjust the instructions of 5.x so that the Graphics window will display the results of this modeling calculation in the desired units.

Click > Temperature (ht) twistie in the Model Builder window.

Click > Line Graph.

In the Settings – Line Graph window,

Under Y-Axis Data, Click > Unit pull-down menu > Select > degF.

Click > the Plot button at the top of the Settings – Line Graph window.

Figure 2.16 shows the Graphics window of the Calculated Results for a Multipane Window in degrees Fahrenheit.

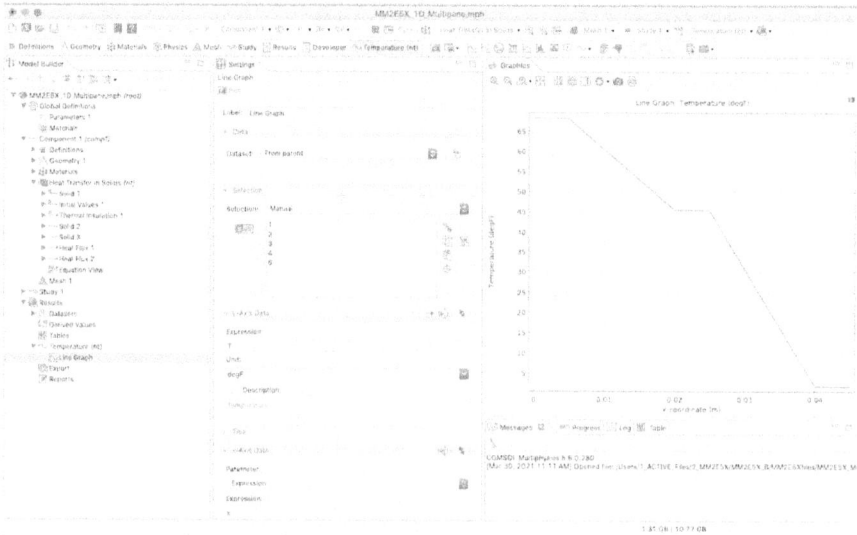

FIGURE 2.16 Calculated results for a multipane window in degrees fahrenheit.

The results of this model that demonstrates the material property entry techniques are very similar to those of the 1D 3 Pane model demonstrated in Chapter 1 using equivalent materials.

REFERENCES

1. *https://www.comsol.com/material-library*

2. *https://en.wikipedia.org/wiki/Unified_numbering_system*

3. *https://en.wikipedia.org/wiki/Deutsches_Institut_für_Normung*

4. *https://www.asminternational.org/materials-resources/online-databases*

5. *https://www.wikipedia.org*

6. *http://www.matweb.com*

CHAPTER 3

0D Electrical Circuit Interface Modeling Using COMSOL Multiphysics 5.x

GUIDELINES FOR ELECTRICAL CIRCUIT INTERFACE MODELING IN 5.x

In this chapter, the modeler is introduced to the development and analysis of electrical and electronic circuit models. Such "0D" models have proven very valuable to the science and engineering communities, both in the past and currently, as first-cut evaluations of possible power supplies, signal-drivers, and control circuit models. Those and other such ancillary circuits are developed and screened early in a project for potential later use in higher-dimensionality (2D, 3D, etc.) field-based (electrical, magnetic, etc.) geometrically dependent models.

<u>NOTE</u>

Once the Model Wizard button has been clicked the modeler then clicks the 0D button. At that point, COMSOL Multiphysics 5.x displays the Add Physics Interface page. The Physics Interfaces that may be selected have no geometric dependence [no (x, y, z) etc.]. That lack of geometric dependence results from the incorporated underlying assumption in 5.x that a 0D model calculation reflects either a homogeneous, isotropic, universal (throughout all model space) reaction or comprises a model formed of a collection of connected lumped-constant devices, such as an electrical or electronic circuit.

For information on the other Physics Interfaces that may be used to implement non-circuit based analyses in 0D, the modeler should consult the literature {3.1, 3.2, 3.3} that accompanies 5.x.

Electrical / Electronic Circuit Considerations

Electrical circuits {3.4} are typically those comprising passive (lumped-constant) devices, such as simple (non-active) resistors, capacitors, and inductors. Electronic circuits {3.5} comprise both passive and active (lumped-constant) devices (e.g. transistors, integrated circuits, thermistors, and many other device types). In this chapter, the 5.x Electrical Circuit Interface will be used to present an introduction to the modeling of the most important basic electrical circuits. Further circuit development by the curious, innovative modeler is recommended and encouraged. That task is left to the modeler to pursue through exploration of the published literature.

A basic understanding of electrical circuits began with the work of Georg Simon Ohm {3.6}. Ohm discovered and published the relationship between current, voltage, and resistance in 1827 {3.7}, approximately four (4) years before the birth of James Clerk Maxwell {3.8}. Ohm's Law, as it is commonly used throughout the science and engineering communities and as a 5.x modeler would typically employ it, is stated as follows:

$$V = IR \qquad (3.1)$$

Where V is the electromotive force in Volts (V).

  I is the current in Amperes (A).

And   R is the resistance in Ohms (Ω).

NOTE

The modeler may find in other sources that either of the symbols E and V have been employed to represent the Electromotive Force {3.9}. In this text, E will be used for Electric Field (Volts per meter). Ohm's Law, as presented in the previous equation, is formulated such that the resistance (R) has a constant nominal value. That nominal value is, to first-order, independent of temperature (T), frequency (f), and all other potential functional dependence variables.

Figure 3.1 shows an example of a typical simple battery-powered circuit with a resistive-load (R).

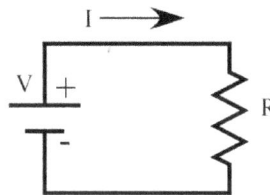

FIGURE 3.1 A typical simple battery-powered circuit with a resistive-load.

NOTE

The modeler should note that, based on a First Principles Analysis, the principle of charge conservation is employed in the analysis that follows. The Law of Charge Conservation {3.10} states that charge is neither created nor destroyed in the absence of sinks or sources, where a sink decreases the magnitude of charge and a source increases the magnitude of charge.

In configuring simple circuits, there are two basic connection sequences that are most commonly employed. In the following examples, resistors are employed for simplicity and clarity. However, other basic electrical components are also later discussed and are also so configured in the basic arrangements.

The first connection sequence discussed herein is that of the series resistive circuit. In the series connection case, two or more resistors are connected in a chain. In this case, the two serially connected resistors form a voltage divider. See Figure 3.2.

Figure 3.2 shows an example of a simple battery-powered series resistive circuit (a voltage divider).

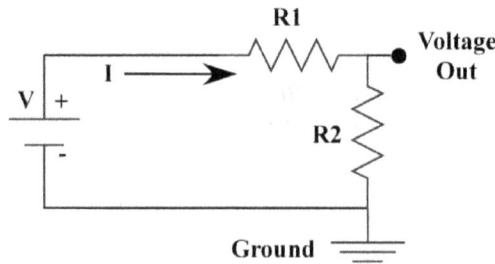

FIGURE 3.2 An example of a simple battery-powered series resistive circuit (a voltage divider).

In order to calculate the ohmic resistance value of the series equivalent resistor (R_{SE}), consider the following argument. Since the two resistors are connected in series, and there are no sinks, sources, or branch paths, the same magnitude of current (I) must flow through both resistors.

Thus:

$$V = V_1 + V_2 = I(R1 + R2) \tag{3.2}$$

Where I is the common circuit current in Amperes (A).

V is the electromotive force in Volts (V).

V_1 is the voltage dropped across R1 by the current I.

V_2 is the voltage dropped across R2 by the current I.

R1 is the resistance in Ohms (Ω) of the first resistor in the series.

And   R2 is the resistance in Ohms (Ω) of the second resistor in the series.

And therefore:

$$R_{SE} = R1 + R2 \tag{3.3}$$

Where R_{SE} is the series equivalent resistance of resistors R1 and R2.

R1 is the resistance in Ohms (Ω) of the first resistor in the series.

And   R2 is the resistance in Ohms (Ω) of the second resistor in the series.

The voltage at the output of this resistive divider is:

$$V_{out} = \frac{V_2}{V} \bullet V = \frac{I \bullet R2}{I(R1 + R2)} \bullet V = \frac{R2}{R1 + R2} \bullet V = \frac{R2}{R_{SE}} \bullet V \qquad (3.4)$$

Where V_{out} is the output voltage of the resistive divider in Volts (V).

V_2 is the voltage dropped across resistor R2 in Volts (V).

And   I is the common circuit current in Amperes (A).

The output voltage of the resistive divider is the product of the source voltage (battery voltage) and the ratio of the ohmic value of the grounded resistor to the ohmic value of the circuit series equivalent resistance.

The next basic connection sequence discussed herein is that of the parallel resistive circuit. In the parallel connection sequence, two or more resistors are connected in a parallel configuration. In this case, the resistors have a common (the same) voltage and the circuit acts as a current divider. See Figure 3.3.

Figure 3.3 shows an example of a simple battery-powered parallel resistive circuit (a current divider).

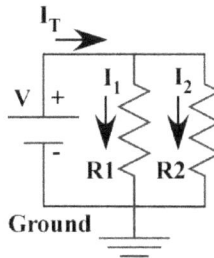

FIGURE 3.3 An example of a simple battery-powered parallel resistive circuit (a current divider).

The value of the equivalent series resistance as viewed from the battery is calculated as follows:

$$I = I_1 + I_2 = \frac{V}{R_1} + \frac{V}{R_2} = V\left(\frac{1}{R_1} + \frac{1}{R_2}\right) = V\left(\frac{R_1 + R_2}{R_1 R_2}\right) = \frac{V}{R_{SE}} \qquad (3.5)$$

Thus:

$$\frac{1}{R_{SE}} = \left(\frac{R_1 + R_2}{R_1 R_2} \right) \;\rightarrow\; R_{SE} = \frac{R_1 R_2}{R_1 + R_2} \tag{3.6}$$

Where R_{SE} is the series equivalent resistance of resistors R1 and R2 in parallel.

R1 is the resistance in Ohms (Ω) of the first resistor connected in parallel.

And   R2 is the resistance in Ohms (Ω) of the second resistor connected in parallel.

NOTE *The modeler can easily see that the calculations for the equivalent series resistance at any point in the circuit can rapidly become lengthy and tedious for even a somewhat complicated circuit.*

In Figure 3.4, resistors are shown connected into a series-parallel configuration.

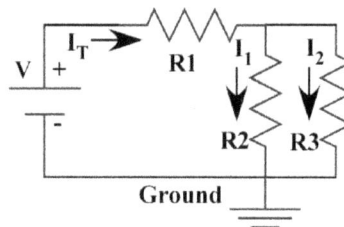

FIGURE 3.4 An example of a simple battery-powered series-parallel resistive circuit.

Figure 3.4 shows an example of a simple battery-powered series-parallel resistive circuit.

The next major expansion in the understanding and analysis of electrical circuits, building on the work of Ohm, occurred through the work of Gustav Robert Kirchhoff {3.11}. Kirchhoff developed two circuit laws, both of which are fundamentally employed in 5.x, one for voltage and one for current. He first described his circuit laws in 1845, when he was still a student {3.12}.

Kirchhoff's first law, the Current Law:

$$\sum_{n=1}^{n} A_{mn} I_{mn} = 0 \tag{3.7}$$

Where I_{mn} is the current flowing into or out of the particular junction (node).

A_{mn} has a value of +1 for each in-flowing nodal current and a value of −1 for each out-flowing nodal current.

m is the index number of the node under analysis.

And n is the index number of a particular branch on the mth node.

Figure 3.5 shows an example of a five-branch node.

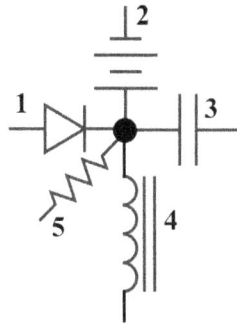

FIGURE 3.5 An example of a five-branch node.

Kirchhoff's second law, the Voltage Law:

$$\sum_{n=1}^{n} B_{mn} V_{mn} = 0 \qquad (3.8)$$

Where V_{mn} is the voltage dropped at the nth element of the mth loop as determined by the direction of the elemental current flow.

B_{mn} has a value of +1 for each non-source element and a value of −1 for each source element.

m is the index number of the loop under analysis.

And n is the index number of a particular element in the mth loop.

Loop 1 of Figure 3.6 shows an example of a four-element loop.

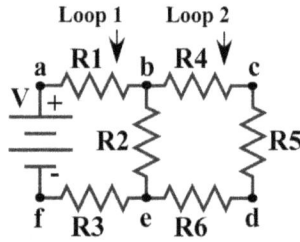

FIGURE 3.6 An example of a four-element loop.

The modeler can easily see that Figure 3.6 Loop 1 shows:

$$V_{ab} + V_{be} + V_{ef} - V_{af} = 0 \tag{3.9}$$

and that:

$$V_{ab} + V_{bc} + V_{cd} + V_{de} + V_{ef} - V_{af} = 0 \tag{3.10}$$

and that Loop 2 shows:

$$V_{bc} + V_{cd} + V_{de} - V_{be} = 0 \tag{3.11}$$

The basics of Ohm's Law and Kirchhoff's Laws have been introduced using resistive circuits. To expand the modeler's analytical capability, he needs to consider more than simply circuit resistance.

There are two other very important lumped-constant electrical components employed in most circuits. They are the capacitor and the inductor. Ewald Georg von Kleist {3.13} first reported the capacitive charge storage effect in 1745. Pieter van Musschenbroek invented the first capacitor {3.14}, the Leiden Jar, in 1746 {3.15}.

Capacitors comprise two conductors, separated by a dielectric (insulating) {3.16} medium (material). The charge stored is as follows:

$$Q = CV \tag{3.12}$$

Where Q is the charge stored on the capacitor.

      C is the capacitance in Farads.

And   V is the electromotive potential measured between the two conductors in Volts.

NOTE *Q =CV is correct for an IDEAL capacitor. For the equivalent circuit of real capacitors, there will generally be some additional terms that include a small series or parallel resistance and a small series inductance, depending upon the operating frequency of the circuit or model. Each of those terms depends upon the physical specification and geometrical configuration of the actual component. (See the manufacturer's specification data sheet.)*

Michael Faraday {3.17} and Joseph Henry {3.18} discovered electromagnetic induction in 1831 {3.19}. The effect of electromagnetic induction is to induce an electromagnetic force into a conductor that is proportional to the rate of change of the magnetic flux. The Maxwell-Faraday formulation is as follows {3.20}:

$$\nabla \times E = -\frac{\partial B}{\partial t} \tag{3.13}$$

Where E is electromagnetic force (electric field) in Volts (V) per meter (m).

B is the magnetic flux in webers per square meter.

And   t is the time in seconds.

These two components are inherently time dependent in their behavior {3.21, 3.22, 3.23}.

The self-inductance of a circuit is defined as follows:

$$V_L = L\frac{dI}{dt} \tag{3.14}$$

Where V_L is the induced voltage in Volts.

L is the inductance in webers per ampere (henries).

I is the current flow in amperes (A).

And   t is the time in seconds.

NOTE *Self-inductance occurs when a voltage is induced into the same circuit by a current flowing within that circuit.*

Mutual-inductance occurs when a voltage is induced into a second circuit by a current flowing in a first circuit (transformers, etc.).

The capacitance of a circuit is defined as follows:

$$V_C = \frac{1}{C} \int i \, dt = \frac{Q}{C} \tag{3.15}$$

Where V_C is the voltage between the terminals of the capacitor in Volts.

   C is the capacitance in Farads (1 Coulomb per Volt).

   i is the current flow in Amperes.

And   t is the time in seconds.

The current flow through a capacitor is defined as:

$$I = C \frac{dV_C}{dt} \tag{3.16}$$

Where V_C is the voltage between the terminals of the capacitor in Volts.

   C is the capacitance in Farads (1 Coulomb per Volt).

   I is the current flow in Amperes.

And   t is the time in seconds.

Utilizing Kirchhoff's Voltage Law for a simple series-resistor-capacitor-inductor circuit yields the following equation:

$$V_R + V_C + V_L - V(t) = 0 \tag{3.17}$$

Where V_R is the voltage-drop across the resistor in Volts.

   V_C is the voltage-drop across the capacitor in Volts.

   V_L is the voltage-drop across the inductor in Volts.

And   V(t) is the time-varying source voltage driving the circuit.

Now that the modeler has been introduced to the underlying physical concepts involved in the functioning of electrical circuits, the modeler will now be shown examples of how to solve for the physical behavior of some of the basic circuits using 5.x. These examples can, of course, be used as guidance to the development of more complex circuits by the modeler at some later time when the need arises.

Simple Electrical Circuit Interface Model Setup Overview

Figure 3.7a and b show the default COMSOL Desktop Displays immediately after startup for a Macintosh (a) and a PC (b). The difference between the two displays is the Application Wizard button, available only on the PC platform.

The default coordinate selection for modeling calculations at startup in 5.x is 3D. In the case of 0D, the underlying equations in the multiphysics model have either no relational dependence on geometrical factors or react in a homogeneous and isotropic manner (effectively geometrically relationally independent).

NOTE

The 0D Space Dimension will be used in this book only to explore electrical and/or electronic circuit model behavior through the use of Ohm's Law and Kirchhoff's Law calculations.

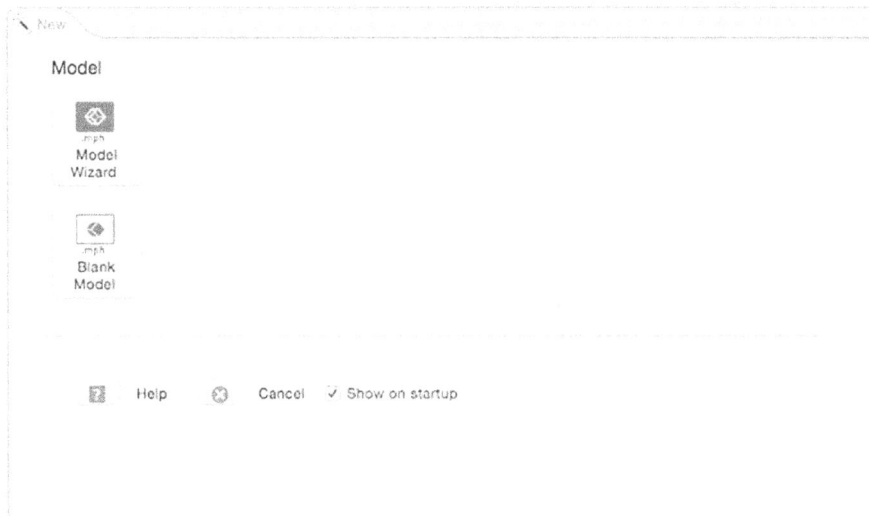

FIGURE 3.7a Default COMSOL desktop display for macintosh.

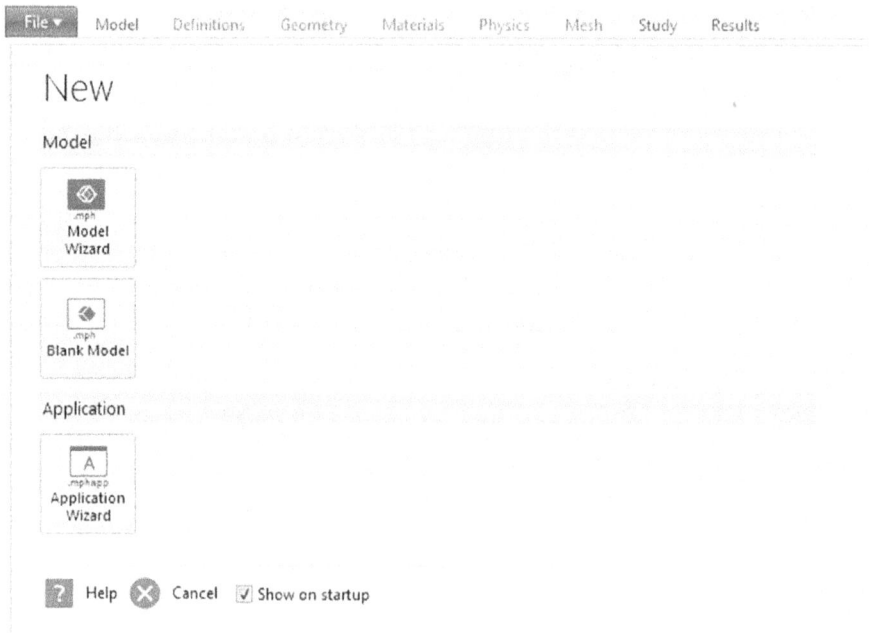

FIGURE 3.7b Default COMSOL desktop display for PC.

Select a Space Dimension

To start the Modeling process, Run 5.x.

For this first demonstration model of the Engineering Circuit Interface, Select (Click) the Model Wizard button, then Click > 0D selection button.

Once the 0D Selection button is clicked, the Desktop Display changes to show the Select Physics window, as shown in Figure 3.8.

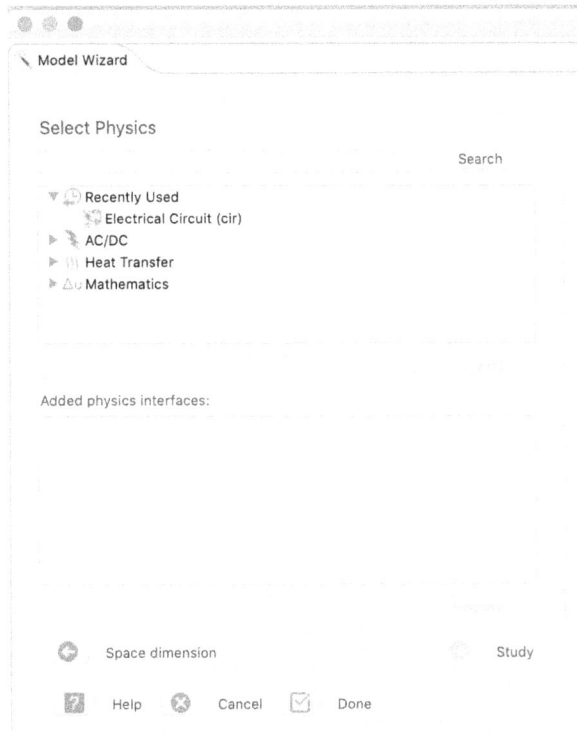

FIGURE 3.8 Desktop display with the select physics window.

Figure 3.8 shows the default COMSOL Desktop Display with the Select Physics window.

Click > Twistie of the AC/DC Physics Interface.

Select > Electrical Circuit (cir).

Figure 3.9a shows the Desktop Display with the Select Physics window after the selection of Electrical Circuit (cir).

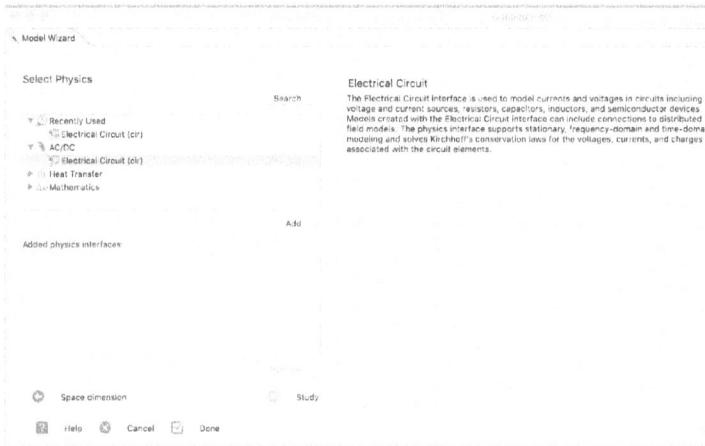

FIGURE 3.9a Electrical circuit (cir) selected.

Figure 3.9a shows the Electrical Circuit (cir) selected.

Click > Add.

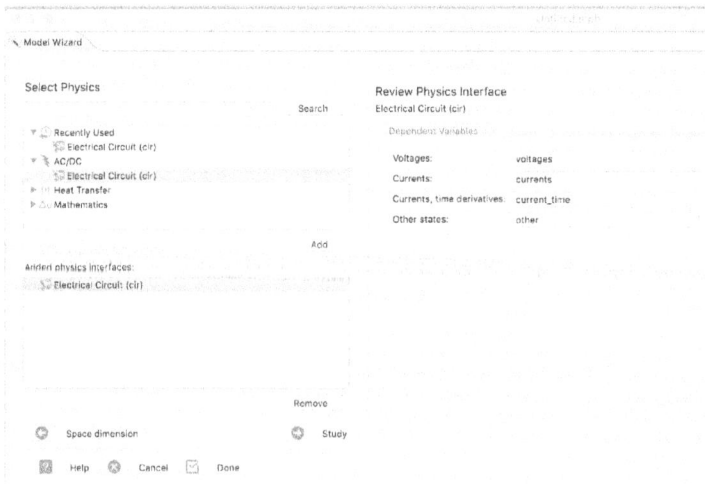

FIGURE 3.9b Electrical circuit (cir) added.

Figure 3.9b shows the Electrical Circuit (cir) Added.

Click > Study button (Right Pointing Arrow).

Select > General Studies > Time Dependent.

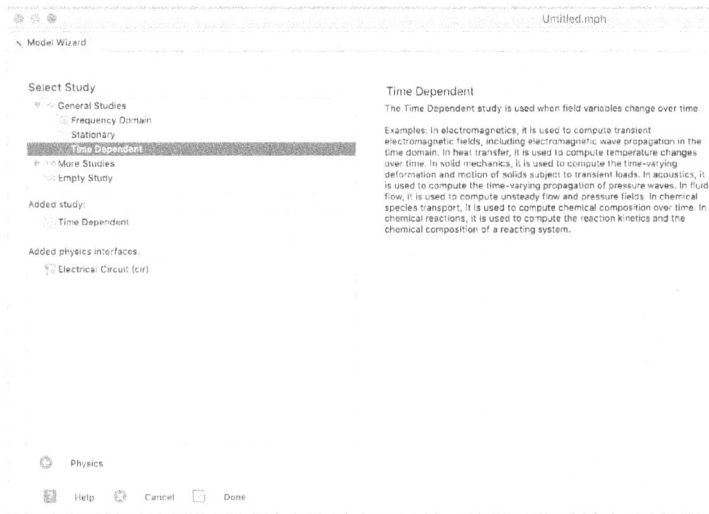

FIGURE 3.9c Time dependent selected.

Figure 3.9c shows the Time Dependent selected.

Click > Done button (Check Mark).

The modeler has now been shown fundamentally how to select the 0D Space Coordinates and how to access the Electrical Circuit (cir) Interface. See Figure 3.10.

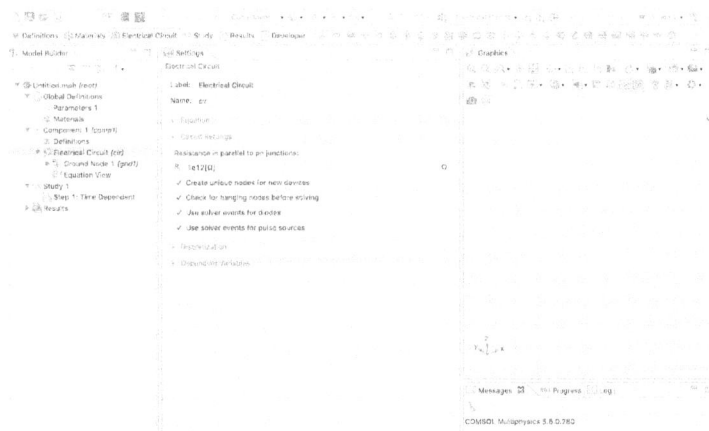

FIGURE 3.10 Desktop display prepared for an electrical circuit (cir) model.

Figure 3.10 shows the Desktop Display prepared for an Electrical Circuit (cir) model.

Now that the initial steps have been demonstrated to start building a new model, the modeler can close down the 5.x application.

If the modeler is using a Macintosh,

Select > COMSOL Multiphysics menu > Quit COMSOL Multiphysics.

Select the Do Not Save option on closing the 5.x Macintosh application.

If the modeler is using a PC,

Select > File > Exit.

Select the Do Not Save option on closing the 5.x PC application.

Basic Problem Formulation and Implicit Assumptions

NOTE

A first-cut problem solution is the equivalent of a back-of-the-envelope or an on-a-napkin solution. Problem solutions of that type are more easily formulated, more quickly built, and typically provide a first estimate of whether the final solution to the full problem will be deemed to be (should possibly be) within reasonable time constraint or budgetary bounds. Creating first-cut solutions will often allow the 5.x modeler to easily decide whether or not it is worth the additional effort and cost needed to build a fully implemented higher-dimensionality model.

The Electrical Circuit Interface models built herein (0D) will start with very simple models and then begin to explore more complicated models. As the modeler knows, there are an infinite number of variations of simple and complex circuits that could be explored. The models built in this chapter will demonstrate the use of the Electrical Circuit Interface as a means to explore the building of possible circuit configurations as a tutorial mechanism.

In later chapters, the Electrical Circuit Interface will be used to develop driver circuits for more complex field-based physical models.

NOTE

In a transient solution model, all of the appropriate variables in the model are a function of time. The model solution builds from a set of suitable initial conditions, through a set of incremental intermediate solutions, to a final solution.

0D BASIC CIRCUIT MODELS

Let us consider first the application of the 5.x Electrical Circuit Interface to some of the basic circuits just presented.

0D Resistor-Capacitor Series Circuit Model

Startup 5.x.

Click > Model Wizard > Select > 0D, on the Select Space Dimension page.

Click > Twistie for AC/DC in the Select Physics window.

Select > Electrical Circuit (cir).

Click > the Add button.

Click > Study (Right Pointing Arrow).

Select > General Studies > Time Dependent in the Select Study window.

Click > Done (Checked Box button).

Click > Save As > MM2E5X_0D_SRC1.mph.

NOTE *In the Electrical Circuit (cir) Interface, the Node Connections specify exactly where each component is connected. Each circuit must have a Ground connection and that Node is 0. The completed circuit must tie back to the Ground Node for completion.*

The completed battery-powered series-resistor-capacitor circuit will be connected as follows:

Ground (0) <> (0) Voltage Source (1) <> (1) Resistor (2) <> (2) Capacitor (0)

NOTE *In this book <> means is electrically connected to.*

See Figure 3.11.

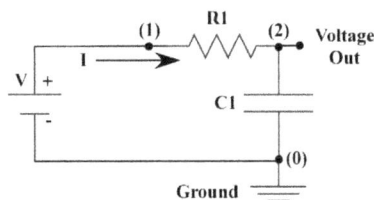

FIGURE 3.11 Circuit diagram for the MM2E5X_0D_SRC1.mph model.

Figure 3.11 shows the circuit diagram for the MM2E5X_0D_SRC1.mph model.

Right-Click in the Model Builder window on Electrical Circuit (cir).

Select > Voltage Source from the Pop-up menu.

See Figure 3.12.

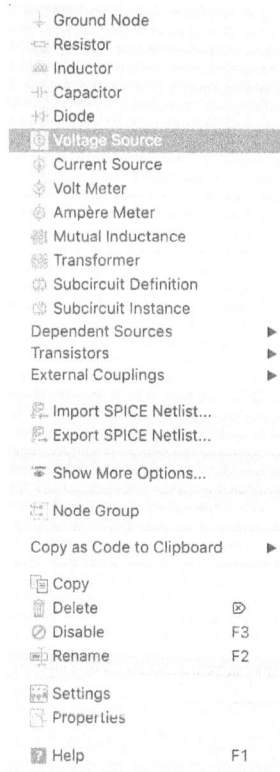

FIGURE 3.12 Electrical circuit (cir) pop-up window.

Figure 3.12 shows the Electrical Circuit (cir) Pop-up window.

In Settings > Voltage Source > Node Connections > Enter n=0, p=1.

Enter 5[V] in the Settings >Voltage Source > Device Parameters > Voltage entry window.

See Figure 3.13.

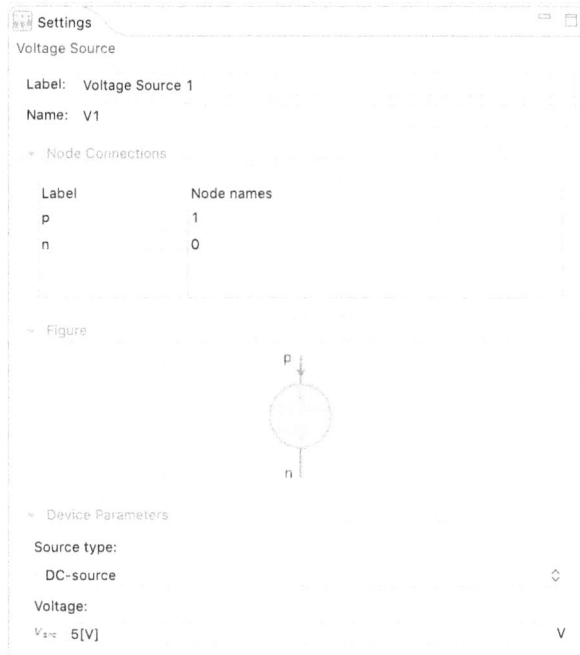

FIGURE 3.13 Desktop display – Settings –Voltage source entry window.

Figure 3.13 shows the Desktop Display – Settings – Voltage Source entry window.

Right-Click in the Model Builder window on Electrical Circuit (cir).

Select > Resistor from the Pop-up menu.

This resistor is the first component in the series circuit.

Enter 1, 2 in the Settings – Resistor – Node Connections – Node names entry windows.

Leave the value in the Resistance entry window (1000[ohm]) set at the default value.

See Figure 3.14.

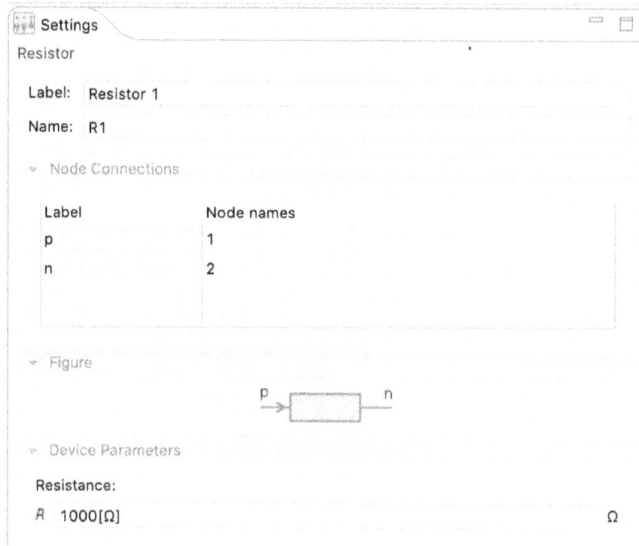

FIGURE 3.14 Settings – Resistor – Node connections – Node names.

Figure 3.14 shows the Settings – Resistor – Node Connections – Node names.

Right-Click in the Model Builder window on Electrical Circuit (cir).

Select > Capacitor from the Pop-up menu.

This capacitor is the second component in the series circuit.

Enter 2, 0 in the Settings – Capacitor – Node Connections – Node names entry windows.

Enter 1000[nF] in the Settings – Capacitor – Device Parameters – Capacitance entry window.

See Figure 3.15.

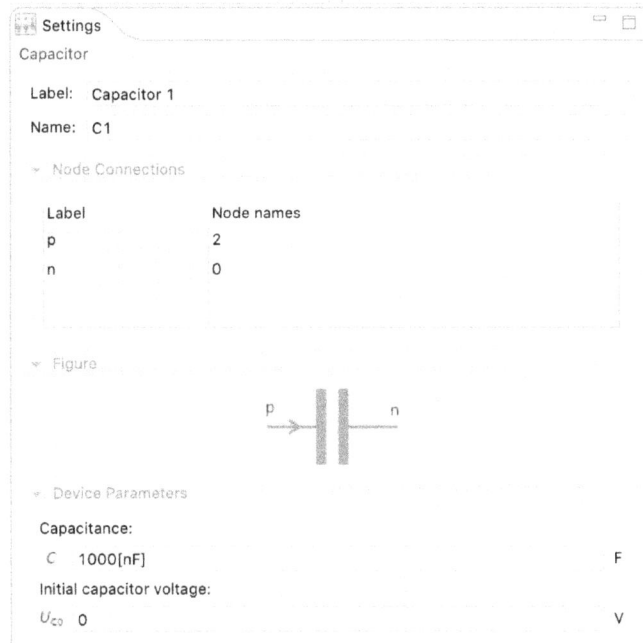

FIGURE 3.15 Settings – Capacitor – Device parameters – Capacitance.

Figure 3.15 shows the Settings – Capacitor – Device Parameters – Capacitance.

Now that the building of the series-resistor-capacitor circuit model is complete, the next step is to prepare to solve the model.

NOTE *The default settings for the Time Dependent Study Type are not appropriate for the solution of this problem and thus need to be modified.*

Click > Study 1 twistie (as needed).

Select > Step 1: Time Dependent.

See Figure 3.16a.

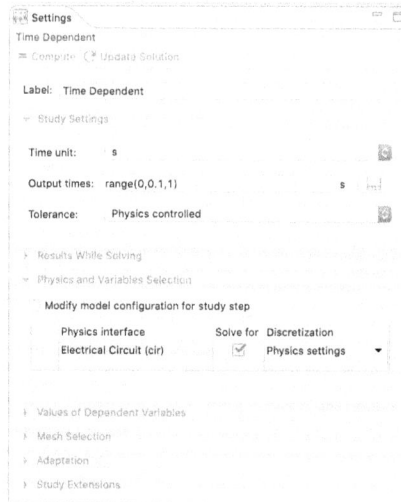

FIGURE 3.16a Settings – Time dependent window.

Figure 3.16a shows the Settings – Time Dependent window.

Click the Range button (at the right of the Times entry window).

Enter > Start = 0.0, Step = 1e-4, Stop = 7e-3.

Click > Replace.

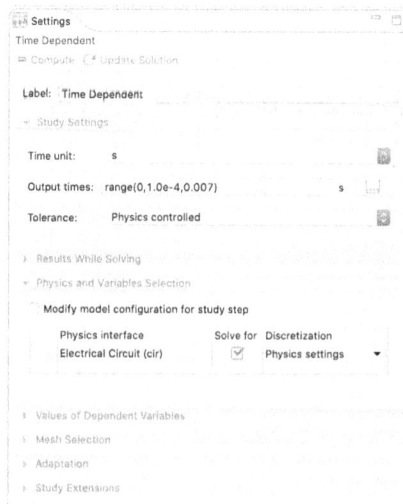

FIGURE 3.16b Settings – Time dependent window modified settings.

Figure 3.16b shows the Settings – Time Dependent window modified settings.

Right-Click > Study 1.

Select > Compute.

NOTE *The default settings for the 1D Plot of the results of this model are not appropriate and thus the settings need to be modified.*

Right-Click > Results > Select >1D Plot Group.

Right-Click > Results > 1D Plot Group 1 > Select > Global.

Click > Settings Global >y-Axis Data > Replace Expression (red-green triangles button).

Select > Component 1 > Electrical Circuit > Devices > C1> comp1.cir.C1_v – Voltage across device C1.

Click > Settings Global > Plot.

Click > 1D Plot Group 1

Settings > 1D Plot Group > Legend > Position > Lower right.

See Figure 3.17 for the computed solution.

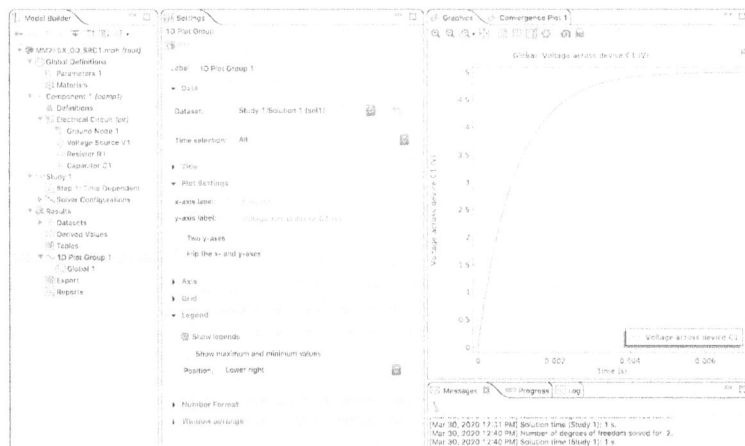

FIGURE 3.17 Voltage across C1 as a function of time.

Figure 3.17 shows the Voltage across C1 as a function of Time.

Be sure to save the just completed model.

0D Inductor-Resistor Series Circuit Model

Startup 5.x.

Click > Model Wizard > Select > 0D, on the Select Space Dimension page.

Click > Twistie for AC/DC in the Select Physics window.

Select > Electrical Circuit (cir).

Click > the Add button.

Click > Study (Right Pointing Arrow).

Select > General Studies > Time Dependent in the Select Study window.

Click > Done (Checked Box button).

Click > Save As > MM2E5X_0D_SLR1.mph.

NOTE *In the Electrical Circuit (cir) Interface, the Node Connections specify exactly where each component is connected. Each circuit must have a Ground connection and that Node is 0. The completed circuit must tie back to the Ground Node for completion.*

The completed battery-powered series inductor-resistor circuit will be connected as follows:

Ground (0) <> (0) Voltage Source (1) <> (1) Inductor (2) <> (2) Resistor (0)

NOTE *In this book <> means is electrically connected to.*

See Figure 3.18.

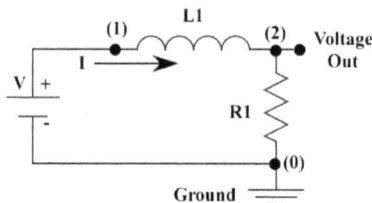

FIGURE 3.18 Circuit diagram for the MM2E5X_0D_SLR1.mph model.

Figure 3.18 shows the circuit diagram for the MM2E5X_0D_SLR1.mph model.

Right-Click in the Model Builder window on Electrical Circuit (cir).

Select > Voltage Source from the Pop-up menu.

See Figure 3.19.

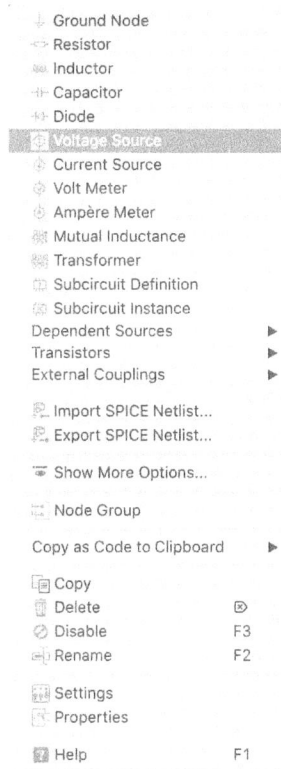

FIGURE 3.19 Electrical circuit (cir) pop-up window.

Figure 3.19 shows the Electrical Circuit (cir) Pop-up window.

In Settings > Voltage Source > Node Connections > Enter n=0, p=1.

Enter 5[V] in the Settings >Voltage Source > Device Parameters > Voltage entry window.

See Figure 3.20.

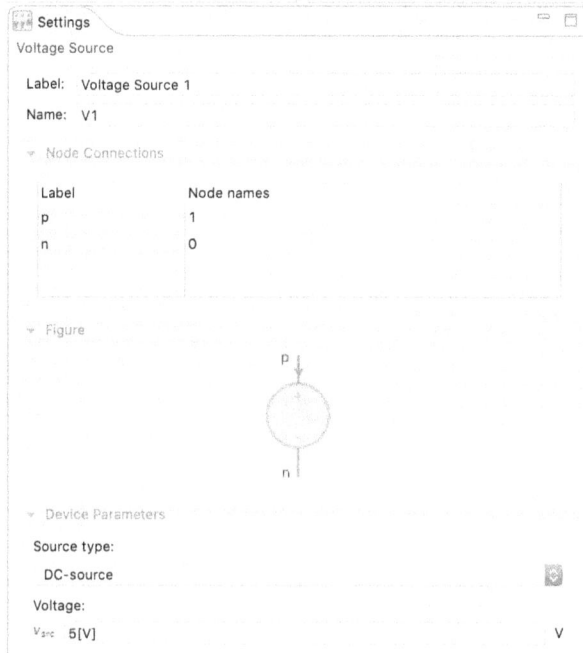

FIGURE 3.20 Desktop display – Settings –Voltage source – Device parameters – Voltage entry window.

Figure 3.20 shows the Desktop Display – Settings – Voltage Source – Device Parameters – Voltage entry window.

Right-Click in the Model Builder window on Electrical Circuit (cir).

Select > Inductor from the Pop-up menu.

This inductor is the first component in the series circuit.

Enter 1, 2 in the Settings – Inductor – Node Connections – Node names entry windows.

Leave the value in the Inductance entry window (1[mH]) set at the default value.

See Figure 3.21.

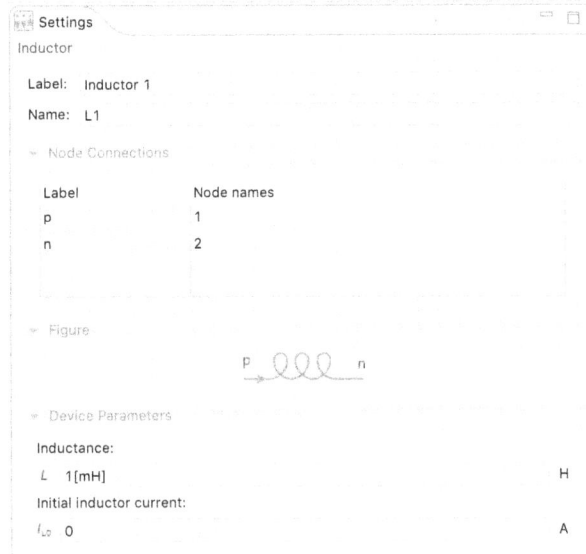

FIGURE 3.21 Settings – Inductor – Node connections – Node names.

Figure 3.21 shows the Settings – Inductor – Node Connections – Node names.

Right-Click in the Model Builder window on Electrical Circuit (cir).

Select > Resistor from the Pop-up menu.

This resistor is the second component in the series circuit.

Enter 2, 0 in the Settings – Resistor – Node Connections – Node names entry windows.

Enter 1[ohm] in the Settings – Resistor – Resistance entry window.

See Figure 3.22.

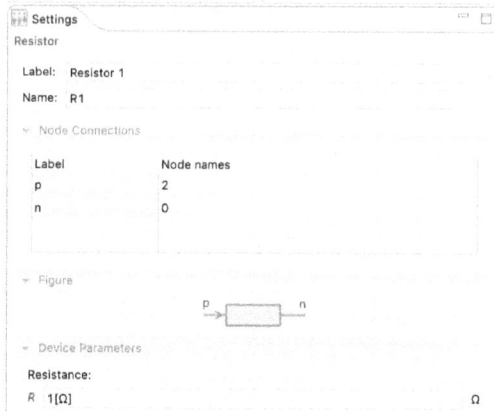

FIGURE 3.22 Settings – Resistor – Device parameters – Resistance.

Figure 3.22 shows the Settings – Resistor – Device Parameters – Resistance.

Now that the building of the series inductor-resistor circuit model is complete, the next step is to prepare to solve the model.

NOTE

The default settings for the Time Dependent Study Type are not appropriate for the solution of this problem and thus need to be modified.

Click > Study 1 twistie (as needed).

Select > Step 1: Time Dependent. See Figure 3.23.

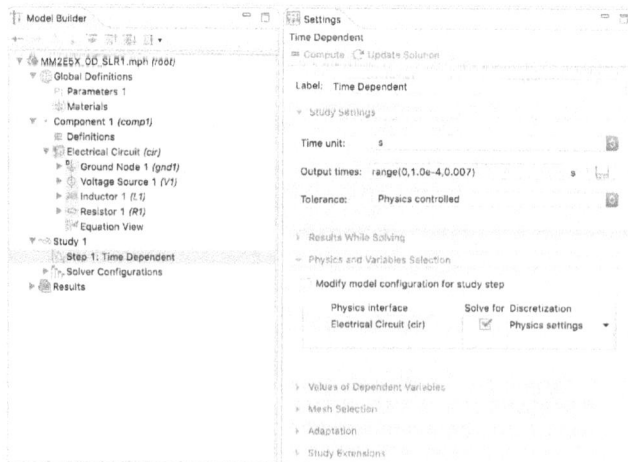

FIGURE 3.23 Settings – Time dependent window.

Figure 3.23 shows the Settings – Time Dependent window.

Click the Range button (at the right of the Times entry window).

Enter > Start = 0.0, Step = 1e-4, Stop = 7e-3.

Click > Replace.

Right-Click > Study 1.

Select > Compute.

NOTE *The default settings for the 1D Plot of the results of this model are not appropriate and thus the settings need to be modified.*

Right-Click > Results > Select >1D Plot Group.

Right-Click > 1D Plot Group 1 > Select > Global.

Click > Settings Global >y-Axis Data > Replace Expression.

Select > Component 1 > Electrical Circuit > Devices > R1> comp1.cir.R1_v – Voltage across device R1.

Click > Settings Global > Plot.

Click > 1D Plot Group > Settings 1D Plot Group > Legend > Position > Lower right.

See Figure 3.24 for the computed solution.

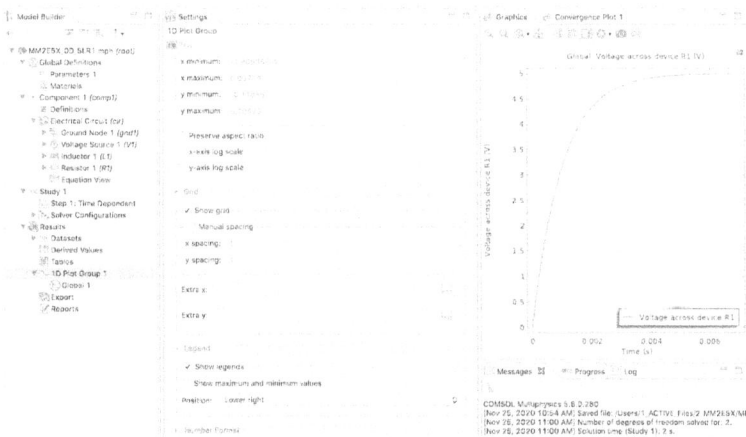

FIGURE 3.24 Voltage across R1 as a function of time.

Figure 3.24 shows the Voltage across R1 as a function of Time.

Be sure to save the just completed model.

0D Series-Resistor Parallel-Inductor-Capacitor Circuit Model

Startup 5.x.

Click > Model Wizard > Select > 0D, on the Select Space Dimension page.

Click > Twistie for AC/DC in the Select Physics window.

Select > Electrical Circuit (cir).

Click > the Add button.

Click > Study (Right Pointing Arrow).

Select > General Studies > Time Dependent in the Select Study window.

Click > Done (Checked Box button).

Click > Save As > MM2E5X_0D_SRPLC1.mph.

NOTE *In the Electrical Circuit (cir) Interface, the Node Connections specify exactly where each component is connected. Each circuit must have a Ground connection and that Node is 0. The completed circuit must tie back to the Ground Node for completion.*

The completed battery-powered series-resistor parallel inductor-capacitor circuit will be connected as follows:

Ground (0) <> (0) Voltage Source (1) <> (1) Resistor (2) <> (2) Inductor (0) <> (2) Capacitor (0)

NOTE *In this book <> means is electrically connected to.*

See Figure 3.25.

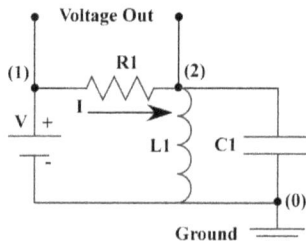

FIGURE 3.25 Circuit diagram for the MM2E5X_0D_SRPLC1.mph model.

Figure 3.25 shows the Circuit Diagram for the MM2E5X_0D_SRPLC1. mph model.

Right-Click in the Model Builder window on Electrical Circuit (cir).

Select > Voltage Source from the Pop-up menu.

See Figure 3.26.

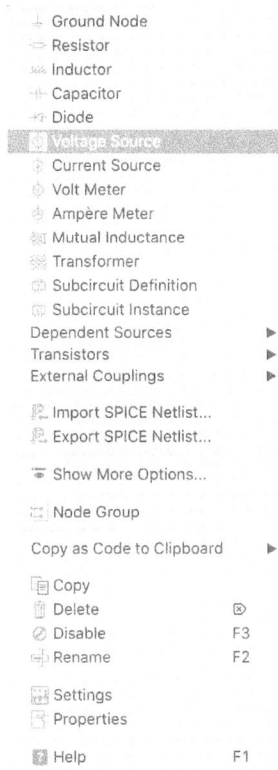

FIGURE 3.26 Electrical circuit (cir) pop-up window.

Figure 3.26 shows the Electrical Circuit (cir) Pop-up window.

In Settings > Voltage Source > Node Connections > Enter n=0, p=1.

Enter 5[V] in the Settings >Voltage Source > Device Parameters > Voltage entry window.

See Figure 3.27.

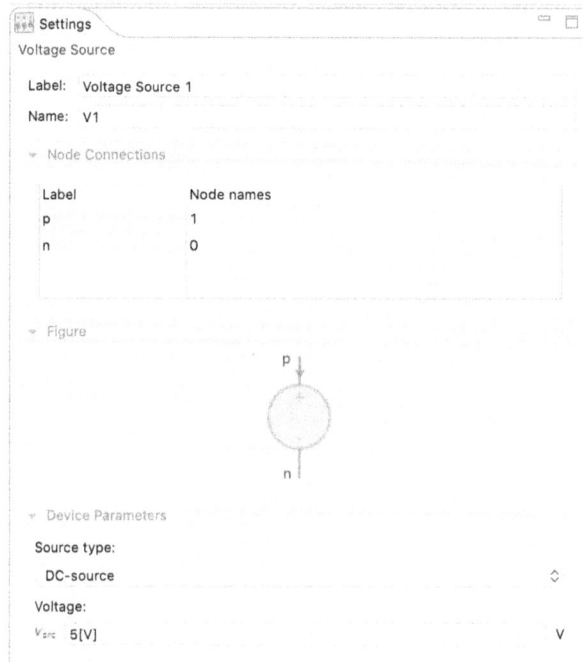

FIGURE 3.27 Desktop display – Settings –Voltage source – Device parameters – Voltage entry window.

Figure 3.27 shows the Desktop Display – Settings – Voltage Source – Device Parameters – Voltage entry window.

Right-Click in the Model Builder window on Electrical Circuit (cir).

Select > Resistor from the Pop-up menu.

This resistor is the first component in the series-parallel circuit.

Enter 1, 2 in the Settings – Resistor – Node Connections – Node names entry windows.

Enter the value in the Resistance entry window (3e3[ohm]).

See Figure 3.28.

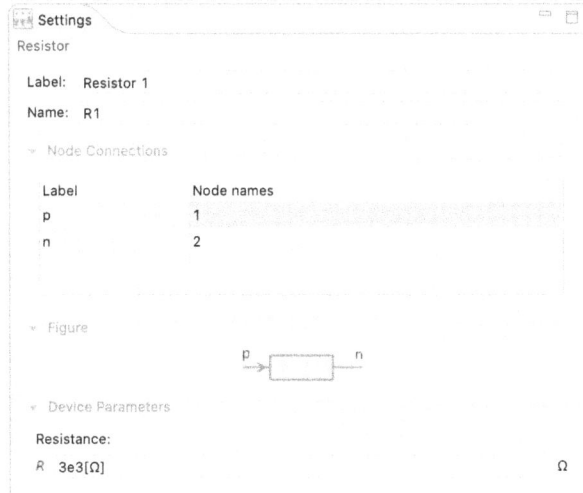

FIGURE 3.28 Settings – Resistor – Node connections – Node names and device parameters – Resistance.

Figure 3.28 shows the Settings – Resistor – Node Connections – Node names and Device Parameters - Resistance.

Right-Click in the Model Builder window on Electrical Circuit (cir).

Select > Inductor from the Pop-up menu.

This inductor is the second component in the series circuit and the first component in the parallel circuit.

Enter 2, 0 in the Settings – Inductor – Node Connections – Node names entry windows.

Enter 100[mH] in the Settings – Inductor– Inductance entry window.

See Figure 3.29.

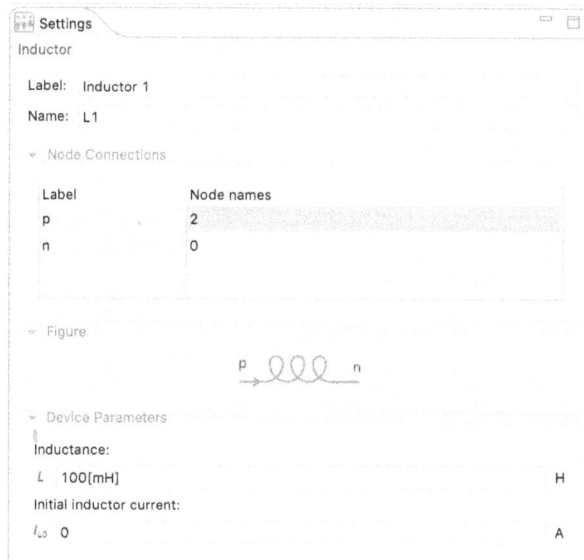

FIGURE 3.29 Settings – Inductor – Node connections – Node names and device parameters – Inductance.

Figure 3.29 shows the Settings – Inductor – Node Connections – Node names and Device Parameters — Inductance.

Right-Click in the Model Builder window on Electrical Circuit (cir).

Select > Capacitor from the Pop-up menu.

This capacitor is the third component in the series circuit and the second component in the parallel circuit.

Enter 2, 0 in the Settings – Capacitor – Node Connections – Node names entry windows.

Enter 1000[nF] in the Settings – Capacitor – Capacitance entry window.

See Figure 3.30.

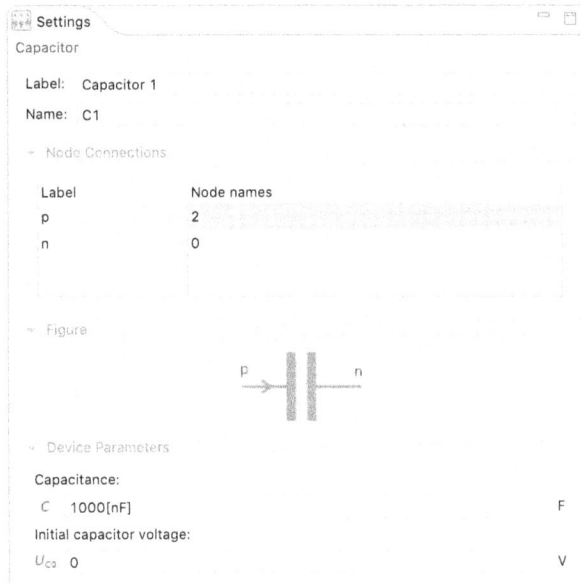

FIGURE 3.30 Settings – Capacitor – Node connections – Node names and device parameters – Capacitance.

Figure 3.30 shows the Settings – Capacitor – Node Connections – Node names and Device Parameters -– Capacitance.

Now that the building of the series-resistor parallel inductor-capacitor circuit model is complete, the next step is to prepare to solve the model.

NOTE *The default settings for the Time Dependent Study Type are not appropriate for the solution of this problem and thus need to be modified.*

Click > Study 1 twistie (as needed).

Select > Step 1: Time Dependent. See Figure 3.31

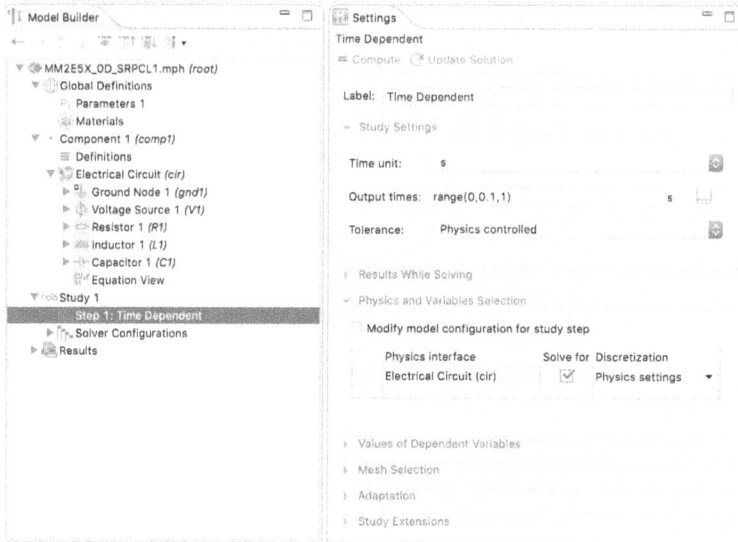

FIGURE 3.31 Settings – Time dependent window.

Figure 3.31 shows the Settings – Time Dependent window.

Click the Range button (at the right of the Times entry window).

Enter > Start = 0.0, Step = 1e-4, Stop = 3.7e-2.

Click > Replace.

Right-Click > Study 1.

Select > Compute.

NOTE

The default settings for the 1D Plot of the results of this model are not appropriate and thus the settings need to be modified.

Right-Click > Results > Select >1D Plot Group.

Right-Click > 1D Plot Group 1 > Select > Global.

Click > Settings Global >y-Axis Data > Replace Expression.

Select > Component 1 > Electrical Circuit > Devices > R1> comp1.cir.R1_v – Voltage across device R1.

Click > Settings Global > Plot.

Click > 1D Plot Group > Settings 1D Plot Group > Legend > Position > Lower right.

See Figure 3.32 for the computed solution.

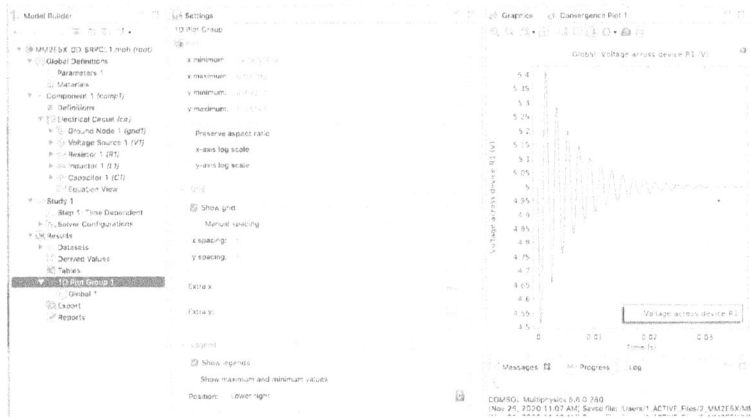

FIGURE 3.32 Damped voltage oscillation across R1 as a function of time.

Figure 3.32 shows the Damped Voltage Oscillation across R1 as a function of Time.

Be sure to save the just completed model.

0D Basic Circuit Models Analysis and Conclusions

The 0D Resistor-Capacitor Series Circuit Model and the 0D Inductor-Resistor Series Circuit Model demonstrate that the addition of inductance or capacitance into a circuit along with resistance results in a finite transient response time. The transient response time of such circuits increases proportionally to the value of either the inductance or the capacitance {3.24}.

The 0D Series-Resistor Parallel-Inductor-Capacitor Circuit Model demonstrates that there is a non-linear interaction that may occur when both inductance and capacitance are incorporated into the same circuit. This potentially useful interaction is known as resonance {3.25}. Resonance can be very useful, somewhat annoying, or very destructive, depending upon whether it is purposely designed into the circuit or comes about inadvertently as a result of a lack of careful design.

Useful resonance is found in designs for oscillators, clocks, mechanical vibrators, etc. Somewhat annoying resonance (ringing) can occur in almost any circuit due to unanticipated (parasitic) capacitances and/or inductances incorporated into the physical circuit inadvertently in the design process. Very destructive resonances are typically indicative of flaws incorporated into the artifact during the design process.

FIRST PRINCIPLES AS APPLIED TO 0D MODEL DEFINITION

First Principles Analysis derives from the fundamental laws of nature. In the case of models in this book or from any other source, the modeler should be able to demonstrate that the calculated results derived from the models are consistent with the laws of physics and the basic observed properties of materials. In the case of this Classical Physics Analysis, the laws of conservation in physics require that what goes in (as mass, energy, charge, etc.) must come out (as mass, energy, charge, etc.) or must accumulate within the boundaries of the model. To do otherwise violates fundamental principles.

In the COMSOL Multiphysics software, the default interior boundary conditions are set to apply the conditions of continuity in the absence of sources (e.g. heat generation, charge generation, molecule generation, etc.) or sinks (e.g. heat loss, charge recombination, molecule loss, etc.).

The careful modeler must be able to determine by inspection of the model that the appropriate factors have been considered in the development of the specifications for the various geometries, for the material properties of each domain, and for the boundary conditions. He must also be knowledgeable of the implicit assumptions and default specifications that are normally incorporated into the COMSOL Multiphysics software model, when a model is built using the default settings.

Consider, for example, the three circuit models developed earlier in this chapter. By choosing to develop those models in the simplest 0D non-geometrical mode, the implicit assumption is that the circuit is isothermal (no heat generation or loss). That assumption essentially eliminates the consideration of thermally related changes (mechanical, electrical, etc.). It also assumes that the materials are homogeneous and isotropic and that there are no thin electrical contact barriers at the electrical junctions. None of these assumptions are typically true in the general case. However, by making such assumptions, it is possible to easily build a 0D First Approximation Model.

> *A First Approximation Model is one that captures all the essential features of the problem that needs to be solved, without dwelling excessively on small details. A good First Approximation Model will yield an answer that enables the modeler to determine if he needs to invest the time and the resources required to build a more highly detailed model.*

NOTE

Also, the modeler needs to remember to name model parameters carefully, as pointed out in Chapter 1.

REFERENCES

1. COMSOL Chemical Reaction Engineering Module Users Guide, pp. 17

2. COMSOL Multiphysics Reference Manual, pp. 108

3. COMSOL Optimization Module Users Guide, pp. 8

4. *https://en.wikipedia.org/wiki/Electrical_network*

5. *https://en.wikipedia.org/wiki/Electronic_circuit*

6. *https://en.wikipedia.org/wiki/Georg_Ohm*

7. *https://en.wikipedia.org/wiki/Ohm%27s_law*

8. *https://en.wikipedia.org/wiki/James_Clerk_Maxwell*

9. *https://en.wikipedia.org/wiki/Electromotive_force*

10. *https://en.wikipedia.org/wiki/Charge_conservation*

11. *https://en.wikipedia.org/wiki/Gustav_Kirchhoff*

12. *https://en.wikipedia.org/wiki/Kirchhoff%27s_circuit_laws*

13. *https://en.wikipedia.org/wiki/Ewald_Georg_von_Kleist*

14. *https://en.wikipedia.org/wiki/Capacitor*

15. *https://en.wikipedia.org/wiki/Pieter_van_Musschenbroek*

16. *https://en.wikipedia.org/wiki/Dielectric*

17. *https://en.wikipedia.org/wiki/Michael_Faraday*

18. *https://en.wikipedia.org/wiki/Joseph_Henry*

19. *https://en.wikipedia.org/wiki/Electromagnetic_induction*

20. *https://en.wikipedia.org/wiki/Maxwell%27s_equations*

21. E.U. Condon and H. Odishaw, "Handbook of Physics", McGraw-Hill, New York, 1958, pp. 4-28 – 4-46

22. *https://en.wikipedia.org/wiki/Inductance*

23. *https://en.wikipedia.org/wiki/Capacitance*

24. J. J. Brophy, "Basic Electronics for Scientists", McGraw-Hill, New York, 1966, pp. 104 - 111

25. *https://en.wikipedia.org/wiki/Resonance*

SUGGESTED MODELING EXERCISES

1. Build, mesh, and solve the 0D Resistor-Capacitor Series Circuit Model problem presented earlier in this chapter.

2. Build, mesh, and solve the 0D Inductor-Resistor Series Circuit Model problem presented earlier in this chapter.

3. Build, mesh, and solve the 0D Series-Resistor Parallel-Inductor-Capacitor Circuit Model problem presented earlier in this chapter.

4. Change the values of the components comprising the 0D Resistor-Capacitor Series Circuit Model problem. Compute the new solution. Analyze, compare, and contrast the results with the new component values to the results found earlier.

5. Change the values of the components comprising the 0D Inductor-Resistor Series Circuit Model problem. Compute the new solution. Analyze, compare, and contrast the results with the new component values to the results found earlier.

6. Change the values of the components comprising the 0D Series-Resistor Parallel-Inductor-Capacitor Circuit Model problem. Compute the new solution. Analyze, compare, and contrast the results with the new component values to the results found earlier.

1D MODELING USING COMSOL MULTIPHYSICS 5.x

GUIDELINES FOR 1D MODELING IN 5.x

NOTE

In this chapter, the modeler is introduced to the development and analysis of 1D models. 1D models have a single geometric dimension (x). 1D models are typically classed by level of difficulty as being of two types, introductory and advanced. In 5.x, there are two types of 1D geometries: 1D and 1D Axisymmetric.

In this text, three (3) introductory 1D models and one (1) advanced 1D model will be presented. Such 1D models have proven to be very valuable to the science and engineering communities, both in the past and currently, as first-cut evaluations of potential physical behavior under the influence of external stimuli. Those model responses and other such ancillary information are then gathered and screened early in a project for potential later use in building higher-dimensionality (2D, 3D, etc.) field-based (electrical, magnetic, etc.) models.

For information on the development of basic and advanced models using the 1D Axisymmetric coordinate system, the modeler should consult the 5.x literature {4.1}.

1D Modeling Considerations

1D Modeling can potentially be both the least difficult and the most difficult type of model to build, no matter which modeling software is used. In a 1D model, the modeler can only have a single dimension (a single line or a sequence of line segments) as the modeling space. The potentially least difficult aspect of 1D model building arises from the fact that the geometry is dimensionally simple. However, the underlying physics in a 1D model can range from relatively easy (simple) to extremely complicated (complex).

The 1D model implicitly assumes that the modeling properties, such as the energy flow, the materials properties, the environment, and any other unspecified conditions and variables of interest are homogeneous, isotropic, and/or constant throughout the entire domain, both within the model and in the environs of the model. In other words, the properties assigned to the 1D model are representative of the properties of typical adjacent non-modeled regions. Considering that, the modeler needs to ensure that all of the basic modeling conditions and associated parameters have been properly considered, defined, and set to the appropriate value(s).

NOTE

For any exploratory model built, the modeler should be able to anticipate reasonably accurately the expected numerical results of the model calculation. Calculated model solution values that widely deviate from the anticipated (estimated) values or from the comparison values measured in experimentally derived realistic models are probably indicative of one or more modeling errors either in the original model design, in the earlier model analysis, in the understanding of the underlying physics, or are simply due to incorporated human errors.

1D BASIC MODELS

1D KdV Equation Model

The KdV Equation {4.2} is a well-known example of a group of nonlinear partial differential equations {4.3} termed exactly solvable {4.4}. The exactly solvable type of equation has solutions that can be specified with exactness and precision.

NOTE
Nonlinear partial differential equations are extremely important in the formulation of an accurate mathematical description of physical systems {4.5}. Nonlinear partial differential equations are inherently difficult to solve and when solved often require a unique approach for the solution of each different type of equation.

Diederik Korteweg and Gustav de Vries solved the KdV equation in 1895. The KdV equation mathematically describes the propagation of a surface disturbance on a shallow canal. Their effort to solve this wave propagation problem was enabled as a result of the availability of earlier observations (data) by John Scott Russell in 1834 {4.6} and others. Subsequent activity in this mathematical area has led to soliton applications of the KdV equation in magnetics {4.7} and optics {4.8}. Work on using the KdV equation in the solution of soliton propagation problems is currently an active area of research.

The following numerical solution model (MM2E5X_1D_KdV_1.mph) is based on a model (kdv_equation) that is currently developed by COMSOL for distribution with 5.x as an Equation-Based Model. In the first part of this chapter, we will demonstrate how to build a model in 5.x for the solution of the KdV equation.

NOTE
It is important for the modeler to personally build each model presented within the text. There is no substitute for the hands-on experience of actually building, meshing, solving, and viewing the results of a solved model. The inexperienced modeler will many times make and subsequently correct errors, adding to his experience and his fund of modeling knowledge. Solving even the simplest model will expand the modeler's fund of knowledge.

The KdV Equation (as written in standard notation) is:

$$\partial_t u + \partial_x^3 u + 6u\partial_x u = 0 \tag{4.1}$$

In the 5.x Model Library the models.mph.kdv_equation.pdf documentation, the formula is shown as:

$$u_t + u_{xxx} = 6uu_x \quad in \ \Omega = [-8, 8] \tag{4.2}$$

The difference between the two equations is that (4.2) is the negative form of (4.1), which will be adjusted during the results analysis.

The boundary conditions are periodic, as shown in (4.3)

$$u(-8, t) = u(8, t) \quad periodic \tag{4.3}$$

The initial condition for this model is:

$$u(x, 0) = -6 \sec h^2(x) \tag{4.4}$$

Once the modeler builds and solves this model, it will be seen that the pulse immediately divides into two soliton pulses, with different width and propagation speeds.

NOTE   *5.x does not evaluate third derivatives directly. Thus the original equation (4.2) needs to be rewritten as a system of two variables so that it can be solved in 5.x.*

Rewriting (4.2):

$$u_{1t} + u_{2x} = 6u_1 u_{1x} \tag{4.5}$$

and

$$u_2 = u_{1xx} \tag{4.6}$$

and

$$u(x, t) = \frac{1}{2} \lambda \sec h^2 \left(\frac{\sqrt{\lambda}}{2} (x - \lambda t - a) \right) \tag{4.7}$$

Where

λ = phase velocity.

a = arbitrary constant.

x = position.

t = time.

Building the 1D KdV Equation Model

Startup 5.x.

Click > Model Wizard > Select > 1D, on the Select Space Dimension page.

Click > Twistie for Mathematics in the Select Physics window.

Click > Twistie for PDE Interfaces > General Form PDE (g).

Click > Add button.

In the Model Wizard > Dependent variables > Number of dependent variables edit window.

Enter 2. Press Return.

See Figure 4.1a.

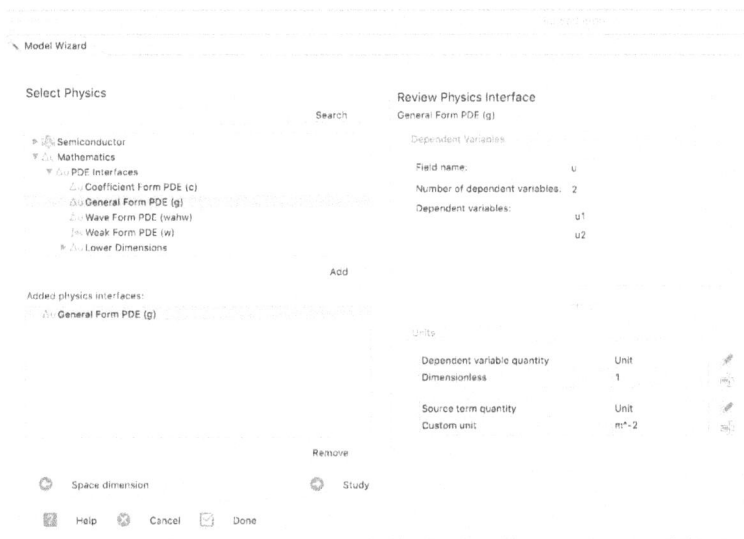

FIGURE 4.1a Desktop display – Model wizard – Dependent variables – Number of dependent variables edit window.

Figure 4.1a shows the Desktop Display – Model Wizard – Dependent variables – Number of dependent variables edit window.

Click > Study (Right Pointing Arrow).

Select > General Studies > Time Dependent in the Select Study window. See Figure 4.1b.

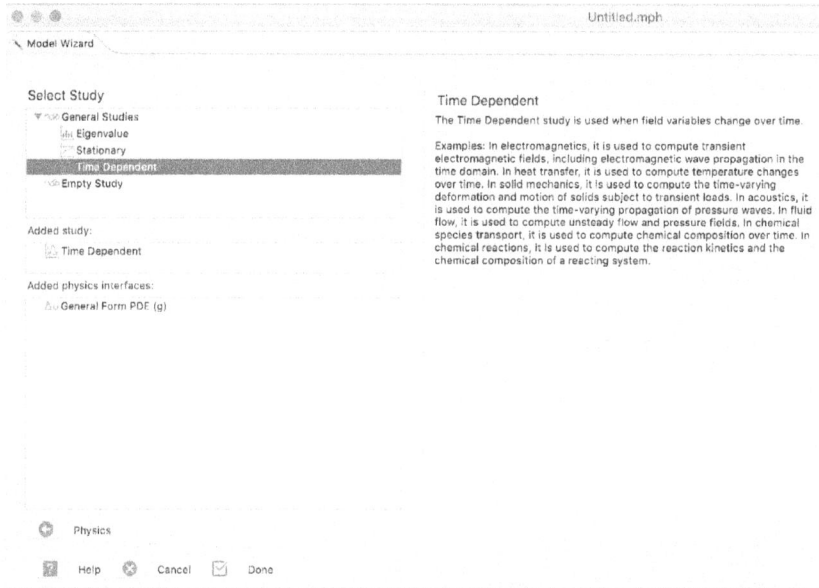

FIGURE 4.1b Desktop display – Model wizard – Select study – Time dependent.

Figure 4.1b shows the Desktop Display – Model Wizard – Select Study – Time Dependent.

Click > Done (Checked Box button).

Click > Untitled.mph (*root*).

Click > Settings Untitled.mph > Unit System.

Click > Unit System Twistie (if needed).

Select > None from the Pull-down Menu.

See Figure 4.2a

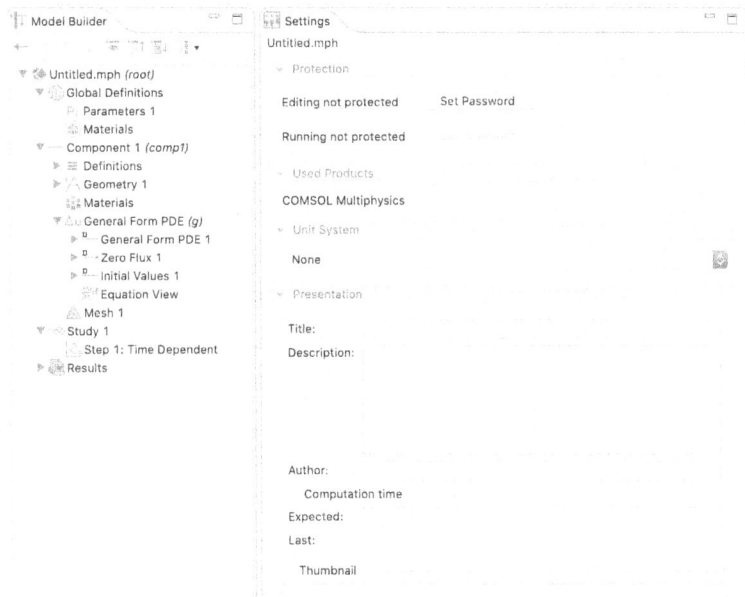

FIGURE 4.2a Desktop display for settings – Root – Unit system – None.

Figure 4.2a shows the Desktop Display for Settings – Root – Unit System – None.

Click > Save As > MM2E5X_1D_KdV_1.mph.

See Figure 4.2b.

FIGURE 4.2b Desktop display for the MM2E5X_1D_KdV_1.mph model.

Figure 4.2b shows the Desktop Display for the MM2E5X_1D_KdV_1.mph model.

Right-Click in the Model Builder window on Geometry 1.

Select > Interval from the Pop-up menu.

In the Settings – Interval window > Click > Interval > Specify: Pull-down menu > Select > Interval lengths.

Enter –8 in the Settings – Interval – Left endpoint entry window.

In the Settings – Interval window > Click > Interval > Length source Pull-down menu > Select > Vector.

Enter 16 in the Settings – Interval – Lengths: entry window.

Click > Build Selected.

See Figure 4.3.

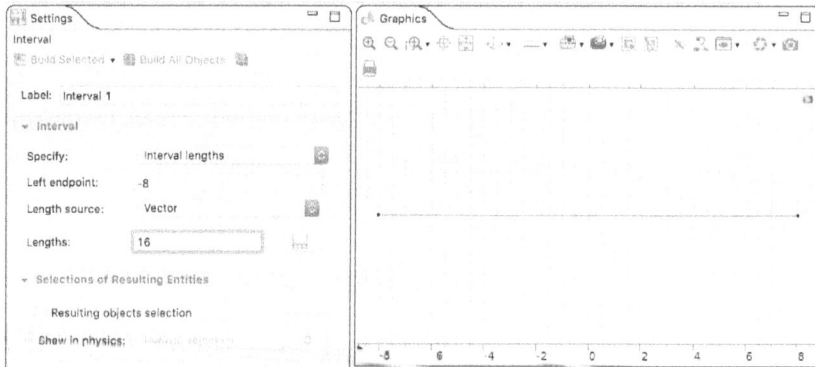

FIGURE 4.3 Desktop display – Settings – Interval entry windows.

Figure 4.3 shows the Desktop Display – Settings – Interval entry windows.

Right-Click in the Model Builder window on Component 1 (*comp1*) > Δu General Form PDE (g).

Select > Periodic Condition from the Selection Pop-up menu.

Select > Settings – Periodic Condition – Boundary Selection > All boundaries from the Pop-up menu.

See Figure 4.4.

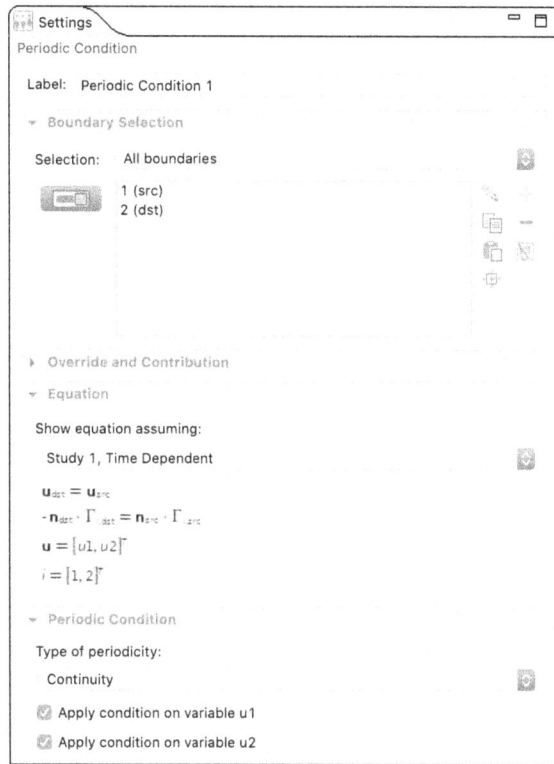

FIGURE 4.4 Desktop display – Settings – Periodic condition 1 – Boundary selection window.

Figure 4.4 shows the Desktop Display – Settings – Periodic Condition 1 – Boundary Selection window.

Click in the Model Builder window on Component 1 (*comp1*) > Δu General Form PDE (g) > General Form PDE 1.

Enter u2 in the first row of the Settings – General Form PDE – Conservative Flux – Γ edit-field array.

Enter u1x in the second row of the Settings – General Form PDE – Conservative Flux – Γ edit-field array.

Enter 6*u1*u1x in the first row of the Settings – General Form PDE – Source Term – *f* edit-field array.

Enter u2 in the second row of the Settings – General Form PDE – Source Term – *f* edit-field array.

Enter 0 in the second column - second row of the Settings – General Form PDE – Damping or Mass Coefficient – d_a edit-field array.

See Figure 4.5.

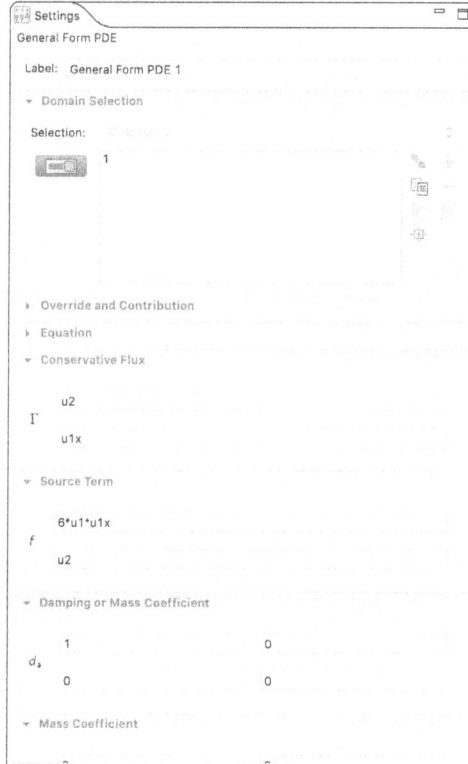

FIGURE 4.5 Desktop display – Settings – General form PDE coefficients.

Figure 4.5 shows the Desktop Display – Settings – General Form PDE coefficients.

Click in the Model Builder window on Component 1 (*comp1*) 1 > Δu General Form PDE (g) > Initial Values 1.

Enter –6*sech(x)^2 in the Settings – Initial Values – Initial value for *u*1.

Enter –24*sech(x)^2*tanh(x)^2+12*sech(x)^2*(1-tanh(x)^2) in the Settings – Initial Values – Initial value for *u*2.

See Figure 4.6.

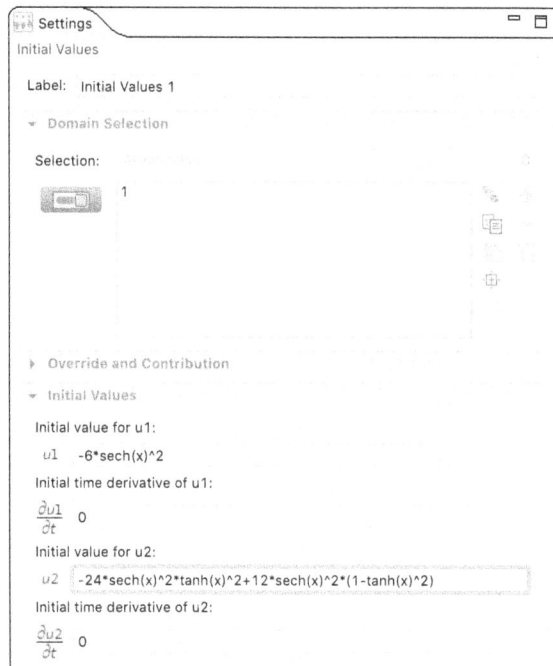

FIGURE 4.6 Desktop display – Settings – Initial values coefficients.

Figure 4.6 shows the Desktop Display – Settings – Initial Values coefficients.

Mesh 1

NOTE *The default settings for the Mesh Type are not appropriate for the solution of this problem and thus need to be modified.*

Right-Click in the Model Builder window on Component 1 (*comp1*) > Mesh 1.

Select > Edge.

Click > Size.

Click > Custom button.

Enter 0.1 in the Settings – Size – Element Size Parameters – Maximum element size edit window.

NOTE *The maximum element size setting (0.1) for the Mesh Type is chosen to ensure adequate resolution of the mesh for use with the hyperbolic trigonometric functions in the solution of this problem.*

See Figure 4.7.

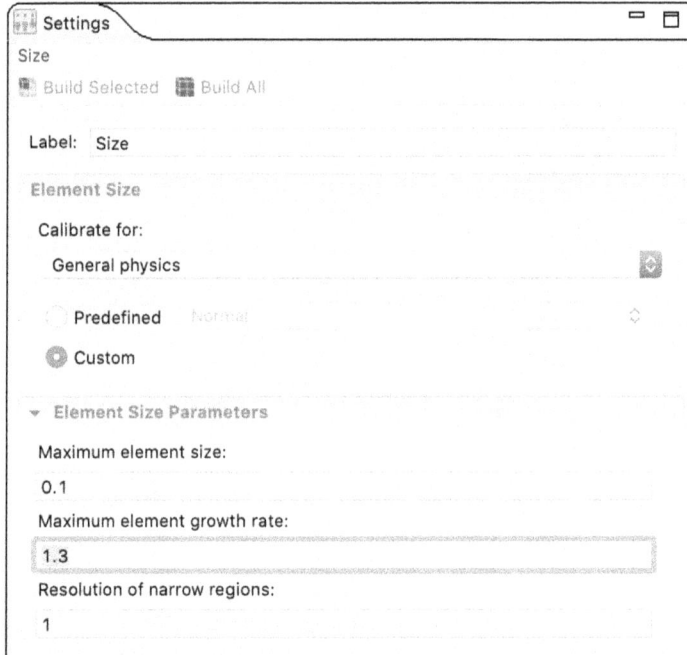

FIGURE 4.7 Desktop display – Settings – Size – Element size parameters – Maximum element size coefficient.

Figure 4.7 shows the Desktop Display – Settings – Size – Element Size Parameters – Maximum element size coefficient.

Click > Build All button.

After meshing, the modeler should see a message in the message window about the number of elements (160 elements) in the mesh.

See Figure 4.8.

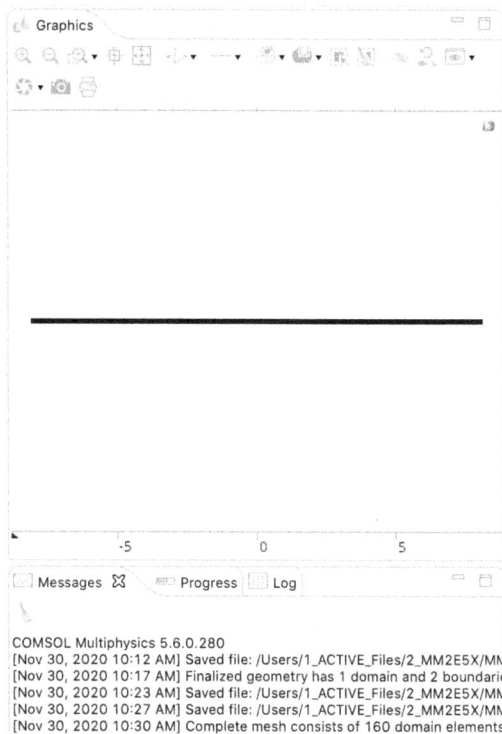

FIGURE 4.8 Desktop display – Graphics – Meshed domain.

Figure 4.8 shows the Desktop Display – Graphics – Meshed Domain.

Study 1

NOTE *The default settings for the Time Dependent Study Type are not appropriate for the solution of this problem and thus need to be modified.*

Click > Study 1 twistie (if needed).

Click > Step 1: Time Dependent.

Click the Range button (at the right of the Times entry window).

Enter > Start = 0, Step = 2.5e-2, Stop = 2.0.

Click > Replace.

Click > Settings-Time Dependent >Study Settings > Tolerance Pull-down menu.

Select > User controlled.

Enter 1.0e-4 in the Relative tolerance edit window.

See Figure 4.9.

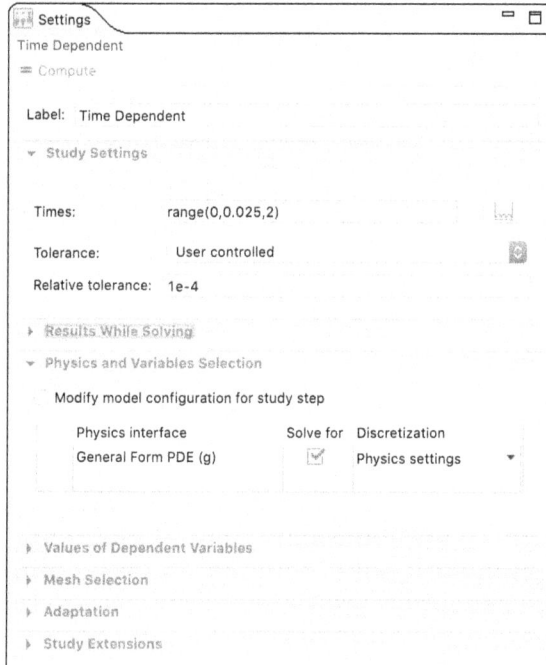

FIGURE 4.9 Desktop display – Settings – Time dependent – Study settings.

Figure 4.9 shows the Desktop Display – Settings – Time Dependent – Study Settings.

Study 1 > *Solver 1*

In Model Builder,

Right-Click > Study 1.

Select > Show Default Solver.

NOTE *The default settings for the default solver for this model are not appropriate and thus the settings need to be modified.*

See Figure 4.10a.

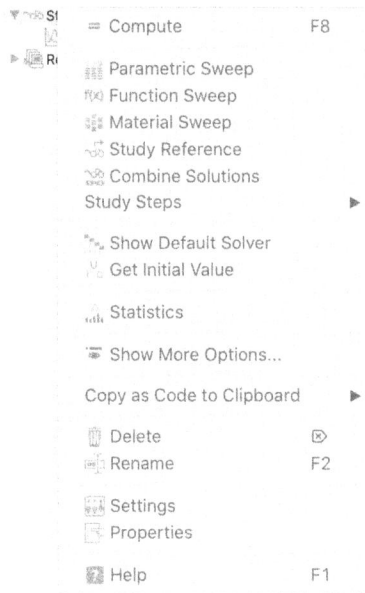

FIGURE 4.10a Desktop display – Study 1 pop-up menu.

Figure 4.10a shows the Desktop Display – Study 1 Pop-up Menu.

Click > Solver Configurations twistie (if needed).

Click > Solution 1 twistie.

Click > Time Dependent Solver 1.

In the Settings – Time Dependent Solver – Time Stepping section,

Select > Method > Generalized alpha from the pull-down menu.

Click > Absolute Tolerance twistie.

Select > Settings-Time Dependent Solver > Absolute Tolerance > Tolerance method Pull-down menu > Manual.

Enter 1e-5 in the Absolute Tolerance – Tolerance edit window.

NOTE

The solver type and the absolute tolerance are chosen to ensure adequate resolution of the convergence process for use with the hyperbolic trigonometric functions in the solution of this problem.

See Figure 4.10b.

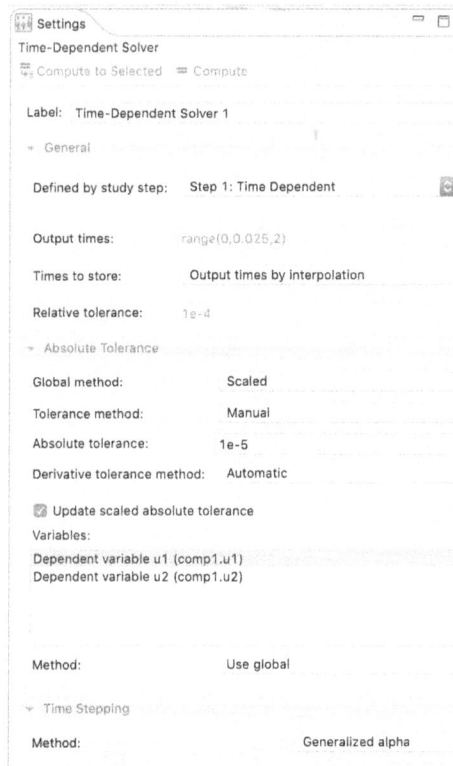

FIGURE 4.10b Desktop display – Settings – Time-dependent solver– Solver settings.

Figure 4.10b shows the Desktop Display – Settings – Time-Dependent Solver– Solver Settings.

In Model Builder, Right-Click Study 1 >Select > Compute.

Results

In Model Builder,

Click > Results > 1D Plot Group 1.

In Settings – 1D Plot Group > Data > Time selection Pull-down menu,

Select: > From list.

In the Times window, locate and Select 0.25.

Click > Plot.

See Figure 4.11.

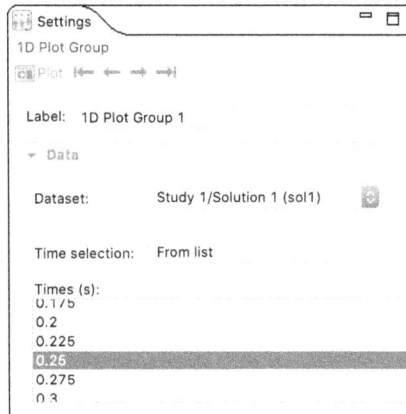

FIGURE 4.11 Desktop display – Settings – 1D plot group – Time data settings.

Figure 4.11 shows the Desktop Display – Settings – 1D Plot Group – Time Data Settings.

See Figure 4.12.

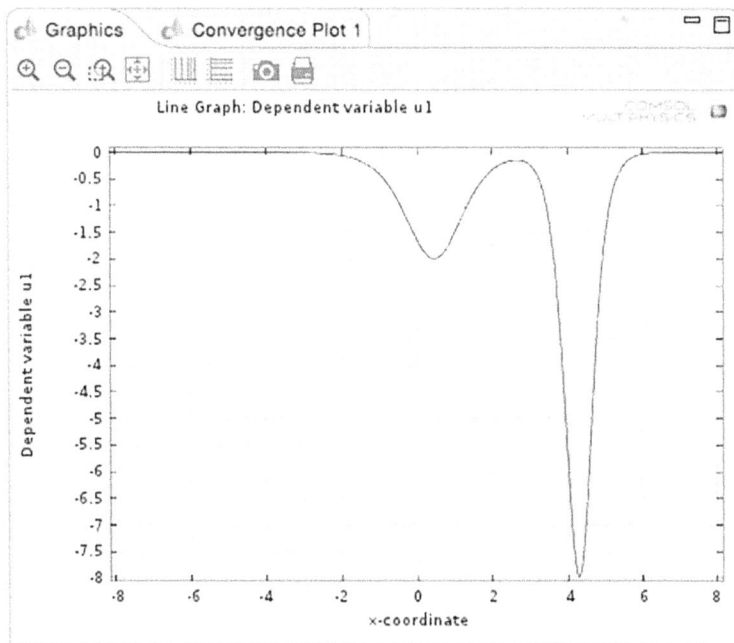

FIGURE 4.12 Desktop display – Graphics – Initial KdV solution plot.

Figure 4.12 shows the Desktop Display – Graphics – Initial KdV Solution Plot.

As mentioned earlier, the plot needs to be inverted.

In Model Builder,

Click > Results > 1D Plot Group 1 twistie.

Click > Results > Line Graph 1.

In Settings – Line Graph – Y-Axis Data – Expression,

Enter -u1.

Click > Plot button.

See Figure 4.13.

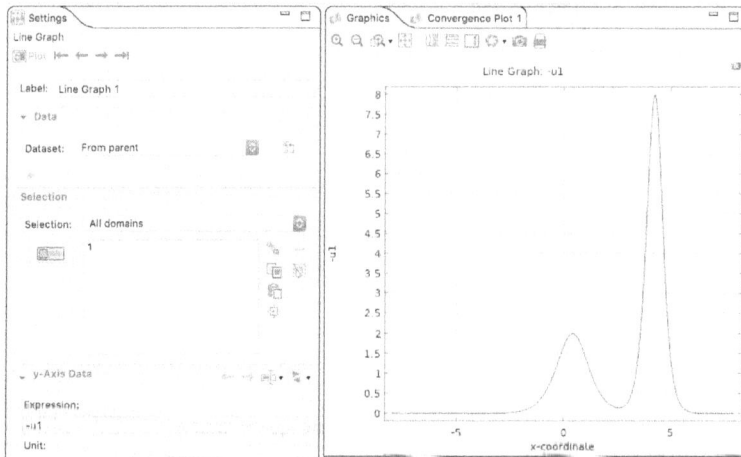

FIGURE 4.13 Desktop display – Graphics – KdV solution plot

Figure 4.13 shows the Desktop Display – Graphics – KdV Solution Plot.

KdV Equation Animation

To demonstrate the propagation of the soliton pulse or pulses in 5.x, the modeler can create an animation and play it in the Graphics window.

In Model Builder,

Right-Click > Results - Export,

Select > Animation > Player.

In Settings-Animation > Scene > Subject Pull-down menu

Select > 1D Plot Group 1 (if needed).

Right-Click > Model Builder > Results > Export > Animation 1: Select > Play.

The modeler can also save the just executed animation to a file.

In Model Builder,

Right-Click > Results – Export.

Select > Animation > File.

In Settings – Animation,

Click > Results – Export – Animation 2.

In Settings – Animation – Output,

Select > Output type Movie from the pull-down menu (if needed).

Select > Format GIF from the pull-down menu (if needed).

NOTE *The detailed procedures for Animation in 5.x on the Macintosh and the PC are different. The Macintosh uses the GIF format. The PC uses the AVI format. If the modeler tries to use AVI on the Macintosh, an error will result.*

In Settings – Animation – Frames,

Click > Lock aspect ratio check box.

In Settings – Animation – Advanced,

Click > twistie (as needed).

Click > Antialiasing check box (as needed).

In the Settings Animation window, on the toolbar,

Click > the Export button.

Select the desired location for saving the movie.

Enter the desired File Name (MM2E5X_1D_KdV_1.gif) in the Save-As edit window.

Click > Save.

Click > Save for the completed MM2E5X_1D_KdV_1.mph KdV Equation model.

1D KdV Equation Model Summary and Conclusions

The 1D KdV Equation model is a powerful tool that can be used to explore soliton wave propagation in many different media (e.g. physical waves in liquids, electromagnetic waves in transparent media, etc.). As has been shown earlier in this chapter, the KdV Equation is easily and simply modeled with a 1D PDE mode model.

1D Telegraph Equation Model

The Telegraph Equation {4.9} was developed by Oliver Heaviside {4.10} and first published in the 1880s. The Telegraph Equation is based on a lumped-constant, four (4) terminal electrical component model, typically with earth (ground) as the return path, as shown in Figure 4.14.

See Figure 4.14.

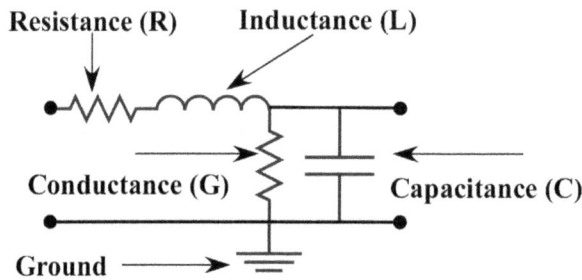

FIGURE 4.14 Telegraph equation lumped-constant circuit.

Figure 4.14 shows the Telegraph Equation Lumped-Constant Circuit.

In this lumped-constant schematic model for telegraph wires (and similar transmission lines), there are four basic electrical components: R (Resistance) per unit of length, L (Inductance) per unit of length, G (Conductance) per unit of length, and C (Capacitance) per unit of length. The differential equations for Voltage (V) and Current (I) have the same form, as shown in equations 4.8 and 4.9.

Equation 4.8 shows the partial differential equation for voltage (V):

$$\frac{\partial^2}{\partial x^2}V = LC\frac{\partial^2}{\partial t^2}V + (RC + GL)\frac{\partial}{\partial t}V + GRV \qquad (4.8)$$

Equation 4.9 shows the partial differential equation for current (I):

$$\frac{\partial^2}{\partial x^2}I = LC\frac{\partial^2}{\partial t^2}I + (RC + GL)\frac{\partial}{\partial t}I + GRI \tag{4.9}$$

Equations 4.8 and 4.9 are similar in form to the equation 4.10 as shown here for the COMSOL Multiphysics Telegraph Equation Model:

$$u_{tt} + (\alpha + \beta)u_t + \alpha\beta u = c^2 u_{xx} \tag{4.10}$$

Where α and β are positive constants, c is the transport velocity and u is the voltage.

Restating equation 4.8 in subscript notation:

$$u_{xx} = LC\,u_{tt} + (RC + GL)u_t + GR\,u \tag{4.11}$$

And rearranging the terms of equation 4.10:

$$u_{xx} = \frac{1}{c^2}u_{tt} + \frac{1}{c^2}(\alpha + \beta)u_t + \frac{1}{c^2}\alpha\beta\,u \tag{4.12}$$

Comparing equations 4.11 and 4.12 yields:

$$LC = \frac{1}{c^2} \tag{4.13}$$

And

$$\alpha + \beta = \frac{(RC + GL)}{LC} \tag{4.14}$$

Also:

$$\alpha\beta = \frac{GR}{LC} \tag{4.15}$$

Solving for α and β:

$$\alpha = \frac{CGL + C^2R - \sqrt{-4CGLR + (-CGL - C^2R)^2}}{2L}$$

and (4.16)

$$\beta = \frac{CGL + C^2R + \sqrt{-4CGLR + (-CGL - C^2R)^2}}{2L}$$

Or:

$$\alpha = \frac{CGL + C^2R + \sqrt{-4CGLR + (-CGL - C^2R)^2}}{2L}$$

and

$$\beta = \frac{CGL + C^2R - \sqrt{-4CGLR + (-CGL - C^2R)^2}}{2L}$$

In the event that:

$$R = G == 0 \tag{4.17}$$

Then, the transmission line is considered lossless and the Telegraph Equation becomes:

$$u_{xx} = LC\, u_{tt} \tag{4.18}$$

Building the 1D Telegraph Equation Model

Startup 5.x.

Click > Model Wizard > Select > 1D, on the Select Space Dimension page.

Click > Twistie for Mathematics in the Add Physics window.

Click > Twistie for PDE Interfaces > Coefficient Form PDE (c).

Click > Add button.

Click > Study (Right Pointing Arrow).

Select > General Studies > Time Dependent in the Select Study Type window.

Click > Done (Checked Box button).

Click > Untitled.mph (*root*).

Click > Settings Untitled.mph > Unit System.

Select > None from the Pull-down Menu.

Click > Save As > MM2E5X_1D_TelE_1.mph.

See Figure 4.15.

FIGURE 4.15 Desktop display for the MM2E5X_1D_TelE_1.mph model.

Figure 4.15 shows the Desktop Display for the MM2E5X_1D_TelE_1.mph model.

In the Model Builder window,

Click > Global Definitions > Parameters 1.

In the Settings – Parameters – Parameters edit window,

Enter the parameters as shown in Table 4.1.

TABLE 4.1 Parameters Window.

Name	Expression	Description
c	1	Transport velocity
alpha	0.25	PDE coefficient parameter alpha
beta	0.25	PDE coefficient parameter beta

The modeler should save the parameters file at this time so that if the model needs to be recovered or the parameters need to be modified, they will be available to be changed or corrected without the need to reenter them.

See Figure 4.16.

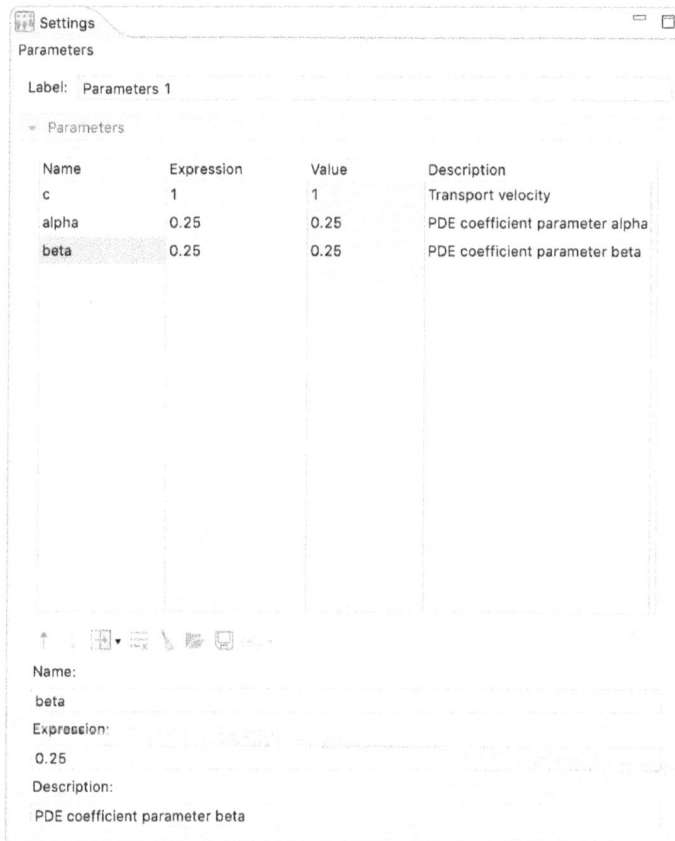

FIGURE 4.16 Desktop display – Settings – Parameters entry windows.

Figure 4.16 shows the Desktop Display – Settings – Parameters entry windows.

Click > Settings – Parameter Save to File button (Disk image).

Enter > MM2E5X_1D_TelE_1_Param.txt in the Save as edit window.

Click > Save button.

Geometry 1

In the Model Builder window,

Right-Click on Component 1 (*comp1*) > Geometry 1.

Select > Interval from the Pop-up menu.

NOTE   *The default settings for the specified interval are adequate for this model.*

In the Model Builder,

Right-Click > Interval 1.

Select > Build Selected.

See Figure 4.17.

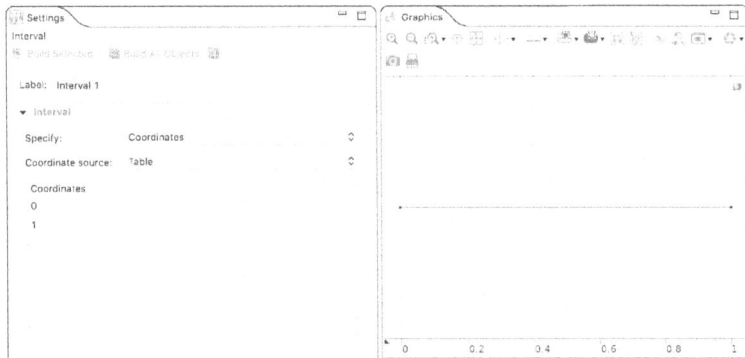

FIGURE 4.17 Desktop display – Graphics – Geometry 1 window.

Figure 4.17 shows the Desktop Display – Graphics – Geometry 1 Window.

PDE

In the Model Builder window,

Click on Component 1 (*comp1*) > Δu Coefficient Form PDE (c) twistie (as needed).

Click > Component 1 (*comp1*) > Δu Coefficient Form PDE (c) >

Coefficient Form PDE 1.

NOTE   *The default settings for the Coefficient Form PDE are not appropriate for the solution of this problem and thus need to be modified. The modeler should note the nominal equation (by Clicking on the Settings – Coefficient Form*

PDE – Equation twistie) and the manner in which the entered coefficients change that equation.

Enter c*c in Settings – Coefficient Form PDE – Diffusion Coefficient – *c* edit window.

Enter alpha*beta in Settings – Coefficient Form PDE – Absorption Coefficient – *a* edit window.

Enter –(alpha+beta)*ut in Settings – Coefficient Form PDE – Source Term – *f* edit window.

Enter 1 in Settings – Coefficient Form PDE – Mass Coefficient – e_a edit window.

Enter 0 in Settings – Coefficient Form PDE – Damping or Mass Coefficient – d_a edit window.

See Figure 4.18.

FIGURE 4.18 Desktop display – Settings – Coefficient form PDE coefficients.

Figure 4.18 shows the Desktop Display – Settings – Coefficient Form PDE coefficients.

Zero Flux 1

The default settings for the Zero Flux 1 set the Neumann Boundary Conditions by default and are appropriate for the solution of this problem. They thus do not need to be modified.

Initial Values 1

Click in the Model Builder window on Component 1 > Δu Coefficient Form PDE (c) > Initial Values 1.

The default settings for the Initial Values are not appropriate for the solution of this problem and thus need to be modified. The entered equation describes a bell-shaped pulse applied to the transmission line.

Enter exp(−3*((x/0.2)−1)^2) in the Settings – Initial Values – Initial value for u.

See Figure 4.19.

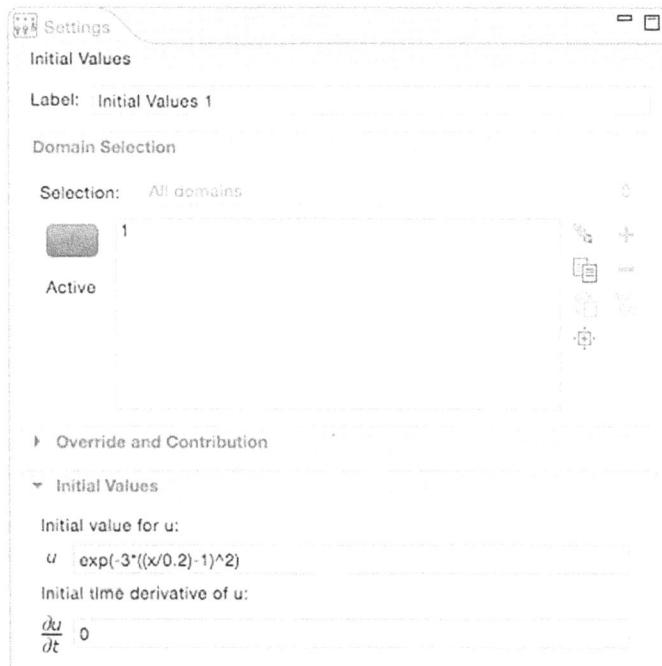

FIGURE 4.19 Desktop display – Settings – Initial values coefficients.

Figure 4.19 shows the Desktop Display – Settings – Initial Values coefficients.

Mesh 1

NOTE *The default settings for the Mesh Type are not appropriate for the solution of this problem and thus need to be modified.*

In the Model Builder window,

Right-Click > Component 1 (*comp1*) > Mesh 1.

Select > Edge.

In the Model Builder window,

Right-Click > Component 1 (*comp1*) > Mesh 1.

Select > Refine.

In the Model Builder window,

Right-Click > Component 1 (*comp1*) > Mesh 1.

Select > Build All.

After meshing, the modeler should see information in the message window about the number of elements (30 domain elements) in the mesh.

See Figure 4.20.

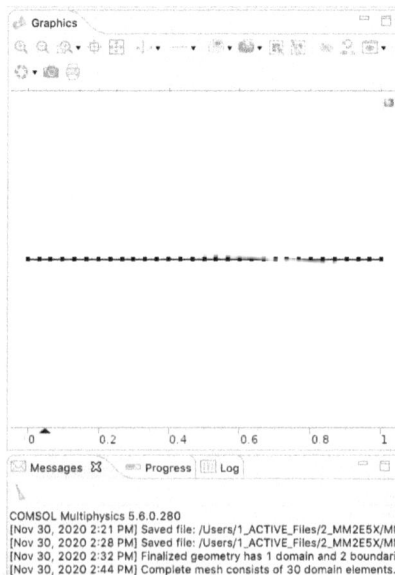

FIGURE 4.20 Desktop display – Graphics – Meshed domain.

Figure 4.20 shows the Desktop Display – Graphics – Meshed Domain.

Study 1

NOTE *A Parametric Sweep study step is added at this point to this model in order to observe the influence of changes in alpha and beta.*

Right-Click > Study 1.

Select > Parametric Sweep.

Click > Settings – Parameter name – Add (Blue Plus Sign).

Click > Active area under Parameter name column title.

Select > alpha (PDE coefficient parameter alpha).

Click > Settings – Parameter name – Add (Blue Plus Sign).

Click > 2nd Active area under Parameter name column title.

Select > beta (PDE coefficient parameter beta).

Enter > 0.25 0.5 1 2 in the Parameter value list edit windows for each of the alpha and beta variables.

NOTE *The individual parametric values for alpha and beta are entered as a string of numbers. Each number is separated from the next number by a single space (e.g. alpha1 alpha2 etc.).*

See Figure 4.21.

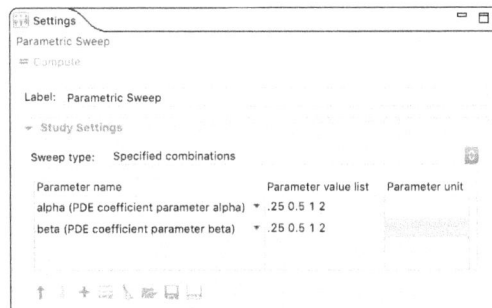

FIGURE 4.21 Desktop display – Settings – Parametric sweep – Study settings edit windows.

Figure 4.21 shows the Desktop Display – Settings – Parametric Sweep – Study Settings edit windows.

Default Solver

Right-Click > Study 1.

Select > Show Default solver.

Click > Component 1 (*comp1*) > Study 1 > Solver Configurations twistie in the Model Builder window (if needed).

Click > Component 1 (*comp1*) > Study 1 > Solver Configurations > Solution 1 twistie in the Model Builder window.

Click > Time-Dependent Solver 1.

Click > Settings – Time-Dependent Solver – Time Stepping twistie.

Click > Settings – Time-Dependent Solver – Time Stepping – Method Pull-down selection menu.

Select > Generalized alpha from the Method pull-down menu.

NOTE *The Generalized alpha ODE solver is usually better for solving this type of wave equation model (d_a=0, e_a=1). It also avoids the need to manually specify time-step size values.*

See Figure 4.22.

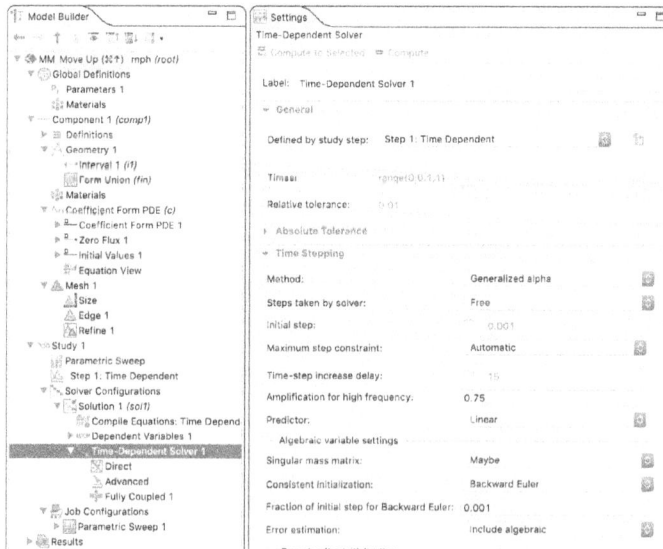

FIGURE 4.22 Desktop display – Settings – Time-dependent solver – Time stepping – Method edit windows.

Figure 4.22 shows the Desktop Display – Settings – Time-Dependent Solver – Time Stepping – Method edit windows.

In Model Builder, Right-Click Study 1 > Select > Compute.

See Figure 4.23a.

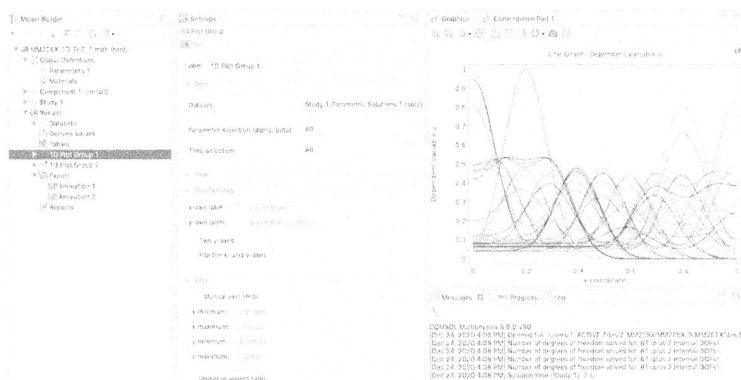

FIGURE 4.23a Desktop display – Calculated parametric sweep results.

Figure 4.23a shows the Desktop Display – Calculated Parametric Sweep Results.

Results

> NOTE
>
> *Upon completion of the Compute step, the Multiphysics default mode displays all of the computed results in the Graphics window. The modeler should note that the shape and amplitude of the pulse change as a function of time and the values of the parameters alpha and beta.*

1D Plot Group 2

In Model Builder,

Right-Click > Results.

Select > 1D Plot Group from the Pop-up menu to create 1D Plot Group 2.

In Settings – 1D Plot Group – Data – Dataset,

Select > Study 1/Parametric Solutions 1 (sol2) from list from the pull-down menu.

In Settings – 1D Plot Group – Data – Parameter selection (alpha, beta),

Select: From list, from the pull-down menu.

In Settings – 1D Plot Group – Data – Parameter values,

Select: 1: alpha=0.25,beta=0.25.

In Settings – 1D Plot Group – Data – Time selection,

Select: Interpolated from the pull-down menu.

In the Times edit window,

Enter > 0 0.5 1.0

See Figure 4.23b.

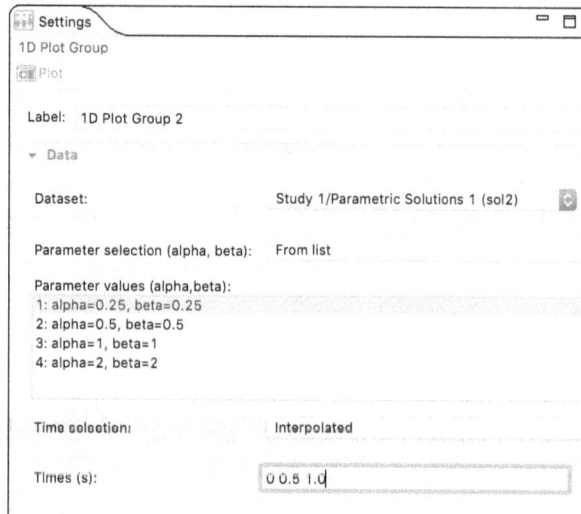

FIGURE 4.23b Desktop display – Settings – 1D plot group – Data edit windows.

Figure 4.23b shows the Desktop Display – Settings – 1D Plot Group – Data edit windows.

Right-Click > Results > 1D Plot Group 2.

Select > Line Graph.

Click > Settings – Line Graph – Selection – Selection.

Select > All domains from the pull-down menu.

Click > Settings – Line Graph – X-Axis Data – Replace Expression (green/red arrows on the far righ)t.

Select > Component 1 (comp1) > Geometry > Coordinate > x - x-coordinate.

See Figure 4.24.

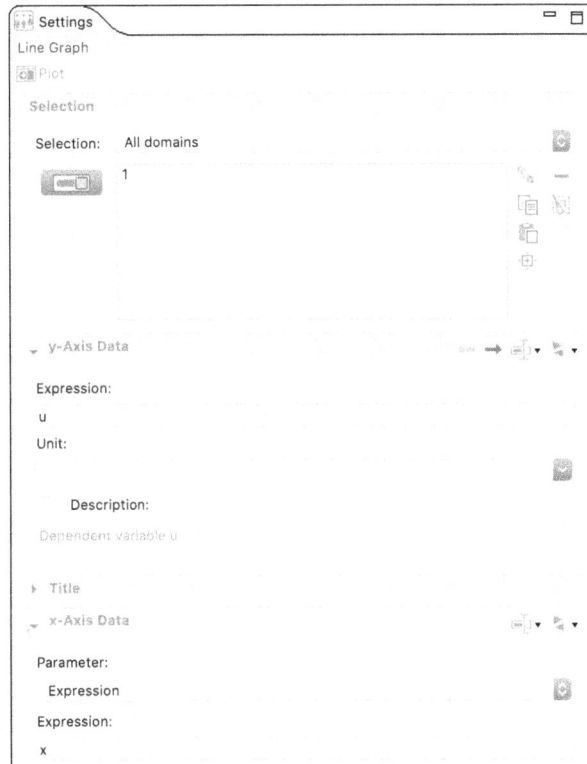

FIGURE 4.24 Desktop display – Settings – Line graph – X-axis data settings.

Figure 4.24 shows the Desktop Display – Settings – Line Graph – X-Axis Data settings.

Click > Plot button.

See Figure 4.25.

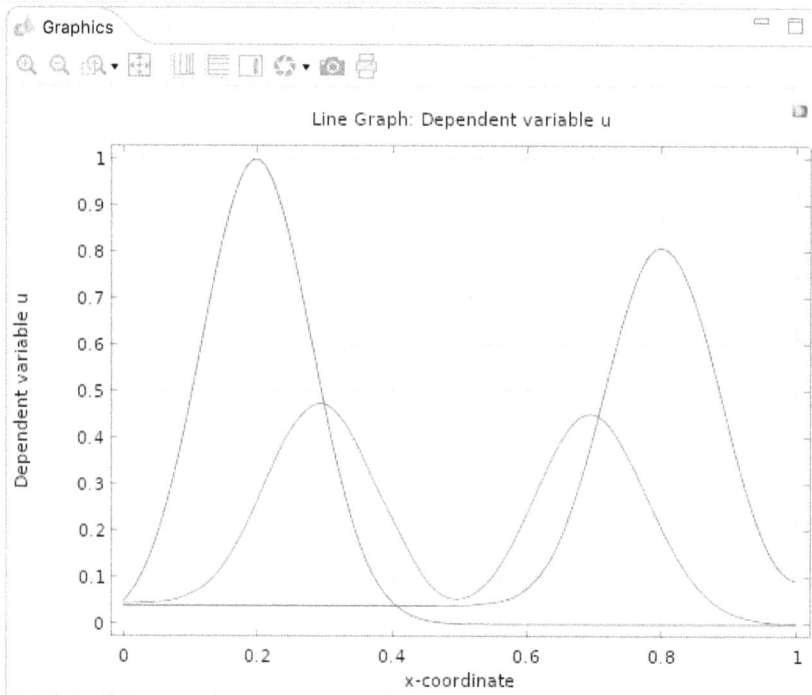

FIGURE 4.25 Desktop display – Graphics – Initial telegraph equation solution plot.

Figure 4.25 shows the Desktop Display – Graphics – Initial Telegraph Equation Solution Plot.

NOTE

The modeler can now note that the shape and amplitude of the pulse change as a function of time and the values of the parameters alpha and beta.

In Model Builder,

Click > Results > 1D Plot Group 2.

In Settings – 1D Plot Group – Data – Parameter values,

Click > 2:alpha=0.5,beta=0.5.

Click > Plot button.

See Figure 4.26.

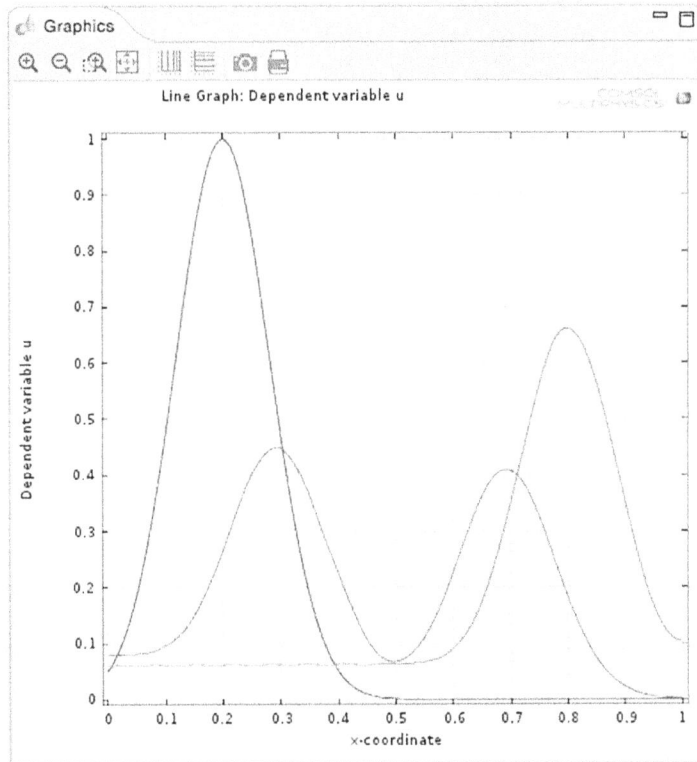

FIGURE 4.26 Desktop display – Graphics – Telegraph equation solution plot (alpha = beta = 0.5).

Figure 4.26 shows the Desktop Display – Graphics – Telegraph Equation Solution Plot (alpha = beta = 0.5).

In Model Builder,

Click > Results > 1D Plot Group 2.

In Settings – 1D Plot Group – Data – Parameter values,

Select: 3: alpha = 1, beta = 1.

Click > Plot button.

See Figure 4.27.

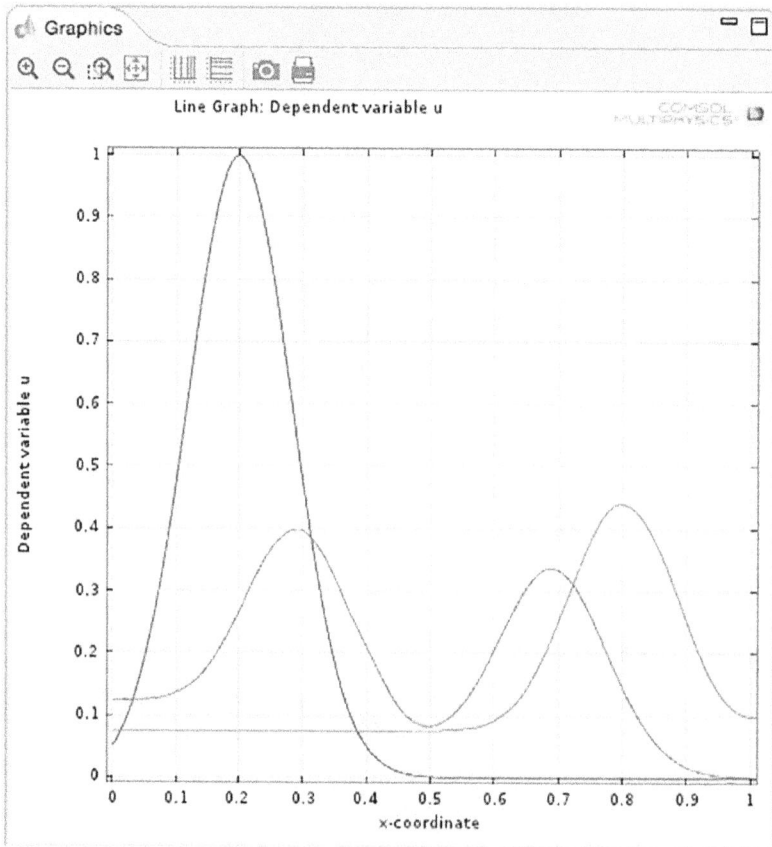

FIGURE 4.27 Desktop display – Graphics – Telegraph equation solution plot (alpha = beta = 1).

Figure 4.27 shows the Desktop Display – Graphics – Telegraph Equation Solution Plot (alpha = beta = 1).

In Model Builder,

Click > Results > 1D Plot Group 2.

In Settings – 1D Plot Group – Data – Parameter values,

Select : 4:alpha=2, beta=2.

Click > Plot button.

See Figure 4.28.

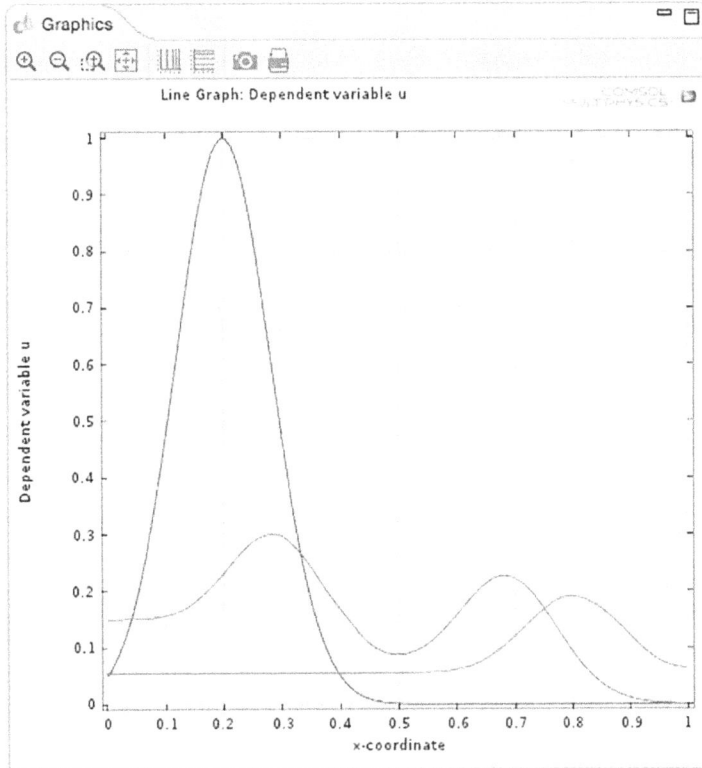

FIGURE 4.28 Desktop display – Graphics – Telegraph equation solution plot (alpha = beta = 2).

Figure 4.28 shows the Desktop Display – Graphics – Telegraph Equation Solution Plot (alpha = beta = 2).

Telegraph Equation Animation

To demonstrate the propagation of the Telegraph Equation pulse or pulses in 5.x, the modeler can create an animation and play it in the Graphics window.

In Model Builder,

Right-Click > Results - Export,

Select > Animation > Player.

In Settings-Animation > Scene > Subject Pull-down menu

Select > 1D Plot Group 1 (if needed).

Right-Click > Model Builder > Results > Export > Animation 1: Select > Play.

The modeler can also save the just executed animation to a file.

In Model Builder,

Right-Click > Results – Export.

Select > Animation > File.

In Model Builder,

Click > Results – Export – Animation 2.

In Settings – Animation – Output,

Select > Output type Movie from the pull-down menu (if needed).

Select > Format GIF from the pull-down menu (if needed).

NOTE *The detailed procedures for Animation in 5.x on the Macintosh and the PC are different. The Macintosh uses the GIF format. The PC uses the AVI format. If the modeler tries to use AVI on the Macintosh, an error results.*

In Settings – Animation – Frames,

Click > Lock aspect ratio check box.

In Settings – Animation – Advanced,

Click > twistie (as needed).

Click > Antialiasing check box (as needed).

In the Settings Animation window,

Click > the Export button (in the Settings Animation toolbar).

Select the desired location for saving the movie.

Enter the desired File Name (MM2E5X_1D_TelE_1.gif) in the Save-As edit window.

Click > Save.

Click > Save for the completed MM2E5X_1D_TelE_1.mph Telegraph Equation model.

1D Telegraph Equation Model Summary and Conclusions

The 1D Telegraph Equation model is a powerful tool that can be used to explore pulse wave propagation in electrical communication cables. As has been shown earlier in this chapter, the Telegraph Equation is easily and simply modeled with a 1D PDE mode model.

1D Spherically Symmetric Transport Model

The fabrication and use of small and microscopic spherical and spheroidal particles for the manufacturing of metal artifacts date to as early as 1200 BC {4.11}. In the present era, spherical and spheroidal particles are also employed in fluidized beds {4.12}, sintering (sintered artifact fabrication) {4.13}, lithium-ion batteries {4.14}, pharmaceuticals {4.15}, combustion {4.16}, and numerous other practical applications.

In this model, by making a few simple assumptions and by converting the model from Cartesian to Spherical coordinates, the modeling calculations are reduced from a 3D model to a 1D model. Figure 4.29 shows both a Cartesian and a Spherical coordinate system parametrically.

See Figure 4.29.

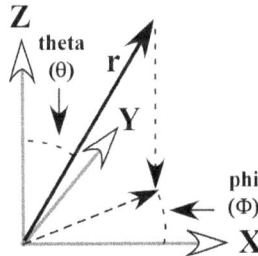

FIGURE 4.29 Both Cartesian and spherical coordinate parameters.

Figure 4.29 shows both the Cartesian and the Spherical coordinate parameters.

A spherical radius vector with magnitude |r|, originating at coordinate (0,0,0) in Cartesian coordinates can be indicated as follows:

$$\vec{r} = \hat{i}x + \hat{j}y + \hat{k}z \qquad (4.19)$$

Where

$$r = \left(x^2 + y^2 + z^2 \right)^{\frac{1}{2}} \qquad (4.20)$$

and where

$$(\hat{i}, \hat{j}, \hat{k})$$

are unit vectors in the X, Y, Z directions, respectively.

The conversion factors for the magnitude of |r| to x, y, z values are {4.17}:

$$x = r \sin \Theta \cos \Phi \qquad (4.21)$$

and

$$y = r \sin \Theta \sin \Phi \qquad (4.22)$$

and

$$z = r \cos \Theta \qquad (4.23)$$

Where theta (Θ) is the angle between the radius vector (r) and the Z-axis (as shown in Figure 4.29).

phi (Φ) is the angle between the projection of the radius vector (r) into the X-Y plane and the X-axis (as shown in Figure 4.29)

NOTE

As mentioned previously, there are a large number of potential applications that are suitable for reduction from 3D to 1D through the use of this technique. The author recommends that the modeler explore those as the opportunity arises. In this text, however, we will explore the case of time-dependent heat transfer by conduction, since that type of model is so broadly applicable to numerous commonly observed different problems in its own right.

Time-Dependent Heat Transfer by Conduction

The time-dependent heat conduction equation in Cartesian coordinates is expressed as follows:

$$\rho c_p \frac{\partial T}{\partial t} + \nabla \cdot (-k \nabla T) = Q \qquad (4.24)$$

Where ρ is the density of the material [kg/m^3].

c_p is the heat capacity of the material [J/(kg×K)].

T is the temperature [T].

k is the thermal conductivity [W/(m×K)].

Q is the internal heat source [W/m^3].

Conversion of the time-dependent heat conduction equation from Cartesian coordinates to Spherical coordinates requires the expression of the gradient operator in Spherical coordinates {4.18}, as shown in equation 4.25:

$$\nabla \cdot \nabla T = \frac{1}{r^2}\frac{\partial}{\partial r}\left[r^2\frac{\partial T}{\partial r}\right] + \frac{1}{r^2}\left[\frac{1}{\sin\Theta}\frac{\partial}{\partial\Theta}\left[\sin\Theta\frac{\partial T}{\partial\Theta}\right] + \frac{1}{\sin^2\Theta}\frac{\partial^2 T}{\partial\Phi^2}\right] \quad (4.25)$$

Converting the Cartesian time-dependent heat conduction equation to a Spherical time-dependent heat conduction equation, yields:

$$\rho c_p\frac{\partial T}{\partial t} - k\left[\frac{1}{r^2}\frac{\partial}{\partial r}\left[r^2\frac{\partial T}{\partial r}\right] + \frac{1}{r^2}\left[\frac{1}{\sin\Theta}\frac{\partial}{\partial\Theta}\left[\sin\Theta\frac{\partial T}{\partial\Theta}\right] + \frac{1}{\sin^2\Theta}\frac{\partial^2 T}{\partial\Phi^2}\right]\right] = Q$$

$$(4.26)$$

NOTE *In order to simplify the 3D problem to a 1D problem, it will be assumed that the sphere is nominally perfectly spherical, of radius R_p, and that T does not vary as a function of the angles Θ and/or Φ.*

Since T does not vary as a function of the angular variables, then:

$$\frac{\partial T}{\partial\Theta} = \frac{\partial T}{\partial\Phi} \equiv 0 \quad (4.27)$$

Based on these assumptions, the time-dependent heat conduction equation is modified as follows:

$$\rho c_p\frac{\partial T}{\partial t} - k\left[\frac{1}{r^2}\frac{\partial}{\partial r}\left[r^2\frac{\partial T}{\partial r}\right]\right] = Q \quad (4.28)$$

In order to eliminate the problems associated with division by zero, the time-dependent heat conduction equation needs to be multiplied by r^2. The following equation results:

$$r^2\rho c_p\frac{\partial T}{\partial t} + \frac{\partial}{\partial r}\left[-kr^2\frac{\partial T}{\partial r}\right] = r^2 Q \quad (4.29)$$

Since k is not a function of r, it can be moved inside the differentiation operation with no adverse effect in equation 4.29.

It is also convenient to solve the 1D problem for a dimensionless radial coordinate, because this allows the modeler to rapidly change the sphere size without changing the model geometry.

The conversion of the time-dependent heat conduction equation to a dimensionless form is as follows:

$$r_a = \frac{r}{R_p} \tag{4.30}$$

where R_p is the radius of the sphere.

Then:

$$\frac{\partial}{\partial r} = \frac{1}{R_p} \frac{\partial}{\partial r_a} \tag{4.31}$$

Substituting the dimensionless operator into equation 4.29, yields:

$$r_a^2 \rho c_p \frac{\partial T}{\partial t} + \frac{\partial}{\partial r_a} \left[\frac{-k r_a^2}{R_p^2} \frac{\partial T}{\partial r_a} \right] = r_a^2 Q \tag{4.32}$$

The dimensionless radius vector for a sphere of nominal size is shown in Figure 4.30.

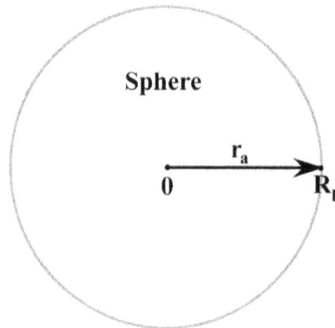

FIGURE 4.30 Radius vector for a dimensionless nominal sphere.

Figure 4.30 shows a Radius Vector for a dimensionless nominal sphere.

Building the 1D Spherically Symmetric Transport Model

Startup 5.x.

Click > Model Wizard > Select > 1D, on the Select Space Dimension page.

Click > Twistie for Mathematics in the Select Physics window.

Click > Twistie for PDE Interfaces > General Form PDE (g).

Click > Add button.

In the Model Wizard – Dependent variables – Dependent variables entry window,

Enter > T.

See Figure 4.31.

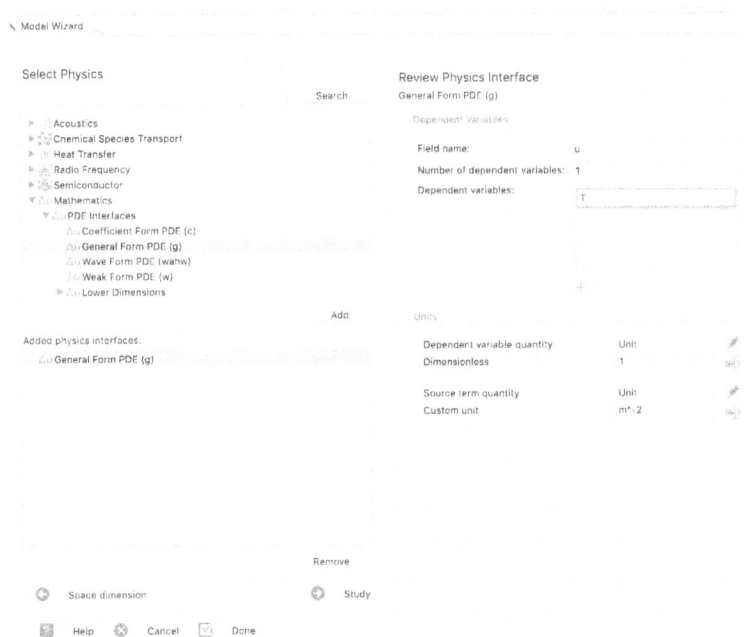

FIGURE 4.31 Dependent variables entry window.

Figure 4.31 shows the dependent variables entry window.

Click > Study (Right Pointing Arrow).

Select > General Studies > Time Dependent in the Select Study window.

Click > Done (Checked Box button).

Click > Save As > MM2E5X_1D_SST_1.mph.

See Figure 4.32a.

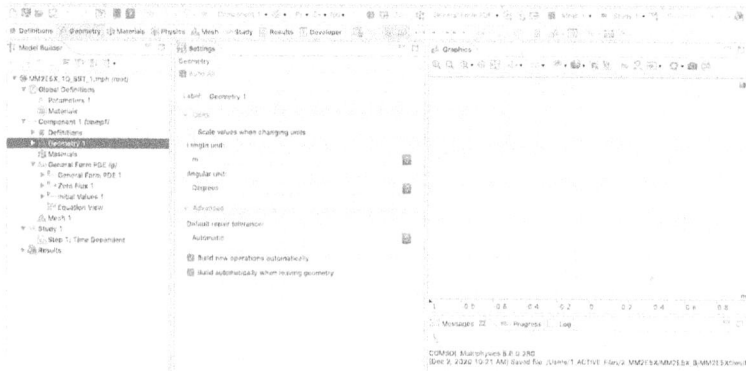

FIGURE 4.32a Desktop display for the MM2E5X_1D_SST_1.mph model.

Figure 4.32a shows the Desktop Display for the MM2E5X_1D_SST_1.mph model.

Click > MM2E5X_1D_SST_1.mph (*root*).

Click > Settings MM2E5X_1D_SST_1.mph > Unit System.

Select: None from the Pull_down Menu.

See Figure 4.32b.

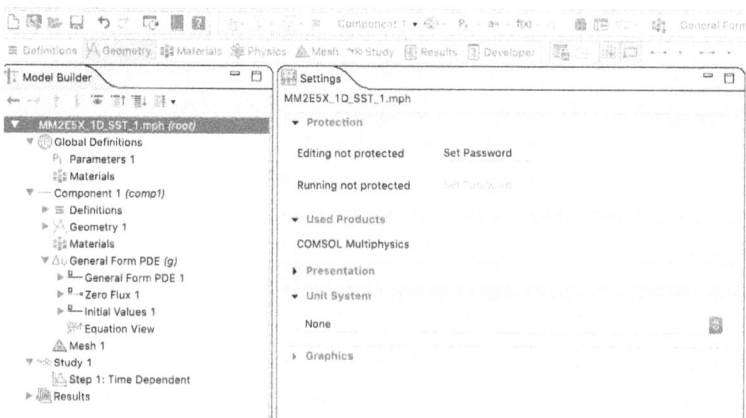

FIGURE 4.32b Desktop display for the MM2E5X_1D_SST_1.mph model.

Figure 4.32b shows the Desktop Display for the MM2E5X_1D_SST_1.mph model.

Global Definitions

In the Model Builder window,

Click > Global Definitions > Parameters 1.

In the Settings – Parameters – Parameters edit window,

Enter the parameters as shown in Table 4.2.

TABLE 4.2 Parameters Window.

Name	Expression	Description
rho	2000 [kg/m^3]	Density
cp	300 [J/(kg*K)]	Heat capacity
k	0.5 [W/(m*K)]	Thermal conductivity
Rp	0.005 [m]	Sphere radius
Qs	0 [W/m^3]	Heat source
hs	1000 [W/(m^2*K)]	Heat transfer coefficient
Text	368 [K]	External temperature
Tinit	298 [K]	Initial temperature

Click > Save to File (Disk image): MM2E5X_1D_SST_1_Param.txt

Click > Save (to save the model).

NOTE

The units are entered for the model parameters, in case the modeler needs to later verify the consistency of the model. The modeler should also save the parameters file at this time so that if the model needs to be recovered or the parameters need to be modified, they will be available to be changed or corrected without the need to reenter them.

See Figure 4.33.

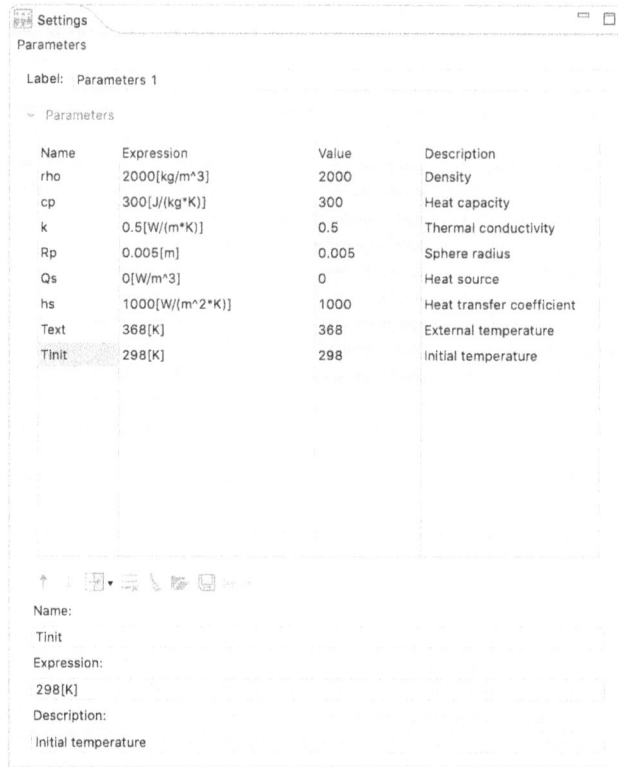

Name	Expression	Value	Description
rho	2000[kg/m^3]	2000	Density
cp	300[J/(kg*K)]	300	Heat capacity
k	0.5[W/(m*K)]	0.5	Thermal conductivity
Rp	0.005[m]	0.005	Sphere radius
Qs	0[W/m^3]	0	Heat source
hs	1000[W/(m^2*K)]	1000	Heat transfer coefficient
Text	368[K]	368	External temperature
Tinit	298[K]	298	Initial temperature

Name:

Tinit

Expression:

298[K]

Description:

Initial temperature

FIGURE 4.33 Desktop display – Settings – Parameters entry windows.

Figure 4.33 shows the Desktop Display – Settings – Parameters entry windows.

Geometry 1

Right-Click in the Model Builder window on Component 1 (*comp1*) > Geometry 1.

Select: Interval from the Pop-up menu.

NOTE *The default settings for the specified interval are adequate for this model.*

Right-Click > Interval 1.

Select: Build Selected.

See Figure 4.34.

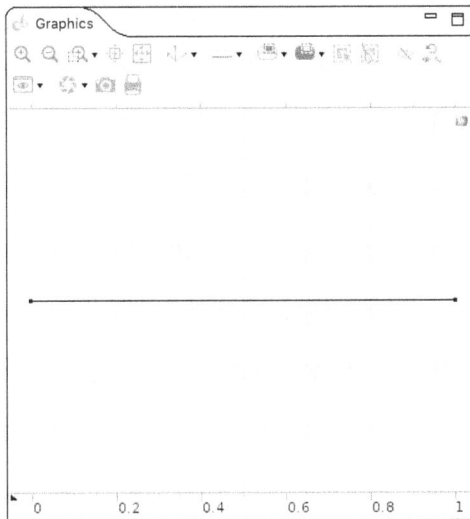

FIGURE 4.34 Desktop display – Graphics – Geometry 1 window.

Figure 4.34 shows the Desktop Display – Graphics – Geometry 1 window.

PDE

Click in the Model Builder window on Component 1 (*comp1*) > Δu General Form PDE (g) twistie.

In the Model Builder window,

Click > Component 1 (comp1) > Δu General Form PDE (g) > General Form PDE 1.

The default settings for the General Form PDE are not appropriate for the solution of this problem and thus need to be modified. The modeler should note the nominal equation (by Clicking on the Settings – General Form PDE – Equation twistie) and the manner in which the entered coefficients change that equation.

NOTE

In Settings – General Form PDE – Conservative Flux – Γ edit window, enter

-k*(x^2/Rp^2)*Tx.

In Settings – General Form PDE – Source Term – *f* edit window, enter x^2*Qs.

In Settings – General Form PDE – Damping or Mass Coefficient – d_a edit window, enter x^2*rho*cp.

See Figure 4.35.

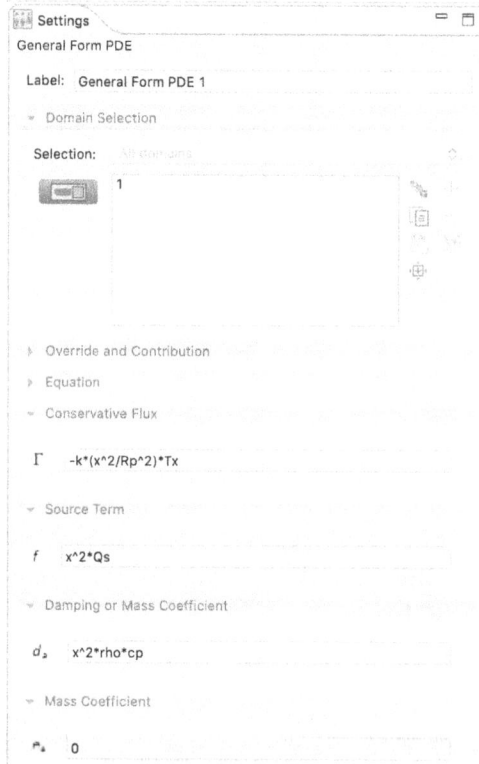

FIGURE 4.35 Desktop display – Settings – General form PDE coefficients.

Figure 4.35 shows the Desktop Display – Settings – General Form PDE coefficients.

Initial Values 1

Click in the Model Builder window on Component 1 (*comp1*) > Δu General Form PDE (g) > Initial Values 1.

NOTE *The default settings for the Initial Values are not appropriate for the solution of this problem and thus need to be modified.*

in the Settings – Initial Values – Initial value for T, enter Tinit.

See Figure 4.36.

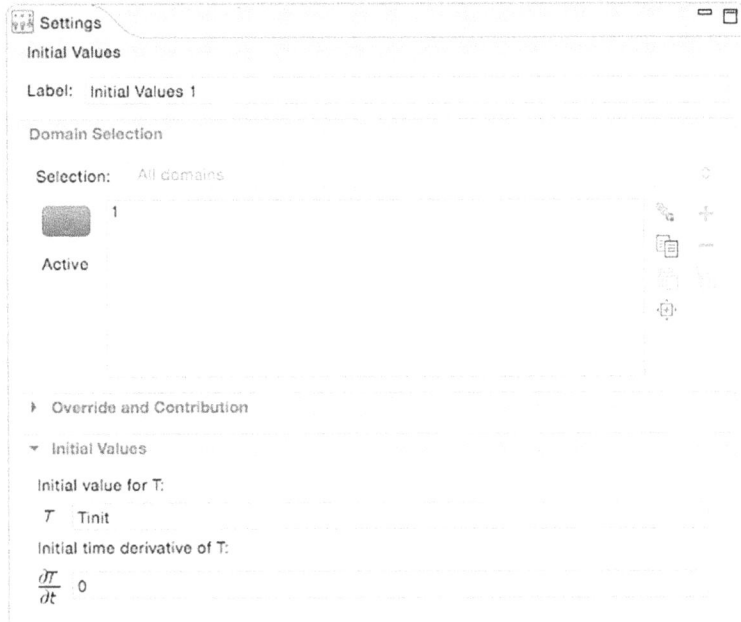

FIGURE 4.36 Desktop display – Settings – Initial values coefficients.

Figure 4.36 shows the Desktop Display – Settings – Initial Values coefficients.

NOTE *The default settings for the Boundary Conditions are not appropriate for the solution of this problem and thus need to be modified.*

In Model Builder,

Right-Click > Component 1 (*comp1*) > Δu General Form PDE (g).

Select: Flux/Source from the Pop-up window.

— Coefficient Form PDE
— General Form PDE
— Weak Form PDE
— Source
— Initial Values
Classical PDEs ▶
More ▶

—• Dirichlet Boundary Condition
—• Constraint
—• Zero Flux
—• Flux/Source
—• Periodic Condition
More ▶
Pairs ▶

Global ▶

☞ Show More Options...

Node Group
Group by Space Dimension

Copy as Code to Clipboard ▶

Copy
Delete   ⊳
Disable   F3
Rename   F2

Settings
Properties

Help   F1

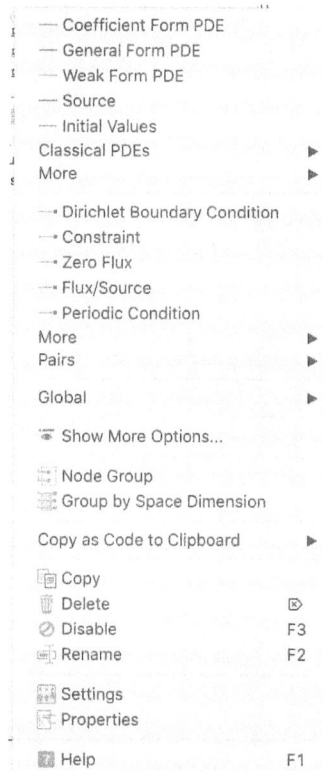

FIGURE 4.37a Desktop display – Pop-up window.

Figure 4.37a shows the Desktop Display – Pop-up-window.

Click > Boundary 1 (left endpoint of the interval) in the Graphics window.

In Model Builder,

Right-Click > Component 1 (*comp1*) > Δu General Form PDE (g).

Select: Flux/Source.

Click > Boundary 2 (right endpoint of the interval) in the Graphics window.

In Settings – Flux/Source – Boundary Flux/Source – *g* edit window,

Enter: (x^2/Rp)*hs*(Text-T).

See Figure 4.37b.

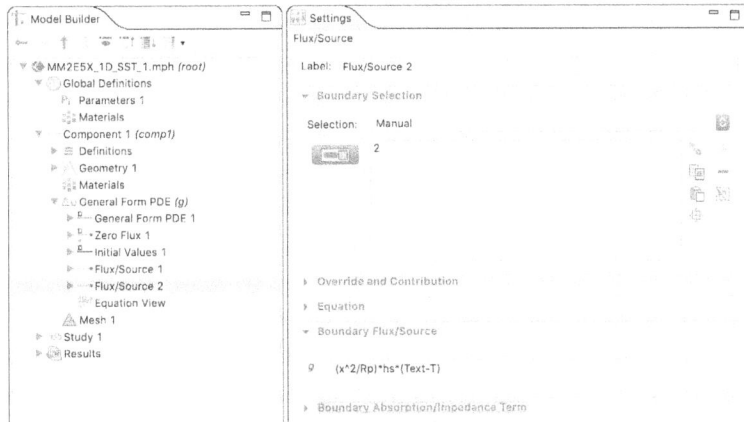

FIGURE 4.37b Desktop display – Settings – Boundary condition.

Figure 4.37b shows the Desktop Display – Settings – Boundary condition.

Mesh 1

NOTE *The default settings for the Mesh Type are not appropriate for the solution of this problem and thus need to be modified.*

Right-Click in the Model Builder window on Component 1 (*comp1*) > Mesh 1.

Select: Scale.

In Settings – Scale – Element size scale edit window,

Enter: 0.4.

Right-Click in the Model Builder window on Component 1 (*comp1*) > Mesh 1.

Select > Edge.

Right-Click in the Model Builder window on Component 1 (*comp1*) > Mesh 1.

Select > Build All.

After meshing, the modeler should see information in the message window about the number of elements (38 elements) in the mesh.

See Figure 4.38.

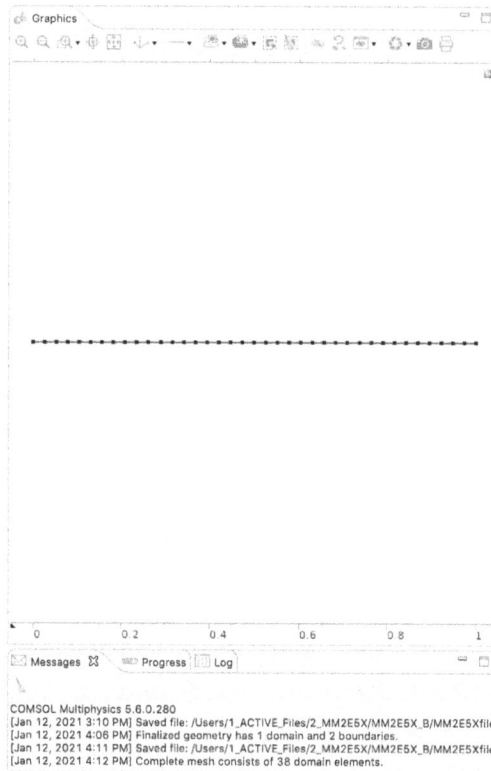

FIGURE 4.38 Desktop display – Graphics – Meshed domain.

Figure 4.38 shows the meshed domain in the Desktop Display – Graphics window.

Study 1

Click > Study 1 twistie.

Click > Step 1: Time Dependent.

In Settings – Time Dependent – Study Settings,

Click > Times Range button.

Enter > Start = 0, Step = 0.25, Stop = 10.

Click > Replace.

In Model Builder, Right-Click Study 1 >Select > Compute.

Results

The as-computed results are shown in Figure 4.39 using the default plot parameters.

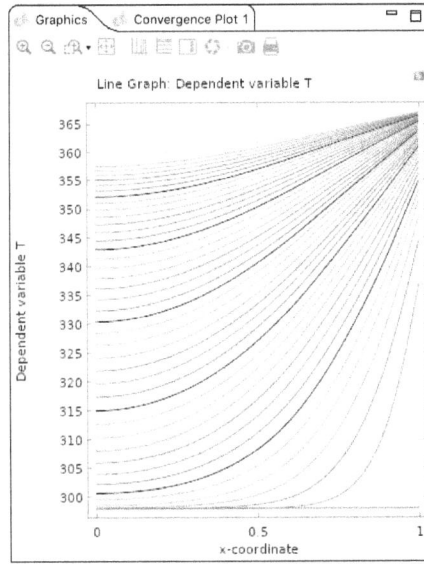

FIGURE 4.39 Desktop display – Graphics – Initial computed radial temperature profile.

Figure 4.39 shows the initial computed radial temperature profile in the Desktop Display – Graphics window.

NOTE *The computed radial temperature profile is plotted using the default plot parameters in Figure 4.39. The modeler can adjust the output plot to better satisfy his particular information display needs as shown below.*

1D Plot Group 1

In Model Builder,

Click > Results – 1D Plot Group 1 (if needed).

In Settings – 1D Plot Group – Click > Title twistie,

Select > Manual from the Title Pull-down menu.

Enter Temperature in the Title edit window.

Click > x-axis label checkbox,

Enter: Dimensionless Radius in x-axis label edit window.

Click > y-axis label checkbox,

Enter: T (K) in the y-axis label edit window.

Click > Plot.

In Model Builder – Results,

Click > 1D Plot Group 1 twistie.

In Model Builder – Results – 1D Plot Group 1,

Click > Line Graph 1.

In Settings – Line Graph,

Click > Legends twistie (scroll down as needed).

Uncheck, as needed, the Show legends checkbox.

On the Graphics Toolbar.

Click > Zoom Extents.

See Figure 4.40.

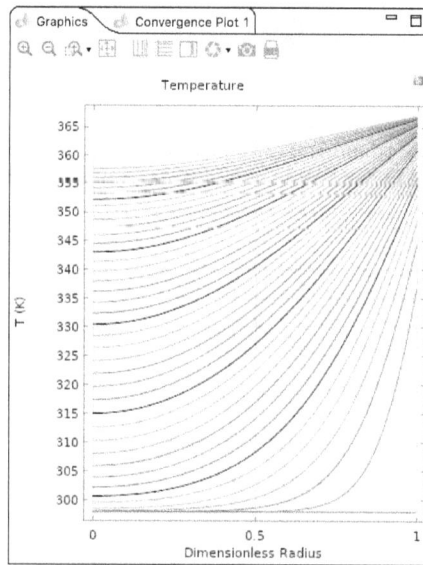

FIGURE 4.40 Desktop display – Graphics –Adjusted plot parameters computed radial temperature profile graph.

Figure 4.40 shows the Desktop Display – Graphics – Adjusted Plot Parameters Computed Radial Temperature Profile Graph.

1D Plot Group 2

In Model Builder,

Right-Click > Results.

Select > 1D Plot Group from the pull-down menu to create 1D Plot Group 2.

Right-Click> 1D Plot Group 2.

Select > Point Graph.

In the Graphics window,

Select > Boundary 1 only (left end boundary).

In Settings – Point Graph,

Click > Legends twistie.

Uncheck, as needed, the Show legends checkbox.

In Model Builder,

Click > 1D Plot Group 2.

In Settings – 1D Plot Group – Click > Title twistie (as needed),

Select > Manual from the Title Pull-down menu.

Enter: Temperature Response at Sphere Center in the Title edit window.

Click > x-axis label checkbox.

Enter: Time (s) in the x-axis label edit window.

Click > y-axis label checkbox.

Enter: T (K) in the y-axis label edit window.

Click > Plot.

See Figure 4.41.

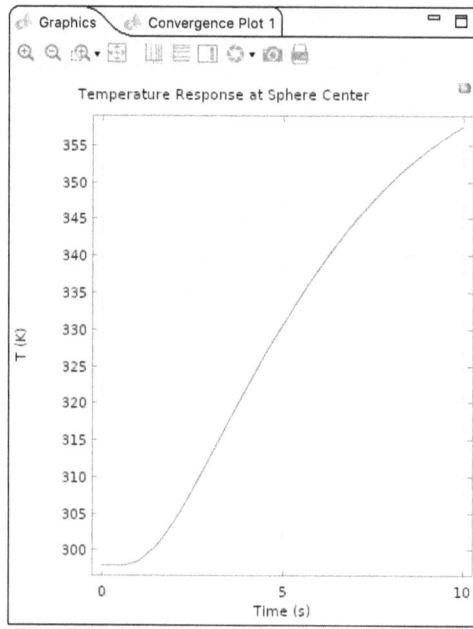

FIGURE 4.41 Desktop display – Graphics – Temperature response at sphere center.

Figure 4.41 shows the Desktop Display – Graphics – Temperature Response at Sphere Center.

1D Spherically Symmetric Transport Model Animation

To demonstrate the propagation of the heat transfer by conduction in 5.x, the modeler can create an animation that runs in the Graphics window.

In Model Builder,

Right-Click > Results - Export,

Select > Animation > Player.

In Settings-Animation > Scene > Subject Pull-down menu

Select > 1D Plot Group 2 (as needed).

Right-Click > Model Builder > Results > Export > Animation 1: Select > Play.

The modeler can also save the just executed animation to a file.

In Model Builder,

Right-Click > Results – Export.

Select > Animation > File.

Click > Results – Export – Animation 2.

In Settings-Animation > Scene > Subject Pull-down menu

Select > 1D Plot Group 2 (as needed).

In Settings – Animation – Output,

Select > Output type Movie from the pull-down menu (if needed).

Select > Format GIF from the pull-down menu (if needed).

NOTE *The detailed procedures for Animation in 5.x on the Macintosh and the PC are different. The Macintosh uses the GIF format. The PC uses the AVI format. If the modeler tries to use AVI on the Macintosh, an error results.*

In Settings – Animation – Frames,

Click > Lock aspect ratio check box.

In Settings – Animation – Advanced,

Click > twistie (as needed).

Click > Antialiasing check box (as needed).

In the Settings Animation window,

Click > the Export button (at the top).

Select the desired location for saving the movie.

Enter the desired File Name (MM2E5X_1D_SST_1.gif) in the Save-As edit window.

Click > Save.

Click > Save for the completed MM2E5X_1D_SST_1.mph 1D Spherically Symmetric Transport Model.

1D Spherically Symmetric Transport Model Summary and Conclusions

The 1D Spherically Symmetric Transport Model is a powerful tool that can be used to reduce the dimensionality of models from 3D to 1D through the assumption of homogeneous, isotropic materials properties and spherical symmetry. As has been shown earlier in this chapter, the 1D Spherically Symmetric Transport Model is easily and simply built with a 1D PDE mode model.

1D ADVANCED MODEL

1D Silicon Inversion Layer Model: A Comparison of the Results obtained from using the Density-Gradient (DG) Theory and the Schrodinger-Poisson (SP) Theory Methodologies

This model is derived from COMSOL Multiphysics Model 73381. The building of this model requires the use of COMSOL Multiphysics and the Semiconductor Module. It compares the calculated electron-density profiles beneath the transistor gate at two different applied voltages, using two different methodologies.

NOTE

Advanced Models are complex and need to be built with great care. If you initially make a mistake (error) in some part of the model, it is best to start over rebuilding the model from the beginning. Then, follow the exact instructions printed in the book and build the model with care.

Detection and amplification of the electronic signals arriving at a particular location, either by conductor (metal wire) and/or through the atmosphere (electromagnetic wave), began with the invention of the Audion tube (thermionic triode) in 1906 {4.19}.

A thermionic triode basically comprises a heated filament (a hot wire), a control element (grid), and a receiving element (plate), within an evacuated glass envelope (tube). The heated filament (cathode) emits electrons into the surrounding vacuum. The next element of the tube, located so that it forms an encompassing sleeve for the filament, is a porous metal structure (the grid). The grid controls the flow of the emitted electrons from the cathode to the plate. Enclosing the grid is a non-porous metal surface (plate) that captures all the emitted electrons that pass from the cathode through the grid.

A variable (AC or DC) voltage (of an amplitude that either adds to or subtracts from the cathode voltage (bias voltage) amplitude as a function of time (signal voltage)) is applied between the grid and the cathode. The current flowing from the cathode through the grid and then to the plate is either increased or decreased by the voltage applied between the grid and the cathode depending upon the relative geometry of the three tube elements and also the relative voltages applied to the tube elements. A small amplitude signal applied to the grid is then replicated in form and becomes larger in amplitude (amplified), at the plate. The signal applied between the grid and the cathode is then replicated in the amplitude and frequency of the current flowing through the plate.

The first working solid state semiconductor amplifying device was developed at Bell Laboratories in 1947 {4.20}. Using a purified crystal of germanium, with selected impurities, and two gold wires, John Bardeen, Walter Brattain and William Shockley created the first working point-contact transistor {4.21}. For their efforts in creating this amplifying semiconductor device, they jointly received the Nobel Prize in 1956. The point-contact transistor was, however, mechanically delicate and subject to malfunction in the presence of mechanical shock.

In order to overcome the inherent mechanical instabilities of the point-contact device, other more stable semiconductor device structures, such as the Bipolar Junction Transistor (BJT) {4.22} and the related surface effect devices were explored at Bell Labs. Those efforts led to the creation of many new configurations of semiconductor surface effect devices (MOSFET) {4.23}. Ultimately, these early discoveries led to the creation of the Integrated Circuit {4.24}, making today's technology possible.

This model demonstrates two of the calculational methodologies that can be used for the comparative determination of the specific calculation of current flow under the gate of one of the typical transistors as used in such an integrated circuit.

1D Silicon Inversion Layer Model using DG and SP Theory Methodologies

It is the purpose of this model to demonstrate the modeling techniques needed to reproduce the calculated inversion layer electron density below the gate oxide as a function of depth curves as published earlier{4.25}and designated as Figure 4 in that paper. The theory underlying this type of calculation is presented and available in another earlier {4.26} paper by the same author and will not be reproduced herein. The parameters used in the Schrodinger-Poisson calculation were obtained from another paper {4.27}. The material properties were obtained from the book by Sze and Ng {4.28}.

Building the 1D Silicon Inversion Layer Model using DG and SP Theory Methodologies

Startup 5.x.

Click > Model Wizard > Select > 1D, on the Select Space Dimension page.

NOTE

In this step, only the Semiconductor physics interface will be added. The Schrodinger-Poisson multiphysics interface will be added later.

Click > Semiconductor Twistie, in the Select Physics window.

Select: Semiconductor (Semi), in the Select Physics window.

Click > Add.

Click > Study (The button with the Right pointing White Arrow in the Blue Circle).

NOTE *It is assumed that in this model, the system is assumed to be in thermal equilibrium.*

Click > In the Select Study window > Preset Studies for Selected Physics Interfaces > Semiconductor Equilibrium.

Click > Done (The button with the Check Box containing the Blue Check).

Click > Save As > MM2E5X_1D_DGSP_SIL_1.mph.

See Figure 4.42.

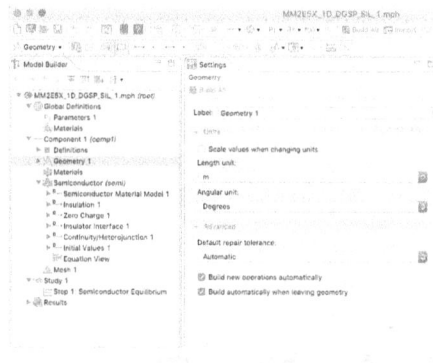

FIGURE 4.42 Desktop display.

Figure 4.42 shows the Desktop.

Geometry 1

NOTE *In this step, a convenient length unit is chosen.*

In the **Model Builder** window, under **Component 1 (compl),**

Click > Geometry 1 (as needed).

See Figure 4.43.

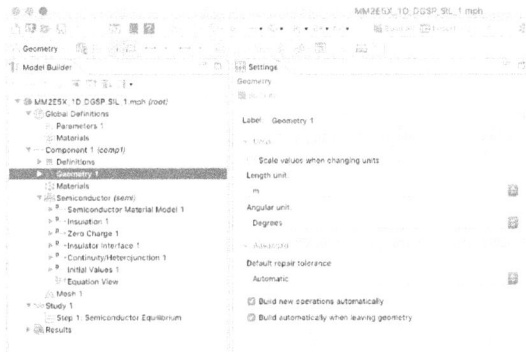

FIGURE 4.43 Model builder > Geometry 1.

Figure 4.43 shows the Model Builder > Geometry 1.

In the Settings window for Geometry,

Click > Units > Length unit > Select nm.

See Figure 4.44.

FIGURE 4.44 Settings geometry window > Units section.

Figure 4.44 shows the Settings Geometry window, Units section.

In the **Model Builder** window, Right-Click > **Component 1 (*comp1*)** > Geometry 1.

Select > Interval.

See Figure 4.45.

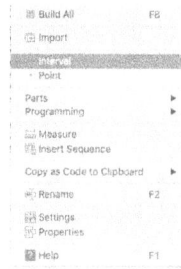

FIGURE 4.45 Geometry 1 interval selection.

Figure 4.45 shows the Geometry 1 Interval selection.

In the Settings Interval window, Interval Section,

Click > Interval > Coordinates source: pull-down,

Select > Vector.

In the Coordinates: text field window,

Enter: 0, 10, 70, 300, 1e3

See Figure 4.46.

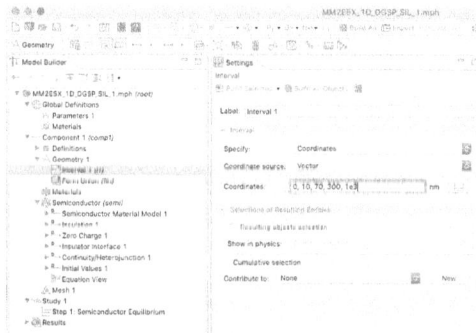

FIGURE 4.46 Geometry 1 interval parameters.

Figure 4.46 shows the Geometry 1 Interval parameters.

Global Definitions

Parameters 1

NOTE *At this point, the Modeler will enter the parameters needed for the Density-Gradient portion of this model.*

In the Model Builder window, Click > Global Definitions > Parameters 1. See Figure 4.47.

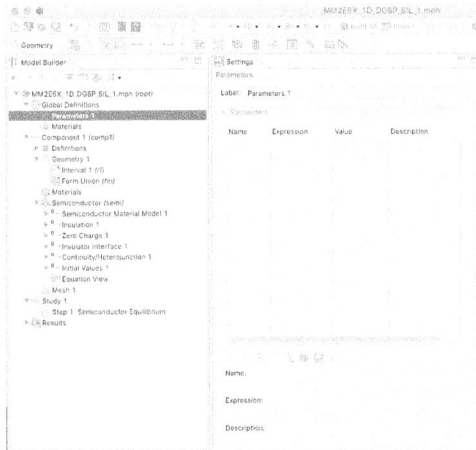

FIGURE 4.47 Global definitions > Parameters 1.

Figure 4.47 shows Global Definitions > Parameters 1.

In the Settings Parameters window, in the Label text field,

Enter > Parameters for Density-Gradient.

TABLE 4.3 Density-Gradient Parameters Window.

Name	Expression	Value	Description
T0	300 [K]	300 [K]	Temperature
mnDG	me_const/3	3.0365E−31 [kg]	Density-gradient effective mass
Na0	3.8e16 [1/cm^3]	3.8E22 [1/m^3]	Doping concentration
Vg	0 [V]	0 [V]	Gate voltage
epsr0x	3.9	3.9	Oxide dielectric constant
d0x	3.1 [nm]	3.1E−9 [m]	Oxide thickness
Phi0	4.01 [V]	4.01 [V]	Gate metal work function

Next, Enter in the Parameters for Density-Gradient text window, the parametric values, as shown in Table 4.3.

See Figure 4.48.

FIGURE 4.48 Density-gradient parameters.

Figure 4.48 shows Density-Gradient Parameters.

NOTE *At this point, the Modeler will add a blank material and then fill in the material parameter values.*

Materials

Material 1 (mat1)

In the Model Builder window, Right-Click > Component 1 (comp1) > Materials Select > Blank Material.

See Figure 4.49.

FIGURE 4.49 Blank materials selection panel.

Figure 4.49 shows Blank Materials Selection Panel.

In the Settings Material window, locate the Material Contents tab entry section.

Click > Material Contents twistie (as needed),

Enter the Material Properties Values found in Table 4.4 below.

TABLE 4.4 Material Parameters Values.

Property	Variable	Value	Unit	Property Group
Electron mobility	mun	1450 [cm^2/(V*s)]	m^2/(V*s)	Semiconductor material
Hole mobility	mup	500 [cm^2/(V*s)]	m^2/(V*s)	Semiconductor material
Relative permittivity	epsilonr_iso; epsilonrii = epsilon_iso, epsilonrij = 0	11.9	1	Basic
Band gap	Eg0	1.12 [V]	V	Semiconductor material
Electron affinity	chi0	4.05 [V]	V	Semiconductor material
Effective density of states, conduction band	Nc	2.80e19 [1/cm^3]	1/m^3	Semiconductor material
Effective density of states, valence band	Nv	2.65e19 [1/cm^3]	1/m^3	Semiconductor material

See Figure 4.50.

FIGURE 4.50 Settings material window with values.

Figure 4.50 shows Settings Material window with Values.

NOTE *At this point, the Modeler will set up the physics formulations and the parameters needed for the next portion of this model.*

SEMICONDUCTOR (SEMI)

In the Model Builder window, Click > Component 1 (comp1) > Semiconductor (semi).

In the Settings Semiconductor window, Click > Model Properties twistie (if needed).

From the Carrier statistics pull-down list, Select Fermi-Dirac.

Click > Discretization tab twistie (if needed),

Select > Finite element density-gradient (quadratic shape function), from the Formulation pull-down list.

See Figure 4.51.

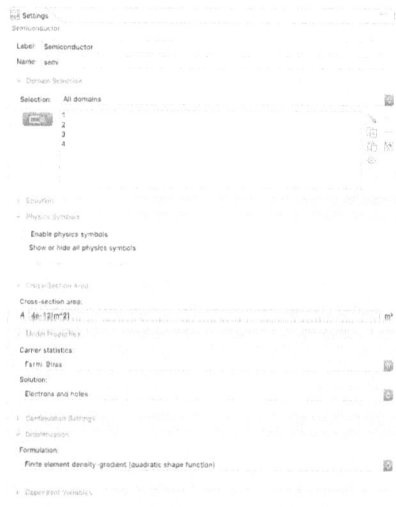

FIGURE 4.51 Settings semiconductor window with selected values.

Figure 4.51 shows Settings Semiconductor window with Selected Values.

Semiconductor (SEMI)

Semiconductor Material Model 1

In the Model Builder window, Click > Component 1 (comp1) > Semiconductor (semi) twistie (if needed) > Semiconductor Material Model 1.

In the Settings Semiconductor Material Model window, Click > Model Input twistie (if needed).

Enter T0 in the Temperature entry text field.

See Figure 4.52.

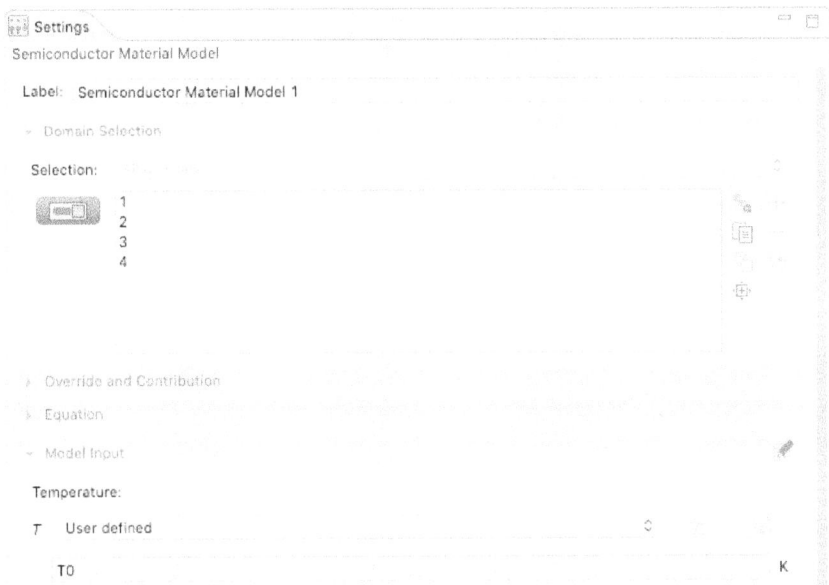

FIGURE 4.52 Settings semiconductor material model window with temperature value.

Figure 4.52 shows Settings Semiconductor Material Model window with Temperature Value.

In the Settings Semiconductor Material Model window, Click > Material Properties, Density-Gradient twistie (if needed).

Enter mnDG in the Electron effective mass, density-gradient ($\mathbf{m_e^{DG}}$) text field.

NOTE

Holes are not treated quantum mechanically in this calculation, as per the reference material. An arbitrarily large effective mass is used herein to minimize the effect of the density-gradient contribution by holes.

Enter 10*me_const in the Hole effective mass, density-gradient ($\mathbf{m_h^{DG}}$) text field.

See Figure 4.53.

FIGURE 4.53 Settings semiconductor material model window for electron and hole effective masses.

Figure 4.53 shows Settings Semiconductor Material Model window with for Electron and Hole Effective Masses.

Analytic Doping Model 1

In the Physics toolbar, Click > Domains and Select > Analytic Doping Model from the pop-up list.

In the Settings Analytic Doping Model window, Click > Domain Selection > Selection: pull-down list.

Select > All domains.

Click > Impurity twistie (as needed).

Enter Na0 in the Acceptor concentration (N_{A0}) text field.

See Figure 4.54.

FIGURE 4.54 Settings analytic doping model window domain and impurity values.

Figure 4.54 shows Settings Analytic Doping Model window Domain and Impurity Values.

Metal Contact 1

In the Physics toolbar, Click > Boundaries and Select > Metal Contact from the pop-up list.

In the Settings Metal Contact window, Click > Boundary Selection > Selection: pull-down list.

Select > Manual (if needed), Paste (Clipboard Symbol) 5.

See Figure 4.55.

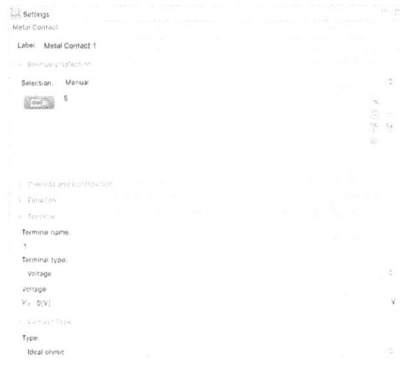

FIGURE 4.55 Settings metal contact model window boundary value.

Figure 4.55 shows Settings Metal Contact Model window Boundary Value.

Thin Insulator Gate 1

In the Physics toolbar, Click > Boundaries and Select > Thin Insulator Gate from the pop-up list.

In the Settings Thin Insulator Gate window, Click > Boundary Selection > Selection: pull-down list.

Select > Manual (if needed),

Paste: (Clipboard Symbol) 1.

In the Settings Thin Insulator Gate window, Click > Terminal twistie (if needed),

Enter Vg in the V_0 text field.

In the Settings Thin Insulator Gate window, Click > Gate Contact twistie (if needed),

Enter epsr0x in the e_{ins} text field,

Enter d0x in the d_{ins} text field,

Enter Phi0 in the Φ text field,

NOTE *The Schrodinger-Poisson computation, earlier referenced, assumes an infinite barrier height at the oxide interface. The modeler can use a large (~very large) barrier height for electrons to make rough first approximation to the "infinite" barrier height. Alternatively, the modeler can use a "zero" barrier height for holes (missing electrons) to suppress the density-gradient formulation quantum confinement effect.*

In the Settings Thin Insulator Gate window, Click > Density-Gradient twistie (if needed),

Click > Formulation pull-down list > Select Potential barrier.

Enter 1e4[V] in the Φ_n^{0x} text field,

Enter 0[V] in the Φ_p^{0x} text field,

See Figure 4.56.

FIGURE 4.56 Settings thin insulator gate window values.

Figure 4.56 shows Settings Thin Insulator Gate window Values.

Mesh 1

Size

In the Model Builder window, Right-Click > Component 1 (comp1) > Mesh 1, Select: Edit Physics-Induced Sequence from the pop-up window.

See Figure 4.57.

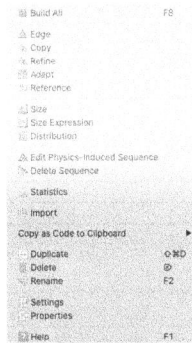

FIGURE 4.57 Mesh 1 pop-up window.

Figure 4.57 shows Mesh 1 pop-up window.

Click > Size.

In the Settings Size window,

In the tab Element Size, Click > Predefined pull-down list,

Select > Fine.

See Figure 4.58.

FIGURE 4.58 Mesh 1 size window settings.

Figure 4.58 shows Mesh 1 Size window Settings.

NOTE *The next step removes the artificial interior boundaries from the interior auto-generated list for mesh refinement.*

Size 1

In the Model Builder window, Click > Mesh 1> Size 1,

Remove: Boundaries 2, 3, 4, (Click > 2, Click > Minus operator, Then do the same for 3 and 4).

Leave: Boundary 5 only.

See Figure 4.59.

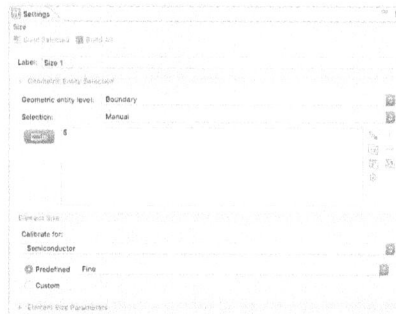

FIGURE 4.59 Mesh 1 size 1 window boundary settings.

Figure 4.59 shows Mesh 1 Size 1 window Boundary Settings.

NOTE *The next step specifies a small mesh element size at the oxide interface to facilitate the resolution of the very large gradient in that location.*

Size 2

In the Model Builder window, Click > Mesh 1> Size 2,

In the Settings Size window, on the Element Size tab,

Click the Custom button.

Select the Maximum element size Check box,

Enter 1e-4 in the Maximum element size

See Figure 4.60.

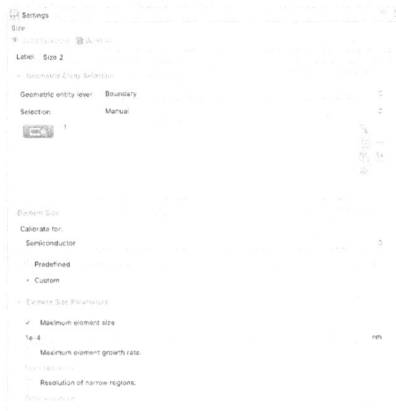

FIGURE 4.60 Mesh 1 size 2 window maximum element size settings.

Figure 4.60 shows Mesh 1 Size 2 window Maximum Element Size Settings.

NOTE *The next step adds a Size node to enhance the Mesh resolution in the regions of interest.*

Size 3

In the Model Builder window, Right-Click > Mesh 1> Select: Size,

Right-Click > Size 3, Select: Move Up from the pop-up window.

In the Settings Size window, on the Geometric Entity Selection tab,

Click > Geometric entity level pop-up list, Select: Domain.

Enter Domains 1-3 only,

Click > Element Size > Calibrate for: pull-down list,

Select > Semiconductor.

Click > Element Size > Predefined pull-down list,

Select > Finer.

Click > Custom button,

Select the Element Size Parameters > Maximum element size Check box,

Enter 0.5 in the Maximum element size

See Figure 4.61.

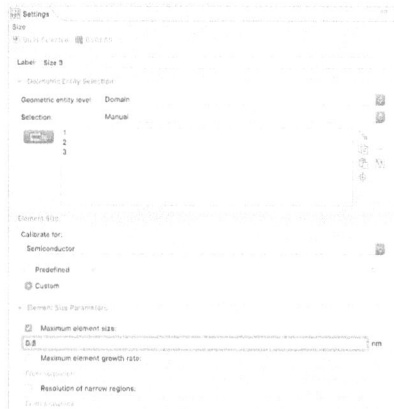

FIGURE 4.61 Mesh 1 size 3 window settings.

Figure 4.61 shows Mesh 1 Size 3 window Settings.

Study 1: Density-Gradient

In the Model Builder window, Click > Study 1.

In the Settings Study window, Label: text edit field,

Enter: Study 1: Density-Gradient

See Figure 4.62.

FIGURE 4.62 Study 1 settings.

Figure 4.62 shows Study 1 Settings.

Step 1: Semiconductor Equilibrium

In the Model Builder window, Click > Study 1: Density-Gradient twistie.

Click > Step 1: Semiconductor Equilibrium

In the Settings Semiconductor Equilibrium window, Click > Study Extensions twistie.

Select: the Auxiliary sweep Check box.

Click > Add (blue +) (Select: the appropriate listed Parameter Name (Vg(Gate voltage), as shown in Table 4.5 Auxillary Sweep Parameters values) by clicking on the pull-down list).

NOTE *The next step starts the voltage sweep at a value that is close to the semiconductor flat band condition so that the model will converge more readily.*

In the Auxiliary sweep table, enter the following settings (be sure that the – sign is in front of the first 1 (–1)):

TABLE 4.5 Auxiliary Sweep Parameters Values.

Parameter Name	Parameter Value List	Parameter Unit
Vg (Gate voltage)	–1 0.2 1	V

See Figure 4.63.

FIGURE 4.63 Semiconductor equilibrium auxiliary sweep settings.

Figure 4.63 shows Semiconductor Equilibrium Auxiliary Sweep Settings.

In the Home toolbar (the first toolbar located at the top of the desktop), Click > Compute.

See Figure 4.63a.

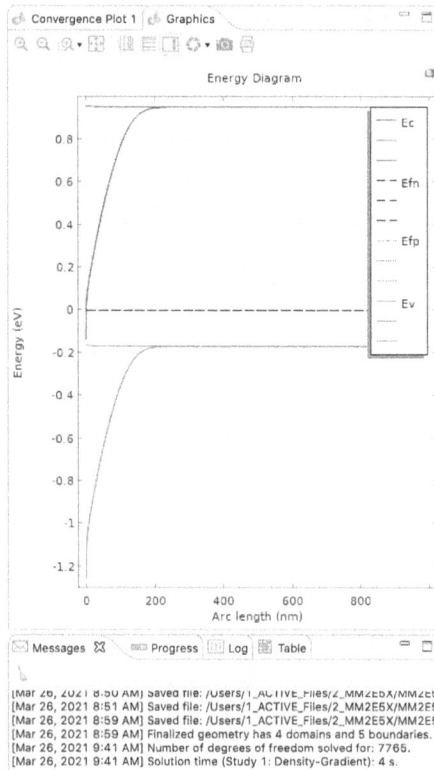

FIGURE 4.63a Computed semiconductor equilibrium energy levels.

Figure 4.63a shows Computed Semiconductor Equilibrium Energy Levels.

> **NOTE** *The next step duplicates the plot group Carrier Concentration to plot the electron density for comparison to external references.*

Results

Electron Concentration Comparison (ECC)

In the Model Builder window, Right-Click > Carrier Concentrations (semi)> Select: Duplicate.

In the Settings 1D Plot Group window, Enter: Electron Concentration Comparison in the Label: text field.

On the Data tab, Click > Parameter selection (Vg) pull-down list,

Select: Manual.

Enter 2 3 in the Parameter indices (1-3) text field.

See Figure 4.64.

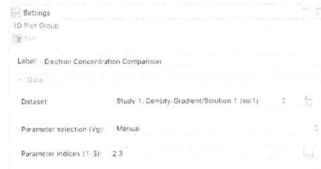

FIGURE 4.64 ECC settings.

Figure 4.64 shows ECC Settings.

Hole Concentration

In the Model Builder window, Click > Electron Concentration Comparison twistie (as needed).

In the Model Builder window, Right-Click > Hole Concentration > Select: Delete.

NOTE *The next step chooses the desired region to be plotted by selecting the correct domains.*

Density-Gradient

In the Model Builder window, Click > Electron Concentration.

In the Settings Line Graph window,

Click > Activate Selection toggle button, Remove: Domains 3 and 4 (Click on a Domain and then use the Minus (–) button), Keep: Domains 1 and 2.

Enter: Density-Gradient, in the Label text field.

Click > Coloring and Style twistie (as needed).

Click > Line: pull-down list, Select: Dashed

Click > Legends twistie (as needed),

Click > Legends: pull-down list, Select: Automatic

In the Include subsection, Enter DG in the Prefix text field.

See Figure 4.65.

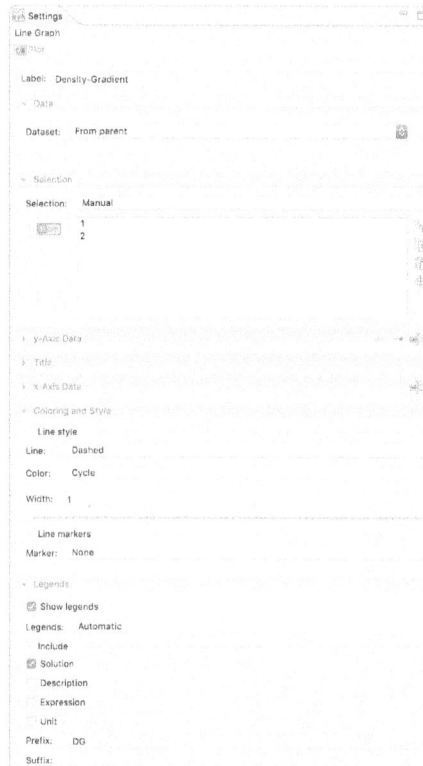

FIGURE 4.65 Electron concentration plot settings.

Figure 4.65 shows Electron Concentration Plot Settings.

Electron Concentration Comparison 1 (ECC1)

In the Model Builder window, Right-Click > Electron Concentration Comparison,

Select > Duplicate.

NOTE *The next step adds the Schrodinger-Poisson (SP) study. In this study, we employ the same hole and ionized dopant concentrations as were used in the Density-Gradient (DG) study. The only difference between the DG and the SP studies is the treatment of the electron distribution. Thus, various results from the DG study will be used as inputs to the SP study, as required.*

NOTE *The next step adds the Schrodinger-Poisson Equation multiphysics interface.*

Add Physics

In the Home toolbar, Click > Add Physics.

In the Add Physics window, in the Physics interfaces in study section,

Uncheck the Solve check box for Study 1: Density-Gradient.

See Figure 4.66.

FIGURE 4.66 Add physics density-gradient uncheck box.

Figure 4.66 shows Add Physics Density-Gradient uncheck box.

In the Add Physics window, Click > Semiconductor twistie.

Click > Schrodinger-Poisson Equation

Click > Add to Component 1 in the Add Physics window toolbar.

Click > Add Physics in the Home toolbar to close the Add Physics window.

Schrodinger Equation (SCHR)

NOTE *The next step adds the parameters used by the Schrodinger-Poisson Equation multiphysics interface.*

Global Definitions

In the Home toolbar, Click > P_i Parameters and Select Add Parameters from the pop-up menu.

In the Settings Parameters window Label: text field, Enter: Parameters for Schrodinger-Poisson.

In the Settings Parameters window, Enter all the Parametric Values as shown in Table 4.6.

TABLE 4.6 Schrodinger-Poisson Parameters Values.

Name	Expression	Value	Description
nv	2	2	Valley degeneracy
fm3	0.916	0.916	Longitudinal effective mass factor
fm1	0.190	0.19	Transverse effective mass factor 1
fm2	0.190	0.19	Transverse effective mass factor 2
m3	fm3*me_const	8.3442E-31 kg	Longitudinal effective mass
m1	fm1*me_const	1.7308E-31 kg	Transverse effective mass 1
m2	fm2*me_const	1.7308E-31 kg	Transverse effective mass 2
md	sqrt(m1*m2)	1.7308E-31 kg	Density of states effective mass

NOTE *The next step restricts the Schrodinger-Poisson Equation domain selection to a reasonable range in proximity to the oxide interface.*

Schrodinger Equation (SCHR)

In the Model Builder window, under Component 1 (comp1), Click > Schrodinger Equation (schr),

Select > Domains 1-3 only.

Next, Click > Electrostatics (es),

Select > Domains 1-3 only.

Effective Mass 1

In the Model Builder window, under Component 1 (comp1), Click > Schrodinger Equation (schr) twistie (as needed),

Click > Effective Mass 1.

In the Settings Effective Mass window, Enter: m3 in the $\mathbf{m}_{eff,e,11}$ text field.

See Figure 4.67.

FIGURE 4.67 Effective mass 1 settings.

Figure 4.67 shows Effective Mass 1 Settings.

Electron Potential Energy 1

In the Model Builder window, under Component 1 (comp1), Click > Schrodinger Equation (schr) twistie (as needed),

Click > Electron Potential Energy 1.

In the Settings Electron Potential Energy window,

Click > V_e pull-down menu

Select > User defined,

Enter: 0 in the associated pop-up text field.

See Figure 4.68.

FIGURE 4.68 Electron potential energy 1 settings.

Figure 4.68 shows Electron Potential Energy 1 Settings.

NOTE *The next step establishes the Zero Probability boundary condition at the oxide interface, to approximate the limiting condition of an infinite barrier height.*

Zero Probability 1

In the Physics toolbar, Click > Boundaries,

Select > Zero Probability.

In the Settings Zero Probability window, Select only Boundary 1.

See Figure 4.69.

FIGURE 4.69 Zero probability 1 settings.

Figure 4.69 shows Zero Probability 1 Settings.

NOTE *The next step uses the D-G Result as the initial estimates for the electron density distribution and the (fixed) space charge contribution from holes and from ionized dopants.*

Electrostatics (ES)

In the Model Builder window, under Component 1 (comp1), Click > Electrostatics (es),

Space Charge Density 1: initial electron density

In the Physics toolbar, Click > Domains,

Select: Space Charge Density.

Enter : Space Charge Density 1 : initial electron density in the Label text field.

Click > Domain Selection > Selection: pull-down list,

Select > All domains.

Enter: -e_const*semi.N in the ρ_v text field.

See Figure 4.70.

FIGURE 4.70 Space charge density 1: initial electron density settings.

Figure 4.70 shows Space Charge Density 1: initial electron density Settings.

Space Charge Density 2: holes and ionized dopants

Right-Click > Space Charge Density 1: initial electron density,

Select: Duplicate from the pop-up list.

Enter: Space Charge Density 2: holes and ionized dopants in the Label text field.

Enter: e_const*(semi.P+semi.Ndplus-semi.Naminus) in the ρ_v text field.

See Figure 4.71.

FIGURE 4.71 Space charge density 2: holes and ionized dopants settings.

Figure 4.71 shows Space Charge Density 2: holes and ionized dopants Settings.

NOTE *The next step uses the D-G Result as the initial estimates for the electrostatics boundary condition.*

Electric Potential 1

In the Physics toolbar, Click > Boundaries

Select: Electric Potential from the pop-up list.

In the Settings Electric Potential window,

Select: Boundary 4 only (Enter 4 using the Clipboard).

Enter: -semi.Ec in the V_0 text field.

See Figure 4.72.

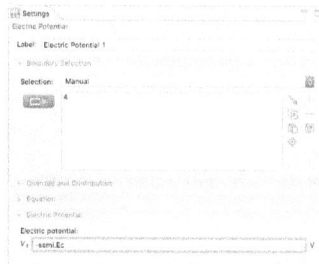

FIGURE 4.72 Electric potential boundary settings.

Figure 4.72 shows Electric Potential Boundary Settings.

Electric Displacement Field 1

In the Physics toolbar, Click > Boundaries,

Select: Electric Displacement Field from the pop-up list.

In the Settings Electric Displacement Field 1 window,

Enter: Boundary 1 only (use the Clipboard tool).

In the D_0 text field, Enter: epsilon0_const*epsr0x*(V_g-(Phi0-semi.chi_semi)-V2)/d0x.

See Figure 4.73.

FIGURE 4.73 Electric displacement field settings.

Figure 4.73 shows Electric Displacement Field Settings.

NOTE *The next step uses the D-G Result to set-up the multiphysics coupling.*

Multiphysics

Schrodinger-Poisson Coupling 1 (schrp1)

In the Model Builder window, under Component 1 (comp1), Click > Multiphysics twistie (as needed),

Click > Schrodinger-Poisson Coupling 1 (schrp1).

In the Settings Schrodinger-Poisson Coupling window, in the Model Input tab,

Enter: T0 in the T text field.

NOTE *The next step uses the **withsol** operator to retrieve the value of the variable for the equilibrium Fermi level.*

In the Settings Schrodinger-Poisson Coupling window, in the Particle Density Computation tab,

In the E_f text field,

Enter: e_const*withsol ('sol1', semi.Ef_0, setval(Vg,Vg)).

Enter: md in the m_d text field.

Enter: nv in the g_i text field.

See Figure 4.74.

FIGURE 4.74 Schrodinger-poisson coupling settings.

Figure 4.74 shows Schrodinger-Poisson Coupling Settings.

NOTE *The next step uses a Stationary study step to solve **only** the electrostatics physics. This methodology establishes a good initial condition for the fully coupled problem.*

In the Home toolbar, Click > Add Study.

In the Add Study window, Select > General Studies > Stationary.

In the Add Study window, clear the Solve check boxes for Semiconductor (semi), Schrodinger Equation (schr) and for the Schrodinger-Poisson Coupling 1 (schrp1).

See Figure 4.75.

FIGURE 4.75 Electrostatics (es) settings.

Figure 4.75 shows Electrostatics (es) Settings.

Click > Add Study.

In the Home toolbar, Click > Add Study, to close the Add Study window.

NOTE *Now use a Parametric Sweep node to set the gate voltage of 0.2V for both subsequent study steps (the first follows immediately).*

Study 2

Parametric Sweep

In the Study toolbar,

Click > Parametric Sweep.

In the Settings Parametric Sweep window, in the Study Settings tab,

Click > Add (blue + sign).

Select: Vg (Gate voltage) from the pop-up list,

Enter: 0.2 in the Parameter value list text field.

See Figure 4.76.

FIGURE 4.76 Parametric sweep settings.

Figure 4.76 shows Parametric Sweep Settings.

NOTE *In the next step, use the Values of variables not solved for section to inherit the solution from the previous study. Make sure to select the solution corresponding to the same gate voltage of 0.2 V specified in the Parametric Sweep node earlier.*

Step 1: Stationary

In the Model Builder under Study 2, Click > Step 1:Stationary.

In the Settings Stationary window,

Click > Values of Dependent Variables twistie (as needed).

Click > Values of variables not solved for pull-down list,

Select: User controlled.

Click > Method pull-down list,

Select: Solution.

Click > Study pull-down list,

Select: Study 1: Density-Gradient, Semiconductor Equillibrium.

Click > Parameter value(Vg(V)) pull-down list,

Select > 0.2 V.

See Figure 4.77.

FIGURE 4.77 Stationary settings.

Figure 4.77 shows Stationary Settings.

In the Model Builder window, Click > Study 2.

In the Settings Study window, Enter: Study 2: Schrodinger-Poisson Vg=0.2V in the Label text field.

In the Settings Study window, Clear the Generate default plots check box.

See Figure 4.78.

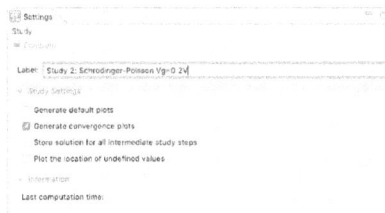

FIGURE 4.78 Study 2 settings.

Figure 4.78 shows Study 2 Settings.

NOTE *In the next step, the dedicated Schrodinger-Poisson study step is added. In the case of a completely new problem, it is often necessary to use the default Manual search option in order to determine the range of the eigenenergies. Once that range is found, the model can be switched to the Region search option, with appropriate value settings for the range and the number of eigenvalues. This practice ensures that all the significant eigenstates will be found by the solver.*

Schrodinger-Poisson

In the Study toolbar, Click > Study Steps,

Select: Eigenfrequency > Schrodinger-Poisson.

In the Settings Schrodinger-Poisson window, Study Settings,

Click > Eigenfrequency search method pull-down list,

Select > Region.

Enter: 100 in the Approximate number of eigenfrequencies text field.

Enter: 300 in the Maximum number of eigenfrequencies text field.

In the Search region > Unit,

Enter: −0.1 in the Smallest real part text field.

Enter: 1 in the Largest real part text field.

Enter: −1e-7 in the Smallest imaginary part text field.

Enter: 1e-7 in the Largest imaginary part text field.

Clear the Semiconductor (semi) Solve for check box, in the Physics and Variables Selection tab of the Settings Schrodinger-Poisson window.

See Figure 4.79.

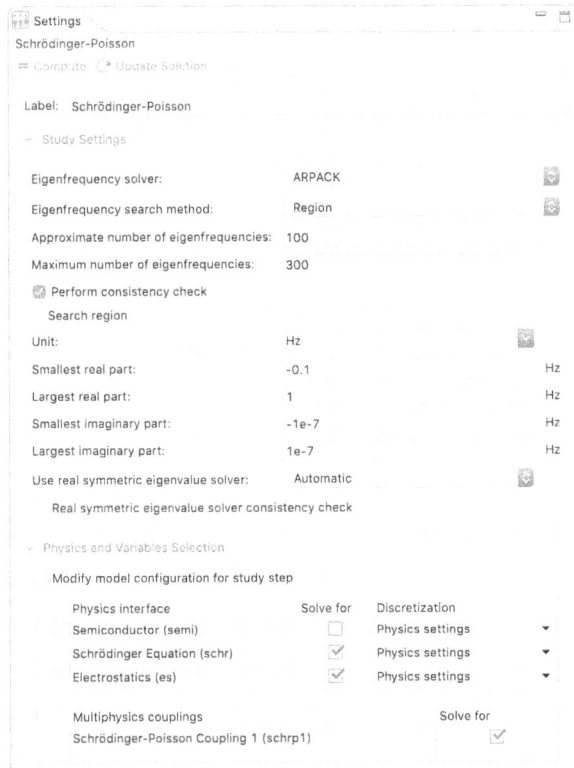

FIGURE 4.79 Schrodinger-poisson settings.

Figure 4.79 shows Schrodinger-Poisson Settings.

NOTE

The next step disables the space charge density contribution from the initial estimate of the electron concentration.

Select: (by clicking), the Modify model configuration for study step Check box.

In the Modify model configuration for study window,

Click > Electrostatics (es) > Space Charge Density 1: initial electron density.

Click > the Disable icon (blue circle with a diagonal line, located immediately below the Physics and Variables Selection window.)

See Figure 4.80.

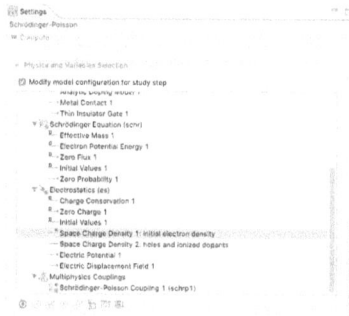

FIGURE 4.80 Physics and variables selection modified settings.

Figure 4.80 shows Physics and Variables Selection modified Settings.

On the Iterations tab,

Click > Termination method pull-down list,

Select > Minimization of global variable.

In the Global variable text field,

Enter: schrp1.global_err.

In the Absolute tolerance text field, Enter: 1e-6.

Click the Study Extensions twistie (as needed),

Check (by clicking) the Auxiliary sweep check box.

Click > Add (blue plus sign),

Select: nv (Valley degeneracy) from the pop-up list,

Enter: 2 4 in the Parameter value list text field.

Click > Add (blue plus sign),

Select: fm3 (Longitudinal effective mass factor) from the pop-up list,

Enter: 0.916 0.190 in the Parameter value list text field.

Click > Add (blue plus sign),

Select: fm2 (Transverse effective mass factor 2) from the pop-up list,

Enter: 0.190 0.916 in the Parameter value list text field.

See Figure 4.81.

FIGURE 4.81 Schrodinger-poisson modified settings.

Figure 4.81 shows Schrodinger-Poisson modified Settings.

In the Study toolbar, Click > Compute.

NOTE *The next step adds these results to the electron concentration comparison plots.*

Results

Schrodinger-Poisson Vg=0.2V

In the Model Builder, Click > Results twistie (as needed).

In the Model Builder window, Right-Click > Electron Concentration Comparison 1 (click twistie as needed) > Density-Gradient,

Select: Duplicate.

In Settings Line Graph,

Click > Dataset pull-down list,

Select: Study 2: Schrodinger-Poisson Vg0.2V/Solution 2 (sol2).

Enter: Schrodinger-Poisson Vg=0.2V in the Label text field.

Enter: schrp1.n_sum in the y-Axis Data Expression text field.

Click > Coloring and Style twistie (as needed),

Click > Line pop-up list,

Select: Solid.

Click > Color pop-up list,

Select: Red.

Click > Legends twistie (as needed),

Select: Manual from the pull-down list.

Enter: SP 0.2 V in the Legends text field.

See Figure 4.82.

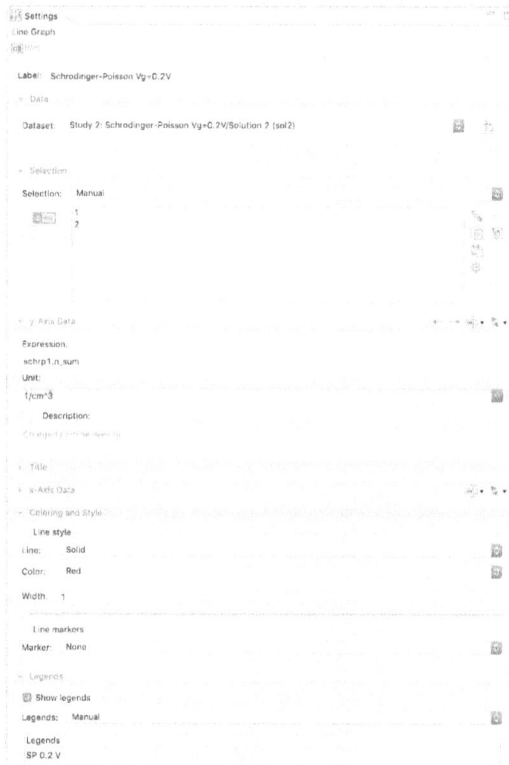

FIGURE 4.82 Schrodinger-poisson modified settings.

Figure 4.82 shows Schrodinger-Poisson modified Settings.

In the Model Builder window, Click > Electron Concentration Comparison 1, See Figure 4.83.

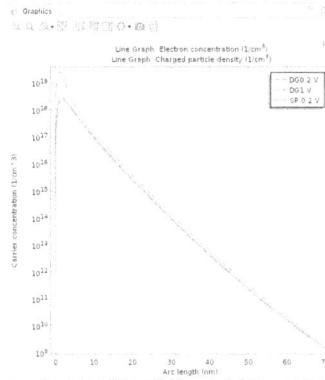

FIGURE 4.83 DG SP comparison graphs.

Figure 4.83 shows the DG SP Comparison Graphs.

1D Silicon Inversion Layer Model: A Comparison of the Results obtained from using the Density-Gradient (DG) Theory and the Schrodinger-Poisson (SP) Theory Methodologies Summary and Conclusions

The 1D Silicon Inversion Layer Model is a powerful tool that can be used to reduce the dimensionality of device layer models from 3D to 1D through the assumption of homogeneous, isotropic materials properties and geometrical symmetry.

FIRST PRINCIPLES AS APPLIED TO 1D MODEL DEFINITION

First Principles Analysis derives from the fundamental laws of nature. In the case of models in this book or from any other source, the modeler should be able to demonstrate the calculated results derived from the models are consistent with the laws of physics and the basic observed properties of materials. In the case of models using this Classical Physics Analysis approach, the laws of conservation in physics require that what goes in (as mass, energy, charge, etc.) must come out (as mass, energy, charge, etc.) or must accumulate within the boundaries of the model. To do otherwise violates fundamental principles.

NOTE

In the COMSOL Multiphysics software, the default interior boundary conditions are set to apply the conditions of continuity in the absence of sources (e.g. heat generation, charge generation, molecule generation, etc.) or sinks (e.g. heat loss, charge recombination, molecule loss, etc.).

The careful modeler must be able to determine by inspection of the model that the appropriate factors have been considered in the development of the specifications for the various geometries, for the material properties of each domain, and for the boundary conditions. He must also be knowledgeable of the implicit assumptions and default specifications that are normally incorporated into the COMSOL Multiphysics software model when a model is built using the default settings.

Consider, for example, the four 1D models developed earlier in this chapter. In these models, it is implicitly assumed there are no thermally related changes (mechanical, electrical, etc.). It is also assumed the materials are homogeneous and isotropic and there are no thin electrical contact barriers at the electrical junctions. None of these assumptions are typically true in the general case. However, by making such assumptions, it is possible to easily build a 1D First Approximation Model.

NOTE

A First Approximation Model is one that captures all the essential features of the problem that needs to be solved, without dwelling excessively on small details. A good First Approximation Model will yield an answer that enables the modeler to determine if he needs to invest the time and the resources required to build a more highly detailed model.

Also, the modeler needs to remember to name model parameters carefully as pointed out in Chapter 1.

REFERENCES

1. COMSOL Multiphysics Reference Manual, p. 464.

2. *https://en.wikipedia.org/wiki/Korteweg-de_Vries_equation*

3. *https://en.wikipedia.org/wiki/List_of_nonlinear_partial_differential_equations*

4. *https://en.wikipedia.org/wiki/Integrable_system#Exactly_solvable_models*

5. *https://en.wikipedia.org/wiki/Nonlinear_system*

6. *https://en.wikipedia.org/wiki/John_Scott_Russell*

7. *https://en.wikipedia.org/wiki/Soliton*

8. *https://en.wikipedia.org/wiki/Soliton_(optics)*

9. *https://en.wikipedia.org/wiki/Telegrapher%27s_equations*

10. *https://en.wikipedia.org/wiki/Oliver_Heaviside*

11. *https://en.wikipedia.org/wiki/Powder_metallurgy*

12. *https://en.wikipedia.org/wiki/Fluidized_bed*

13. *https://en.wikipedia.org/wiki/Sintering*

14. *https://en.wikipedia.org/wiki/Nanoarchitectures_for_lithium-ion_batteries*

15. *https://en.wikipedia.org/wiki/Granulation_(process)*

16. *https://en.wikipedia.org/wiki/Fluidized_bed_combustion*

17. *https://en.wikipedia.org/wiki/Spherical_coordinate_system*

18. E.U. Condon and H. Odishaw, *Handbook of Physics*, McGraw-Hill, New York, 1958, pp. 1–109.

19. *https://en.wikipedia.org/wiki/Audion*

20. *https://en.wikipedia.org/wiki/Transistor*

21. *https://en.wikipedia.org/wiki/Point-contact_transistor*

22. *https://en.wikipedia.org/wiki/Bipolar_junction_transistor*

23. *https://en.wikipedia.org/wiki/MOSFET*

24. *https://en.wikipedia.org/wiki/Integrated_circuit*

25. M. G. Ancona, "Equations of State for Silicon Inversion Layers," *IEEE Trans. Elec. Dev.*, vol. 47, no. 7, July 2000, pp. 1449–1456.

26. M. G. Ancona, "Density-gradient theory: a macroscopic approach to quantum confinement and tunneling in semiconductor devices," *J. Comput. Electron.* vol. 10, 2011, pp. 65–97.

27. F. Stern, "Self-Consistent Results for n-Type Si Inversion Layers," *Phys. Rev. B,* vol. 5, no. 12, p. 4891, 1972.

28. S. M. Sze and K. K. Ng, *Physics of Semiconductor Devices*, 3rd ed., Wiley, ISBN 978-0-471-14323-9.

SUGGESTED MODELING EXERCISES

1. Build, mesh and solve the 1D KdV Equation Model problem presented earlier in this chapter.

2. Build, mesh and solve the 1D Telegraph Equation Model problem presented earlier in this chapter.

3. Build, mesh and solve the 1D Spherically Symmetric Transport Model problem presented earlier in this chapter.

4. Change the values of the parameters in the 1D KdV Equation Model problem. Compute the new solution. Analyze, compare and contrast the results with the new parameter values to the results found earlier.

5. Change the values of the parameters in the 1D Telegraph Equation Model problem. Compute the new solution. Analyze, compare and contrast the results with the new parameter values to the results found earlier.

6. Change the values of the parameters in the 1D Spherically Symmetric Transport Model problem. Compute the new solution. Analyze, compare and contrast the results with the new parameter values to the results found earlier.

CHAPTER 5

2D MODELING USING COMSOL MULTIPHYSICS 5.X

In This Chapter

- Guidelines for 2D Modeling in 5.x
    - 2D Modeling Considerations
- 2D Basic Models
    - 2D Electrochemical Polishing Model
    - 2D Hall Effect Model
- First Principles as Applied to 2D Model Definition
- References
- Suggested Modeling Exercises

GUIDELINES FOR 2D MODELING IN 5.x

NOTE

In this chapter, the modeler is introduced to the development and analysis of 2D models. 2D models have two geometric dimensions (x, y). In 5.x, there are fundamentally two types of 2D geometries: 2D and 2D Axisymmetric. 2D models, for ease of calculation, are assumed to be homogeneous and isotropic in the third (non-planar) dimension. 2D Axisymmetric models are planar representations of cylindrical 3D objects that are rotationally homogeneous and isotropic. 2D models are typically classed by level of difficulty as being of two types, introductory and advanced.

In this chapter, introductory 2D models will be presented. Such 2D models are more geometrically complex than 1D models. 2D models have proven to be very valuable to the science and engineering communities, both in the past and currently, as first-cut evaluations of potential physical behavior under the influence of external stimuli. The 2D model responses and other such ancillary information are then gathered and screened early in a project for initial evaluation and potentially for later use in building higher-dimensionality (3D) field-based (electrical, magnetic, etc.) models.

Since the models in this and subsequent chapters are more complex to build and more difficult to solve than the models presented thus far, except for the Advanced Semiconductor model in Chapter 4, it is important that the modeler have available the tools necessary to most easily utilize the powerful capabilities of the 5.x software. In order to do that, the modeler should go to the main 5.x toolbar, Click > Options – Preferences – Show More Options. When the Preferences – Show More Options edit window is shown, Select: Equation view checkbox and Equation sections checkbox. Click > OK {5.1}.

For information on the development of the more advanced models using the 2D Axisymmetric coordinate system, the modeler should consult Chapter 6 of this text.

2D Modeling Considerations

2D Modeling can in some cases be less difficult than 1D modeling, having fewer implicit assumptions, and yet potentially, it can still be a very challenging type of model to build, depending on the underlying physics involved. The least difficult aspect of 2D model creation arises from the fact that the geometry is relatively simple. (In a 2D model, the modeler has only a single plane as the modeling space.) However, the physics in a 2D model can range from relatively easy to extremely complex.

In compliance with the laws of physics, a 2D model implicitly assumes that energy flow, materials properties, environment, and all other conditions and variables that are of interest are homogeneous, isotropic, and/or constant, unless otherwise specified, throughout the entire domain of interest both within the model and through the boundary conditions and in the environs of the model.

The modeler needs to bear the above-stated conditions in mind and ensure that all of the modeling conditions and associated parameters (default settings) in each model created are properly considered, defined, verified, and/or set to the appropriate values.

It is always mandatory that the modeler be able to accurately anticipate the expected results of the model and accurately specify the manner in which those results will be presented. Never assume that any of the default values that are present when the model is created necessarily satisfy the needs or conditions of a particular model.

NOTE   *Always verify that any parameters employed in the model are the correct value needed for that model. Calculated solutions that significantly deviate from the anticipated solution or from a comparison of values measured in an experimentally derived realistic model are probably indicative of one or more modeling errors either in the original model design, in the earlier model analysis, in the understanding of the underlying physics, or are simply due to human error.*

Coordinate System

In 2D models, there are two geometric coordinates, space (x), space (y), and the temporal coordinate, time (t). In a steady-state solution to a 2D model, parameters can only vary as a function of position in space (x) and space (y) coordinates. Such a 2D model represents the parametric condition of the model in a time-independent mode (quasi-static). In a transient solution model, parameters can vary both by position in space (x), space (y), and in time (t).

The space coordinates (x) and (y) typically represent a distance coordinate throughout which the model is to calculate the change of the specified observables (i.e., temperature, heat flow, pressure, voltage, current, etc.) over the range of coordinate values ($x_{min} <= x <= x_{max}$) and ($y_{min} <= y <= y_{max}$). The time coordinate (t) represents the range of temporal values ($t_{min} <= t <= t_{max}$) from the beginning of the observation period (t_{min}) to the end of the observation period (t_{max}).

Electrochemical Polishing (Electropolishing) Theory

Electrochemical {5.2} polishing (electropolishing {5.3}) is a well-known process in the metal finishing industry. This process allows the finished surface smoothness of a conducting material to be cleanly controlled to a high degree of precision, using relatively simple processing equipment. The electrochemical polishing technique eliminates the abrasive residue typically present on the polished surface from a mechanical polishing process and also eliminates the need for complex, mechanical polishing machinery.

NOTE

It is presently understood that the science of electricity and consequently that of electrochemistry started with the work of William Gilbert through the study of magnetism. He first published his studies in the year 1600 {5.4}. Charles-Augustin de Coulomb {5.5}, Joseph Priestley {5.6}, Georg Ohm {5.7}, and others made additional independent later contributions that furthered that basic understanding of the nature of electricity and electrochemistry. Those contributions led to the discovery and disclosure by Michael Faraday {5.8} of his two laws of electrochemistry in the year 1832.

The following numerical solution model (Electrochemical Polishing) was originally developed by COMSOL for distribution with earlier versions of the Multiphysics software as a COMSOL Multiphysics Electromagnetics Model. This model introduces two important basic concepts, the first concept in applied physics, and the second concept in applied modeling: (1) Electropolishing and the (2) Deformed Mesh and Deformed Geometry {5.9}. The Electrochemical Polishing model built below has been transformed into an Electromagnetics model that is functional in 5.x.

NOTE

It is important for the new modeler to personally build each model presented within this text. The best method of building an understanding of the modeling process is by obtaining the hands-on creation experience of actually building, meshing, solving, and plotting the results of a model. Many times the inexperienced modeler will make and subsequently correct errors, adding to his experience and fund of modeling knowledge. Even building the simplest model will expand the modeler's fund of knowledge.

The polishing (smoothing) of a material surface, either mechanical or electrochemical, results from the reduction of asperities (bumps), thus achieving a nominally smooth surface profile (uniform thickness $\pm \Delta$ thickness). In a mechanical polishing technique, the reduction of asperities occurs through the use of finer (smaller) and finer grit (abrasive) sizes. The mechanical polishing of many nonuniform surfaces is difficult, if not impossible, due to the complexity and/or physical size of such surfaces. Figure 5.1 shows a simple asperity, as will be modeled in this section of the chapter.

See Figure 5.1.

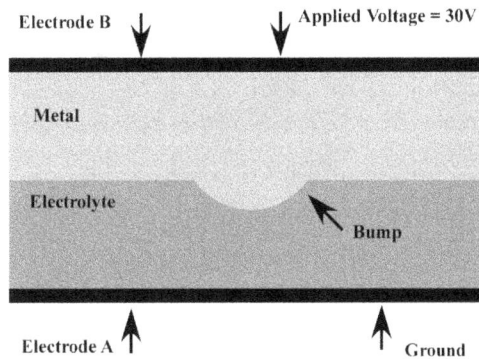

FIGURE 5.1 An asperity (Bump) on a metal electrode.

The surface of the metal electrode, using the electrochemical method, is polished by the differential removal of material from local asperities in selected areas, by the immersion of the nominally rough metal electrode in an electrolyte, and the application of an electrical current (electron bombardment). As shown in Equation 5.1, a good first-order approximation to the experimentally observed material removal process, is that the rate of removal (velocity) of material removal (U) from the electrode surface is proportional to the amplitude of the electrical current and direction of the electrical current **J**, relative to the local surface normal vector **n** (See Figure 5.2).

$$U = -K * \mathbf{J} \bullet \mathbf{n} = -K * J_n \qquad (5.1)$$

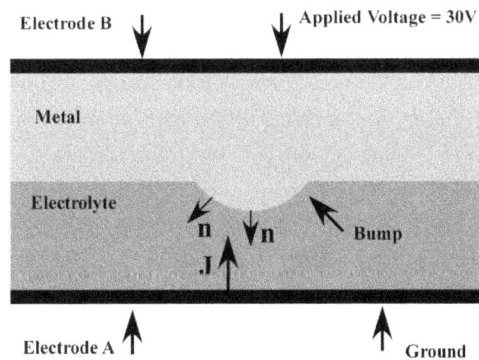

FIGURE 5.2 Surface normal vector n and the current vector J.

NOTE *The electropolishing technique is substantially the inverse of the electroplating process and as a result, the rate of removal of material (velocity = U) from the nominally rough surface of the positive electrode is proportional to the normal current density at the positive electrode surface, as shown in Equation 5.1.*

The exact value of the proportionality constant (K), to first order, in physical applications (e.g., research experiments, processing, etc.), is determined by the electrode material, the electrolyte, and the temperature.

For this model, the proportionality constant is chosen to be:

$$K = 1.0 \, x \, 10^{-11} \; m^3 \, / \, (A * s) \tag{5.2}$$

where K is the proportionality constant in [m^3/(A*s)].

m is in meters [m].

A is amperes [A].

s is seconds [s].

Obviously, since the material is removed from the positive electrode during the electropolishing process, the spacing between the upper and lower electrodes will increase. The time rate of change of the model geometry (electrode spacing) needs to be accommodated somewhere within the model. The Deformed Mesh > Deformed Geometry (dg) interface accommodates that time rate of change, resulting from the normal current (J_n) flowing in the electrolyte during the use of the Electric Currents (ec) Interface {5.10}.

NOTE *The Deformed Mesh > Deformed Geometry (dg) interface allows the modeler to create models in which the physics of the process introduces and controls geometric changes in the model. However, the modeler must be sure to know and to work carefully within the limits of the modeling system. The Deformed Mesh > Deformed Geometry (dg) interface is a powerful tool. However, the calculated Mesh parameters can drift, as the mesh is deformed, and ultimately lead to non-physical, non-convergent results. Avoidance of such non-physical results requires the modeler to understand the basic physics of the modeled problem and to choose the meshing method that yields the best overall results.*

2D BASIC MODELS

2D Electrochemical Polishing Model

Building the 2D Electrochemical Polishing Model

Startup 5.x, Click > Model Wizard button.

Select Space Dimension > 2D.

Click > Twistie for Δu Mathematics > Twistie for Deformed Mesh in the Select Physics window.

Select: Deformed Geometry (dg).

Click > Add.

Click > Twistie for AC/DC in the Select Physics window.

Click > Twistie > Electric Fields and Currents.

Select: Electric Currents (ec).

Click > Add.

Click > Study (Right Pointing Arrow button).

Select > General Studies > Time Dependent in the Select Study window.

Click > Done (Checked Box button).

Click > Save As.

Enter MM2E5X_2D_EP_1.mph.

See Figure 5.3.

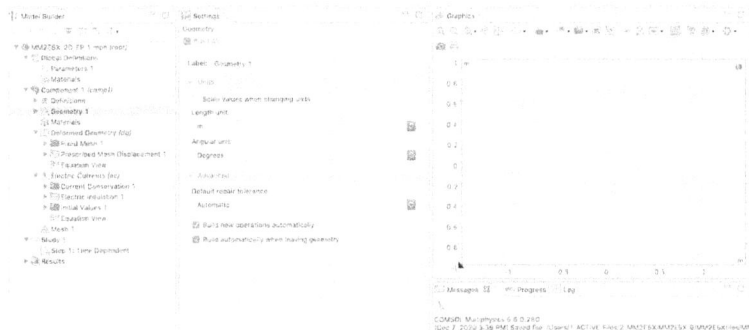

FIGURE 5.3 Desktop display for the MM2E5X_2D_EP_1.mph model.

Figure 5.3 shows the Desktop Display for the MM2E5X_2D_EP_1.mph model.

Parameters

Click > Global Definitions > Parameters 1.

In the Settings – Parameters – Parameters 1 edit window,

Click > The top of the Name column twice.

Enter the following information in the text entry fields below the Parameters 1 table.

Name = K, Expression = 1.0e-11[m^3/(A*s)], Description = Proportionality constant.

See Figure 5.4a.

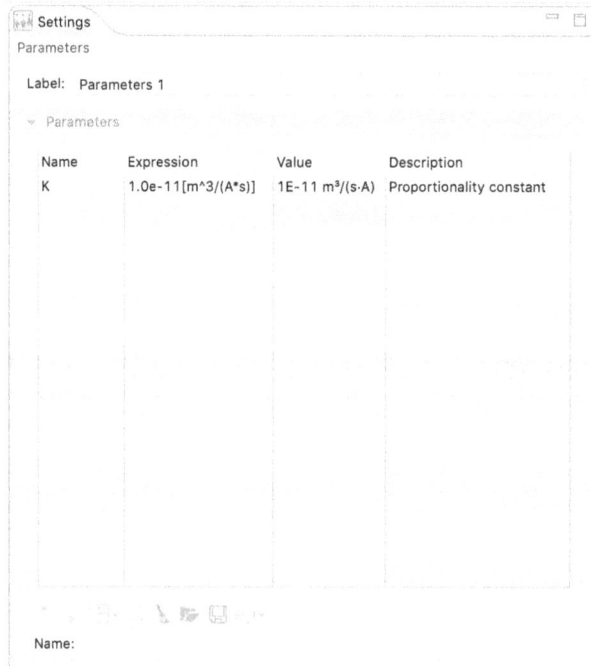

FIGURE 5.4A Settings – Parameters – Parameters 1 edit window filled.

Figure 5.4a shows the Settings – Parameters – Parameters 1 edit window filled.

NOTE *The Deformed Geometry (dg) interface {5.11} requires the specification of the system geometry to enable the calculation of the changing parameters within the model.*

Variables

Right-Click > Global Definitions.

Select > Variables from the Pop-up menu.

In the Settings – Variables – Variables 1 edit window,

Enter > Name = dx, Expression = x-Xg, Description = x-displacement.

Enter > Name = dy, Expression = y-Yg, Description = y-displacement.

See Figure 5.4b.

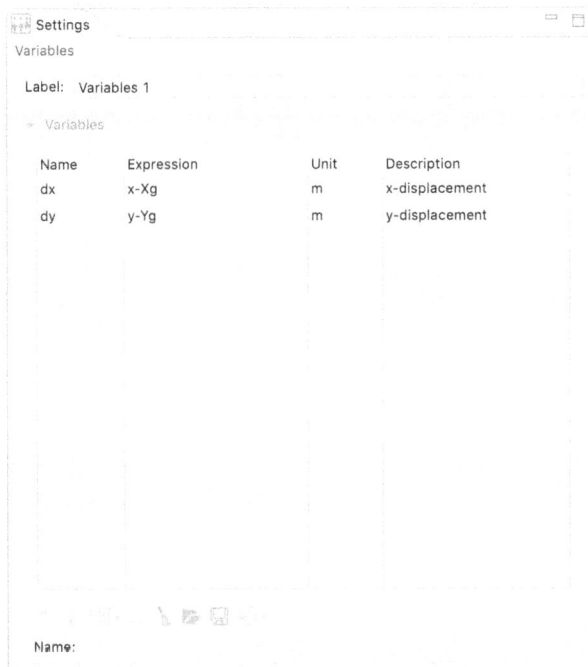

FIGURE 5.4B Settings – Variables – Variables 1 edit window filled.

Figure 5.4b shows the Settings – Variables – Variables 1 edit window filled.

In Model Builder > Component 1 (*comp1*),

Click > Geometry 1.

In Settings – Geometry > Units > Length unit,

Select > mm from the Pull-down edit bar.

In Settings Geometry > Advanced > Select > Default repair tolerance: Pull-down > Relative.

See Figure 5.5.

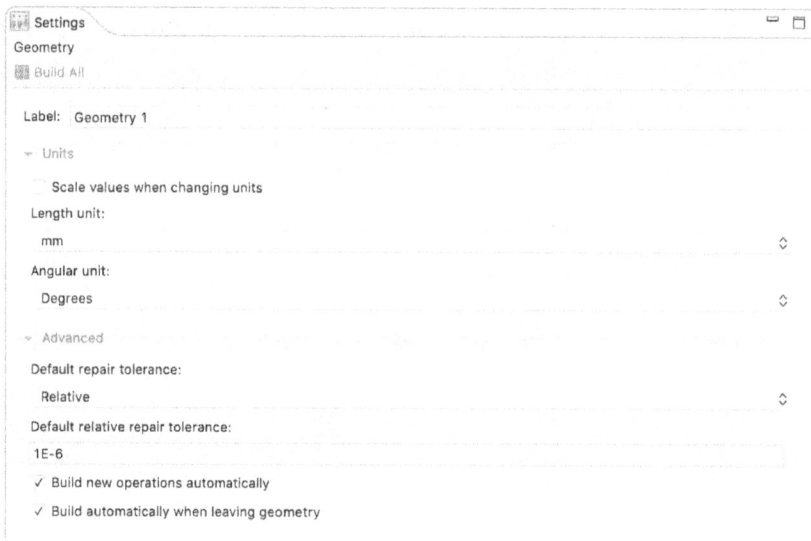

FIGURE 5.5 Model builder – Component 1 (comp1) > Geometry 1 > Settings geometry.

Figure 5.5 shows the Model Builder – Component 1 (*comp1*) > Geometry 1 > Settings Geometry.

Right-Click > Model Builder > Component 1 (*comp1*) > Geometry 1.

Select > Rectangle from the Pop-up menu.

Enter > 2.8 in the Settings – Rectangle – Size – Width entry window.

Enter > 0.4 in the Settings – Rectangle – Size – Height entry window.

Enter > –1.4 in the Settings – Rectangle – Position – x entry window.

Enter > 0 in the Settings – Rectangle – Position – y entry window.
See Figure 5.6.

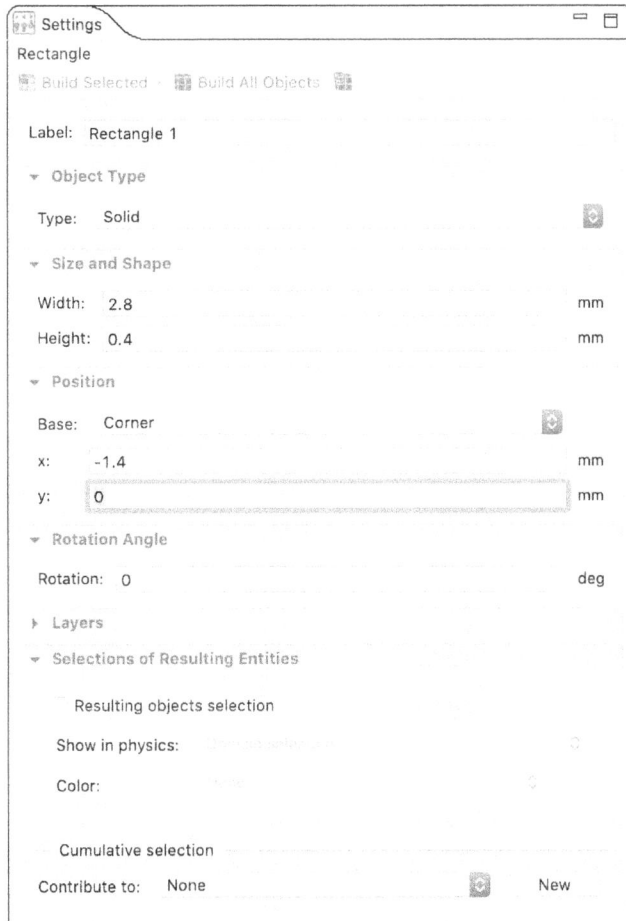

FIGURE 5.6 Settings – Rectangle entry windows.

Figure 5.6 shows the Settings – Rectangle Entry Windows.

Click > Build Selected.

See Figure 5.7.

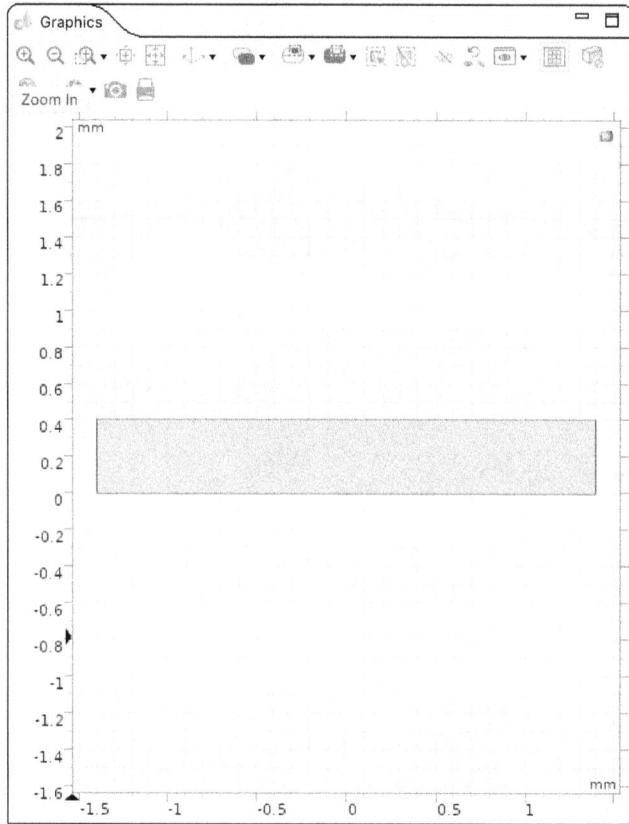

FIGURE 5.7 Graphics window – Rectangle.

Figure 5.7 shows the Graphics window – Rectangle.

Right-Click > Model Builder > Component 1 (*comp1*) > Geometry 1.

Select > Circle from the Pop-up menu.

Enter > 0.3 in the Settings – Circle – Size – Radius entry window.

Enter > –0.6 in the Settings – Circle – Position – x entry window.

Enter > 0.6 in the Settings – Circle – Position – y entry window.

Click > Build All.

See Figure 5.8a.

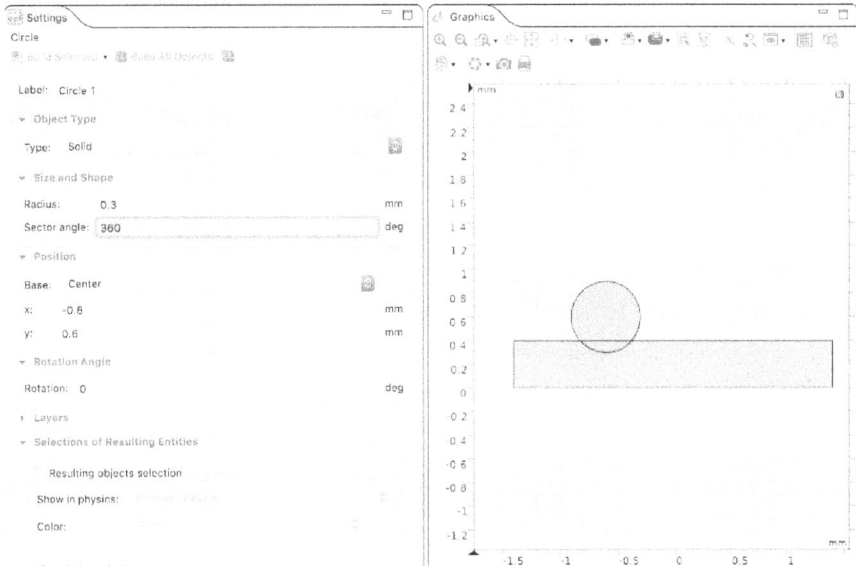

FIGURE 5.8a Settings – Circle entry windows and graphics window.

Figure 5.8a shows the Settings – Circle Entry Windows and Graphics Window.

Right-Click > Model Builder > Component 1 (*comp1*) > Geometry 1.

Select > Circle from the Pop-up menu.

Enter > 0.3 in the Settings – Circle – Size – Radius entry window.

Enter > 0.6 in the Settings – Circle – Position – x entry window.

Enter > 0.6 in the Settings – Circle – Position – y entry window.

Click > Build All.

See Figure 5.8b.

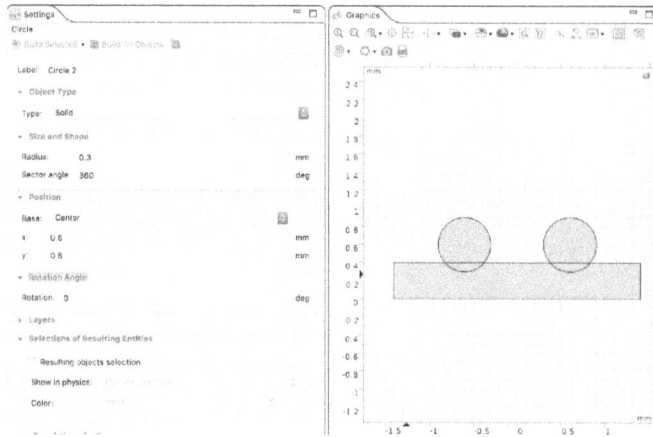

FIGURE 5.8b Settings – Circle entry windows and graphics window.

Figure 5.8b shows the Settings – Circle Entry Windows and Graphics Window.

Right-Click > Model Builder > Component 1 (comp1) > Geometry 1.

Select > Boolean and Partitions – Difference from the Pop-up menu.

Click > r1 in the Graphics window (be sure to click r1 until it turns Blue)

Click > Activate Selection button for the Objects to subtract entry window.

Click > c1 and c2 in the Graphics window.

See Figure 5.9.

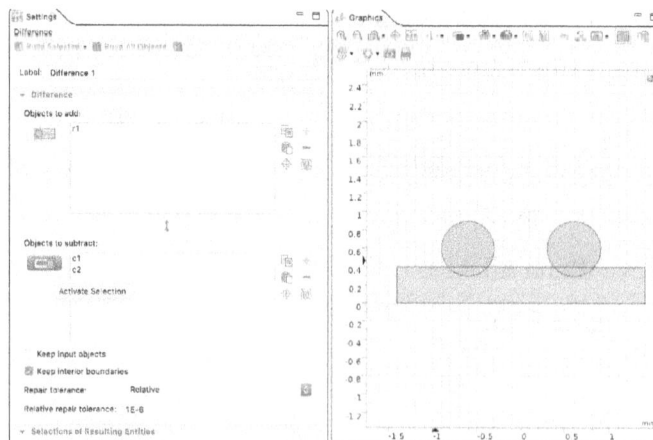

FIGURE 5.9 Settings – Difference – Difference windows.

Figure 5.9 shows the Settings – Difference – Difference windows.

Click > Build All.

See Figure 5.10.

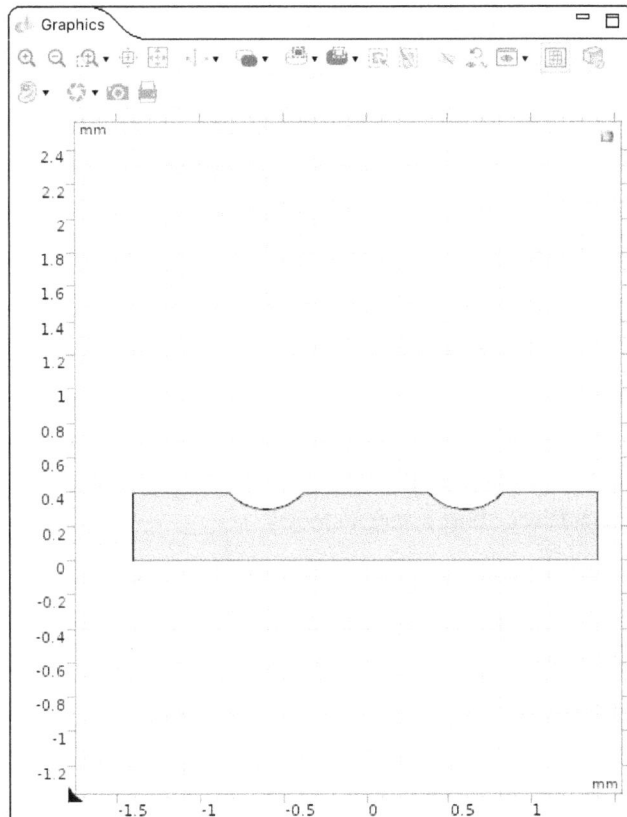

FIGURE 5.10 Model geometry in the graphics window.

Figure 5.10 shows the Model Geometry in the Graphics window.

Deformed Geometry (dg) Interface

Right-Click > Deformed Geometry (dg),

Select > Free Deformation.

Click > Domain 1 in the Graphics window.

See Figure 5.11

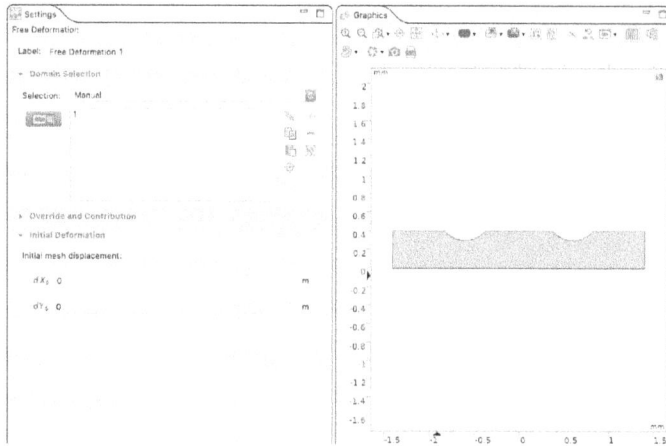

FIGURE 5.11 Selected model geometry in the graphics window.

Figure 5.11 shows the Selected Model Geometry in the Graphics window.

Right-Click > Deformed Geometry (dg),

Select > Prescribed Mesh Velocity.

Click > Boundaries 1 (left end) & 6 (right end) in the Graphics window (be sure the ends turn Blue).

Click and Clear > Prescribed y velocity check box.

See Figure 5.12

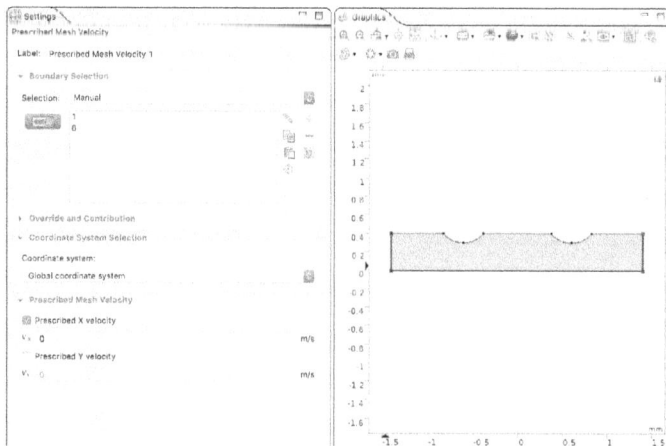

FIGURE 5.12 Prescribed mesh velocity selections.

Figure 5.12 shows the Prescribed Mesh Velocity Selections.

Right-Click > Deformed Geometry (dg),

Select > Prescribed Normal Mesh Velocity.

Click > Boundaries 3, 4, 5, 7, 8, 9, & 10 in the Graphics window (be sure the top surface segments turn Blue).

Enter > -K*(-ec.nJ) in the v_n edit window.

See Figure 5.13.

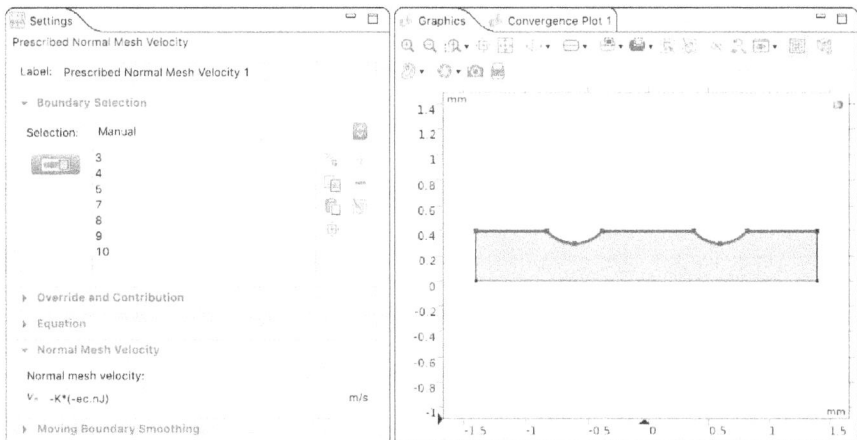

FIGURE 5.13 Prescribed normal mesh velocity selections.

Figure 5.13 shows the Prescribed Normal Mesh Velocity Selections.

Electric Currents (ec) Interface

Click > Twistie for the Electric Currents (ec) interface (as needed).

Click > Electric Currents (ec).

Enter > 2.8e-3[m] in the Settings – Electric Currents –Thickness – Out-of-Plane Thickness d edit window.

See Figure 5.14.

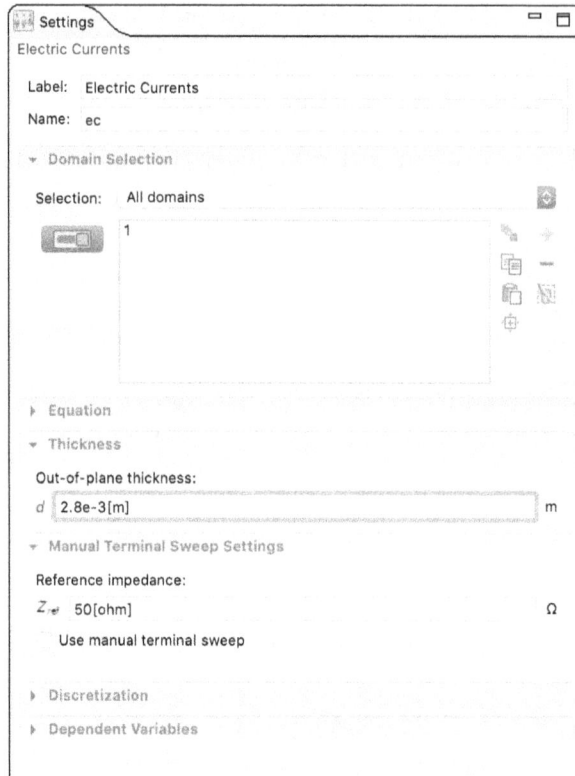

FIGURE 5.14 Model geometry out-of-plane-thickness thickness edit window.

Figure 5.14 shows the Model Geometry Out-of-Plane-Thickness Thickness edit window.

Click > Model Builder – Component 1(*comp1*) – Electric Currents – Current Conservation 1.

Select > User defined from the pull-down menu in Settings – Current Conservation – Constitutive Relation Jc-E – Electrical conductivity.

Enter > 10 in the Settings – Current Conservation – Constitutive Relation Jc-E – Electrical conductivity edit window.

Scroll-down > Find the Settings – Current Conservation – Constitutive Relation D-E – Relative permittivity pull-down menu.

Select > User defined from the pull-down menu in Settings – Current Conservation – Constitutive Relation D-E – Relative permittivity pull-down menu.

Either use the default value of 1 that appears or enter the value 1 in the edit window.

NOTE *In this case, a relative permittivity of 1 is appropriate. If the electrolyte had a different relative permittivity, then the modeler would need to enter that different value there.*

Right-Click > Model Builder – Component 1 (*comp1*) – Electric Currents.

Select > Electric Insulation from the Pop-up menu.

Click > Model Builder – Component 1 (*comp1*) – Electric Currents – Electric Insulation 2.

Click > Boundary 1 (Left end) in the Graphics window.

Click > Boundary 6 (Right end) in the Graphics window.

See Figure 5.15.

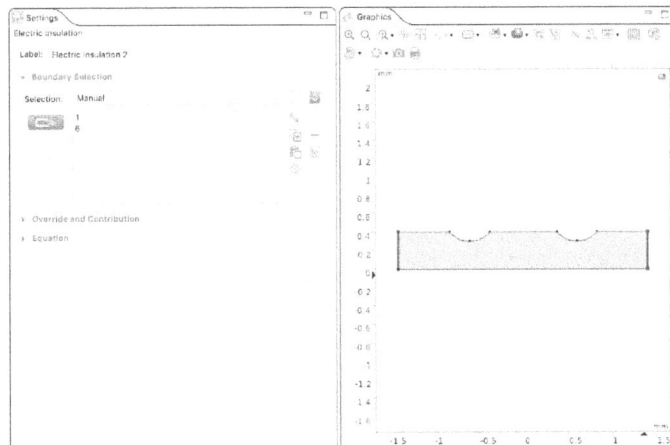

FIGURE 5.15 Model geometry electrically insulation 2 boundaries.

Figure 5.15 shows the Model Geometry Electrically Insulation 2 Boundaries.

Right-Click > Model Builder – Component 1 (*comp1*) – Electric Currents.

Select > Ground from the Pop-up menu.

Click > Model Builder – Component 1 (*comp1*) – Electric Currents – Ground 1.

Click > Boundary 2 (Bottom) in the Graphics window.

See Figure 5.16.

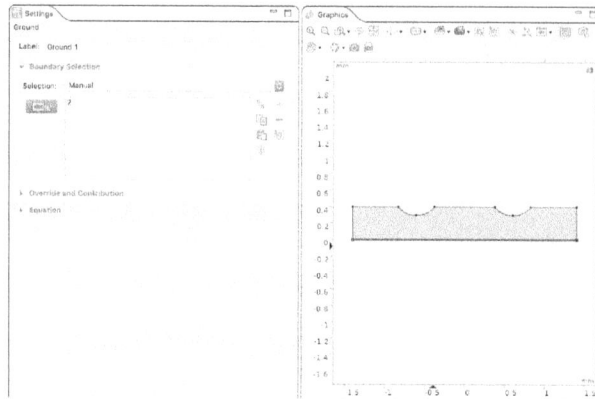

FIGURE 5.16 Model geometry ground 1 boundaries.

Figure 5.16 shows the Model Geometry Ground 1 Boundaries.

Right-Click > Model Builder – Component 1 (*comp1*) – Electric Currents.

Select > Electric Potential from the Pop-up menu.

Click > Model Builder – Component 1 (*comp1*) – Electric Currents – Electric Potential 1.

Click > Boundaries 3, 4, 5, 7, 8, 9, 10 in the Graphics window.

Enter > 30 in the Settings – Electric Potential – Electric Potential – Voltage edit window (V_0).

See Figure 5.17.

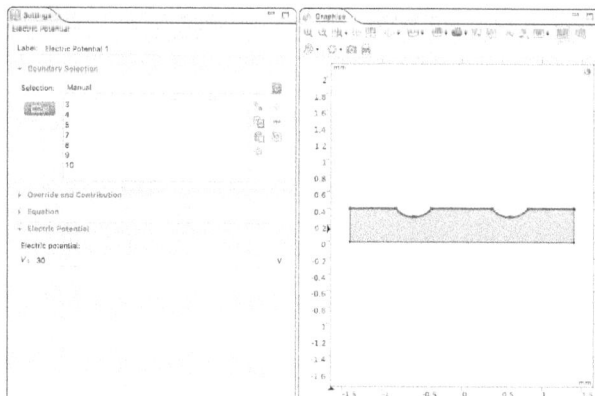

FIGURE 5.17 Model geometry electric potential 1 boundaries.

Figure 5.17 shows the Model Geometry Electric Potential 1 Boundaries.

Mesh 1

The default settings for the Mesh Type are not appropriate for the solution of this problem and thus need to be modified.

Click > Model Builder – Component 1 (*comp1*) – Mesh 1.

Click > Sequence type pull-down menu in Settings – Mesh – Mesh Settings – Sequence type.

Select > Physics-controlled mesh (as needed).

Click > Element size pull-down menu in Settings – Mesh – Physics-Controlled Mesh – Element size.

Select > Finer.

Click > Build All button.

After meshing, the modeler should see a message in the message window about the number of elements (343 domain elements and 71 boundary elements) in the mesh.

See Figure 5.18.

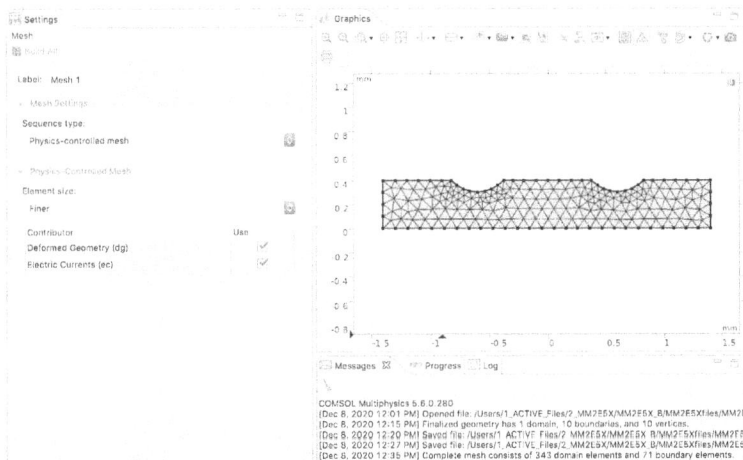

FIGURE 5.18 Desktop display – Graphics – Meshed domain.

Figure 5.18 shows the Desktop Display – Graphics – Meshed Domain.

Study 1

NOTE *The default settings for the Time Dependent Study Type are not appropriate for the solution of this problem and thus need to be modified.*

Click > Study 1 twistie (as needed).

Click > Step 1: Time Dependent.

Click the Range button (at the right of the Times entry window).

Enter > Start = 0, Step = 1, Stop = 10.

Click > Replace.

See Figure 5.19.

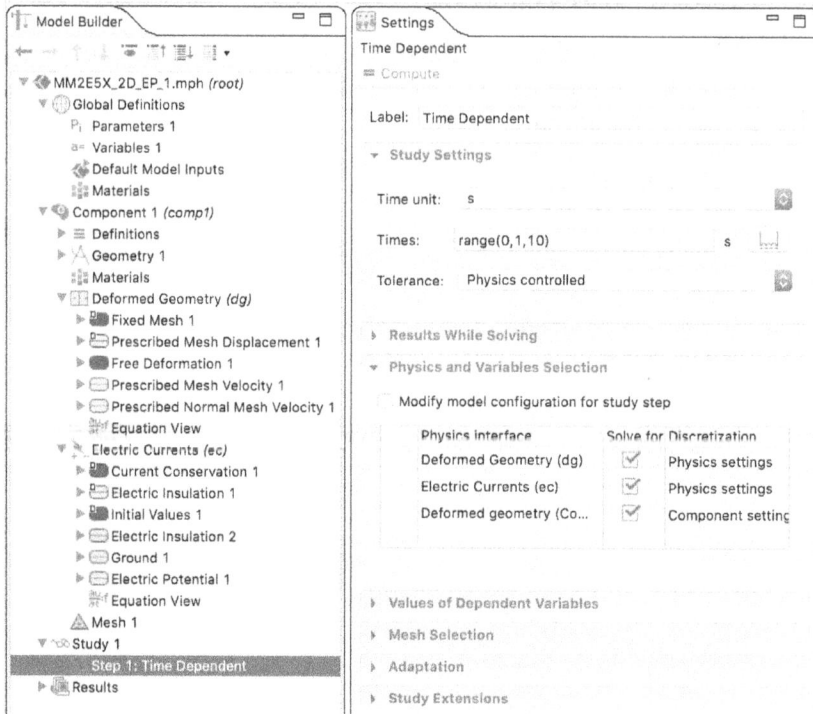

FIGURE 5.19 Desktop display – Settings – Time dependent – Study settings.

Figure 5.19 shows the Desktop Display – Settings – Time Dependent – Study Settings.

Study 1 > **Solver 1**

In Model Builder,

Right-Click Study 1 >Select > Compute.

Computed results, using the default display settings, are shown in Figure 5.20.

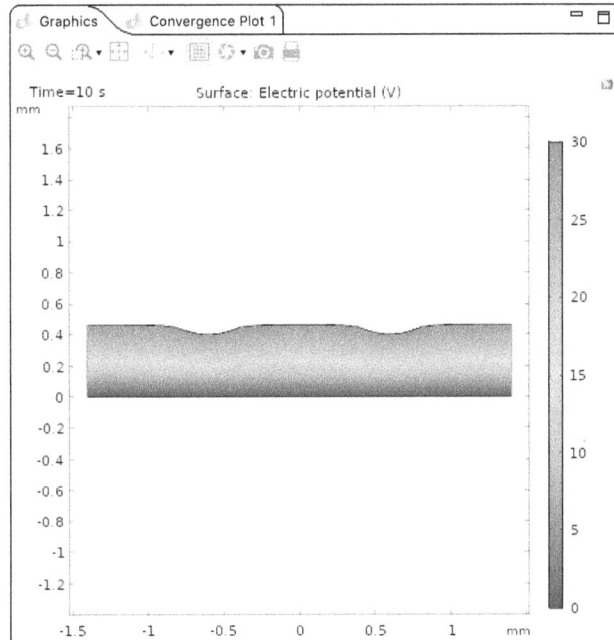

FIGURE 5.20 Computed results, Using the default display settings.

Figure 5.20 shows the computed results, using the default display settings.

Results

In Model Builder,

Right- Click > Results

Select > 2D Plot Group.

Right- Click > Model Builder – Results – 2D Plot Group 2,

Select > Surface.

In Settings – Surface – Expression,

Click > Replace Expression.

Select > Component 1 > Electric Currents > Currents and charge > ec.normJ-Current density norm from the pop-up menus.

Click > Plot.

See Figure 5.21.

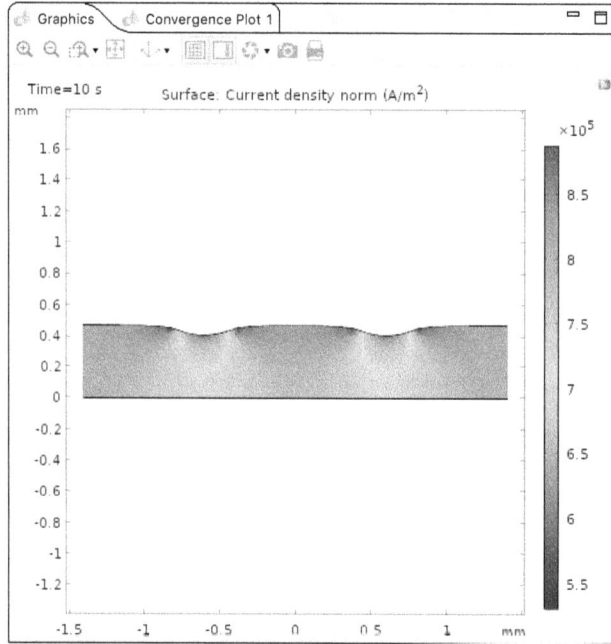

FIGURE 5.21 Graphics display – Surface current density norm.

Figure 5.21 shows the Graphics Display – Surface Current Density norm.

2D Electrochemical Polishing Model Animation

NOTE *It is important that the modeler be able to observe the results of the modeling process. The instructions that follow allow the modeler to precisely do that. The modeler will first build a movie that is viewed in the Graphics window.*

This animation shows the electropolishing process as it progresses in time.

In Model Builder,

Right-Click > Results > Export.

Select > Animation > Player.

In Settings Animation > Scene > Subject: Pull-down, Select > 2D Plot Group 2.

Right-Click > Animation 1: Select > Play.

NOTE *The modeler is now able to observe the results of the modeling process dynamically. The above instructions allowed the modeler to build a movie with the default parameters. The modeler can now build other movies using different parameters that will be viewed in the Graphics window.*

The next set of instructions allow the modeler to export the movie frames to a file.

In Model Builder,

Right-Click > Results – Export.

Select > Animation > File.

In Settings – Animation 2,

In Settings Animation > Scene > Subject: Pull-down, Select > 2D Plot Group 2.

In Settings – Animation – Output,

Select > Output type Movie from the pull-down menu.

Select > Format GIF from the pull-down menu.

NOTE *The detailed procedures for Animation in 5.x on the Macintosh and the PC are different. The Macintosh uses the GIF format. The PC uses the AVI format. If the modeler tries to use AVI on the Macintosh, an error results.*

In Settings – Animation – Frames,

Click > Lock aspect ratio check box.

In Settings – Animation – Advanced (Click > twistie as needed),

Check > Antialiasing check box (as needed).

Click > Settings – Export (located on the Settings toolbar).

Save the Animation as: MM2E5X_2D_EP_1_An.gif

NOTE *The movie has been exported to a file.*

2D Electrochemical Polishing Model Summary and Conclusions

The 2D Electrochemical Polishing Model is a powerful tool that can be used to explore electrochemical polishing techniques in many different conductive media (e.g. metals, semimetals, alloys, graphene films, graphite films, etc.) {5.12}. As has been shown earlier in this chapter, the 2D Electrochemical Polishing Model is easily and simply modeled with a combination of Deformed Geometry (dg) and Electric Currents (ec) in the model.

Save and Close the just generated **2D Electrochemical Polishing Model.**

2D Hall Effect Model Considerations

In 1827, Georg Ohm published {5.13} his now fundamental and famous Ohm's Law:

$$I = \frac{V}{R} \qquad (5.3)$$

Where I = Current in Amperes.

V = Potential Difference in Volts.

R = Resistance in Ohms.

See Figure 5.22.

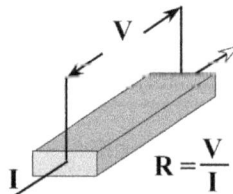

FIGURE 5.22 Ohm's law.

Ohm's Law is an extremely useful phenomenological law. However, in order to better understand conduction in homogeneous, isotropic solid materials, the calculations used herein need to be expanded until they represent the physical behavior (motion) of the fundamental charged particles (electrons, holes).

NOTE *In solid materials (e.g. metals, semiconductors, etc.), there are three potential mobile carriers of charge: electrons (-), holes (+), and ions (charge sign can be either + or -, depending upon the type of ion). Ions in a solid typically have a very low mobility (pinned in position) and thus contribute little to the observed current flow in most solids. Ion flow will not be considered herein.*

In metals, due to the underlying physical and electronic structure, electrons are the only carrier. In semiconductors (e.g. Si, Ge, GaAs, InP), either electrons and/or holes (the absence of an electron) can exist as the primary carrier types. The density of each carrier type (electrons, holes) is determined by the electronic structure of the host material (e.g. Si, Ge, SiGe) and the density and distribution of foreign impurity atoms (e.g. As, P, N, Al) within the host solid material. For further information on the nature of solids and the behavior of impurity atoms in a host matrix see works by Kittel {5.14} and/or Sze {5.15}.

Hall Effect sensors are widely available in a large number of different geometric configurations. They are typically applied in sensing fluid flow, rotating, and/or linear motion, proximity, current, pressure, orientation, etc. In the 2D model presented in the remainder of this chapter, several simplifying assumptions will be made that allow the basic physics principles to be demonstrated without excess complexity.

The resistance of a homogeneous, isotropic solid material R is defined:

$$R = \frac{\rho L}{A} \tag{5.4}$$

Where ρ = resistivity in Ohm-meters (Ω-m).

L = Length of Sample in meters (m).

A = Cross-sectional Area of Sample in meters squared (m²).

See Figure 5.23.

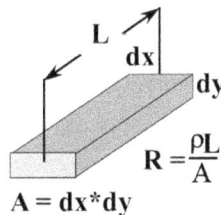

FIGURE 5.23 Resistance.

The resistivity of a homogeneous, isotropic solid material is defined {5.16}:

$$\sigma = \frac{1}{\rho} = n_e |e| \mu_e + n_h |e| \mu_h \qquad (5.5)$$

Where ρ = resistivity Ohm-meters (Ω-m).

σ = conductivity in Siemens per meter (S/m).

n_e = electron density in electrons per cubic meter (N_e/m^3).

n_h = hole density in holes per cubic meter (N_h/m^3).

$|e|$ = absolute value of the charge on an electron (hole) in Coulombs (C).

μ_e = electron mobility in meters squared per volt-second ($m^2/(V^*s)$).

μ_h = hole mobility in meters squared per volt-second ($m^2/(V^*s)$).

The Hall Effect {5.17} was discovered by Edwin Hall in 1879 {5.18}. His discovery was made through making measurements on the behavior of electrical currents in thin gold foils, in the presence of a magnetic field. By introducing a magnetic field into the current flow region of the solid, the measurer effectively adds an anisotropic term into the conductivity of a nominally homogeneous, isotropic solid material. The magnetic field causes the anisotropic conductivity by acting on the carriers through the Lorentz Force {5.19}. The Lorentz Force produces a proportional, differential voltage/charge accumulation between two surfaces or edges of a conducting material orthogonal to the current flow.

The Lorentz Force is:

$$\mathbf{F} = q\,(\mathbf{E} + (\mathbf{v} \times \mathbf{B})) \qquad (5.6)$$

Where $\mathbf{F}$ = Force Vector on the charged particle (electron and/or hole).

q = charge on the particle (electron and/or hole).

$\mathbf{E}$ = Electric Field Vector.

$\mathbf{v}$ = Instantaneous velocity vector of the particle.

$\mathbf{B}$ = Magnetic Field Vector.

The Hall Voltage {5.20} is:

$$V_H = \frac{R_H * I * B}{t} \qquad (5.7)$$

Where V_H = Hall Voltage.

R_H = Hall coefficient.

I = Current.

B = Magnetic Field.

t = thickness of sample.

The Hall coefficient (R_H) is:

$$R_H = -\frac{r}{n_e e} \qquad (5.8)$$

Where R_H = Hall coefficient.

r = $1 \le x \le 2$.

n_e = density of electrons.

e = charge on the electron.

NOTE *In the Hall Effect model presented herein, it is assumed that r =1. That assumption is a valid first approximation. For applied development models, the modeler will need to determine experimentally the best approximation for the value of r for the particular material and physical conditions being modeled.*

For example, in the case that the charge carrier is a "hole," the minus sign (-) in the equation for the Hall coefficient changes to a plus sign (+). In the case of mixed electron/ hole flow, R_H can become zero (0).

The differential voltage/charge accumulation (Hall Voltage (V_H)) that results from the Lorentz Force interaction between any currents (electron and/or hole) flowing through that conducting material and the local magnetic field is shown in Figure 5.24.

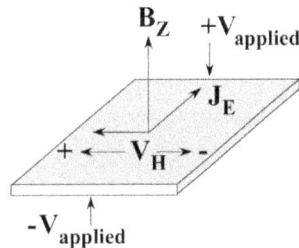

FIGURE 5.24 Hall effect sensor geometry, electron flow.

As can be seen from the introductory material, depending upon the characteristics of the material being modeled, the calculation of the Hall Effect can be very complex. The Hall coefficient (R_H) varies for different materials and has a predominant functional dependence that involves temperature, carrier type, carrier concentration, carrier mobility, carrier lifetime, and carrier velocity. In a dual carrier system, such as semiconducting materials (electrons and holes), under the proper conditions, R_H can become equal to zero. Semiconductor sensors, however, are among the most sensitive magnetic field Hall sensors currently manufactured.

Due to the underlying complexity of the Hall Effect, the model in this section of Chapter 5 requires the use of either the AC/DC module or the MEMS Module, in addition to the basic COMSOL Multiphysics Software. In this model, only a single carrier conduction system, electrons, is employed. For ease of modeling, it is assumed that the system is quasi-static. This model introduces the COMSOL modeling concepts of Pointwise Constraints and Floating Potential {5.21, 5.22}.

2D Hall Effect Model

Building the 2D Hall Effect Model

Startup 5.x, Click > Model Wizard button.

Select Space Dimension > 2D.

Click > Twistie for AC/DC in the Add Physics window.

Click > Twistie for Electric Fields and Currents > Select: Electric Currents (ec).

Click > Add.

Click > Study (Right Pointing Arrow button).

Click the General Studies Twistie (as needed)

Select: Stationary in the Select Study window.

Click > Done (Checked Box button).

Click > Save As.

Enter MM2E5X_2D_HE_1.mph.

See Figure 5.25.

FIGURE 5.25 Desktop display for the MM2E5X_2D_HE_1.mph model.

Figure 5.25 shows the Desktop Display for the MM2E5X_2D_HE_1.mph model.

Click > Global Definitions > Parameters 1.

Enter > In the Settings – Parameters – Parameters edit window

the parameters as shown in Table 5.1.

TABLE 5.1 Parameters Edit Window.

Name	Expression	Description
sigma0	1.04e3[S/m]	Silicon conductivity
Rh	1.25e-4[m^3/C]	Hall coefficient
Bz	0.1[T]	Magnetic field
coeff0	sigma0/(1+(sigma0*Rh*Bz)^2)	Conductivity anisotropy 2
V0	5.0[V]	Applied voltage
t_Si	1.0e-3[m]	Silicon thickness
coeff1	sigma0*Rh*Bz	Conductivity anisotropy 1
s11	coeff0	Conductivity matrix term 11
s12	coeff0*coeff1	Conductivity matrix term 12
s21	-coeff0*coeff1	Conductivity matrix term 21
s22	coeff0	Conductivity matrix term 22
epsilon_Si	11.68	Relative permeability silicon

See Figure 5.26.

Name	Expression	Value	Description
sigma0	1.04e3[S/m]	1040 S/m	Silicon conductivity
Rh	1.25e-4[m^3/C]	1.25E-4 m³/(s·A)	Hall coeffficient
Bz	0.1[T]	0.1 T	Magnetic field
coeff0	sigma0/(1+(sigma0*Rh*Bz)^2)	1039.8 S/m	Conductivity anisotropy 2
V0	5.0[V]	5 V	Applied voltage
t_Si	1.0e-3[m]	0.001 m	Silicon thickness
coeff1	sigma0*Rh*Bz	0.013	Conductivity anisotropy 1
s11	coeff0	1039.8 S/m	Conductivity matrix term 11
s12	coeff0*coeff1	13.518 S/m	Conductivity matrix term 12
s21	-coeff0*coeff1	-13.518 S/m	Conductivity matrix term 21
s22	coeff0	1039.8 S/m	Conductivity matrix term 22
epsilon_Si	11.68	11.68	Relative permeability silicon

Settings
Parameters

Label: Parameters 1

Parameters

Name:

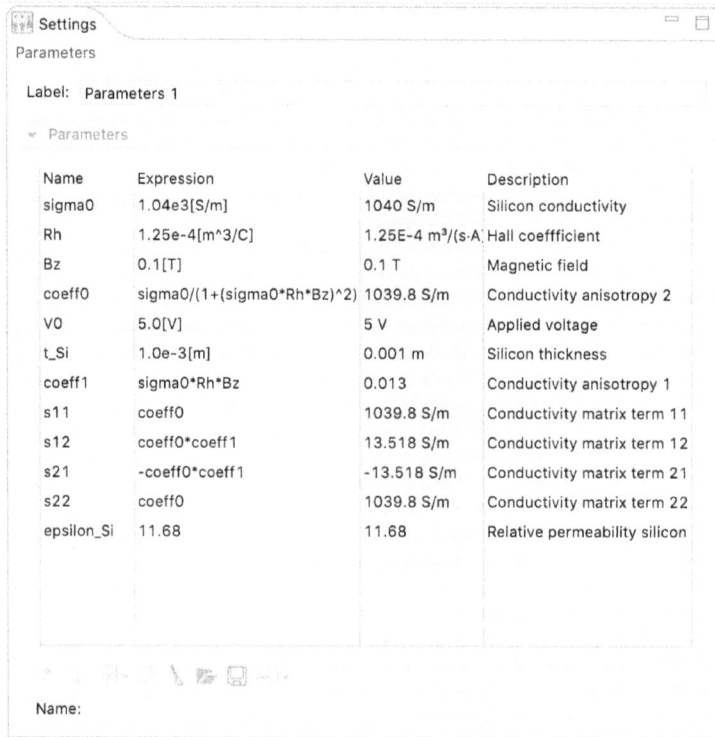

FIGURE 5.26 Settings – Parameters – Parameters edit window filled.

Figure 5.26 shows the Settings – Parameters – Parameters edit window filled.

In Model Builder – Component 1 (*comp1*),

Right-Click > Model Builder > Component 1 (*comp1*) > Geometry 1.

Select > Rectangle from the Pop-up menu.

Enter > 1.8e-2[m] in the Settings – Rectangle – Size – Width entry window.

Enter > 6e-3[m] in the Settings – Rectangle – Size – Height entry window.

Enter > –9e-3[m] in the Settings – Rectangle – Position – x entry window.

Enter > –3e-3[m] in the Settings – Rectangle – Position – y entry window.

Click > Build All.

See Figure 5.27.

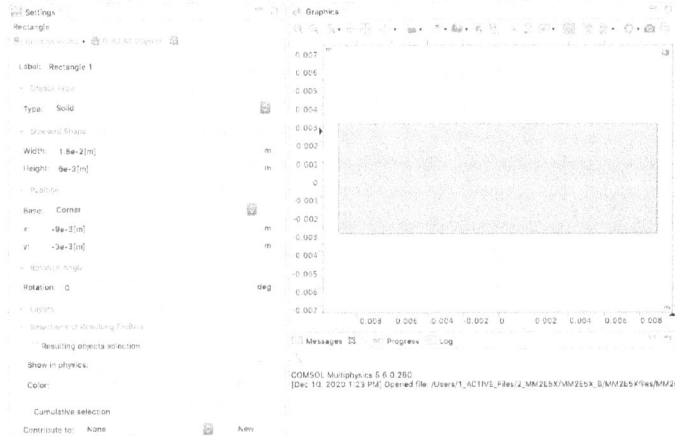

FIGURE 5.27 Settings – Rectangle entry windows.

Figure 5.27 shows the Settings – Rectangle Entry Windows.

Create points at the locations shown in Table 5.2:

TABLE 5.2 Points List.

Point #	x Coordinate	y Coordinate
1	−1e-3[m]	3e-3[m]
2	1e-3[m]	3e-3[m]
3	−1e-3[m]	−3e-3[m]
4	1e-3[m]	−3e-3[m]

Right-Click > Model Builder > Component 1 (*comp1*) > Geometry 1.

Select > Point from the Pop-up menu.

Enter > Point 1 coordinates.

Click > Build All.

Right-Click > Model Builder > Component 1 (*comp1*) > Geometry 1.

Select > Point from the Pop-up menu.

Enter > Point 2 coordinates.

Click > Build All.

Right-Click > Model Builder > Component 1 (*comp1*) > Geometry 1.

Select > Point from the Pop-up menu.

Enter > Point 3 coordinates.

Click > Build All.

Right-Click > Model Builder > Component 1 (*comp1*) > Geometry 1.

Select > Point from the Pop-up menu.

Enter > Point 4 coordinates.

Click > Build All.

See Figure 5.28.

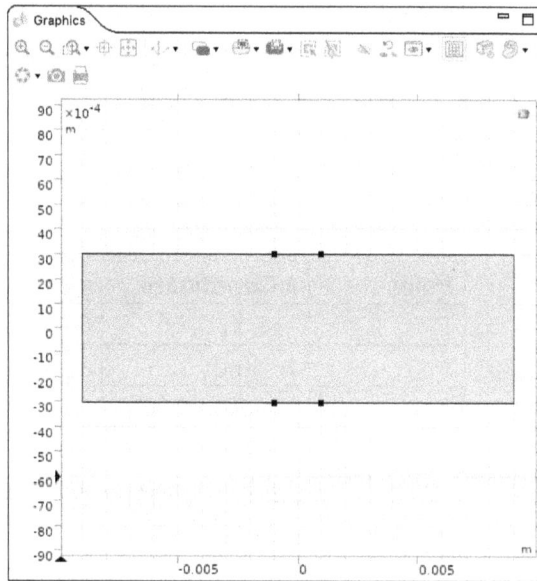

FIGURE 5.28 Rectangle with points in the graphics window.

Figure 5.28 shows the Rectangle with Points in the Graphics Window.

Electric Currents (ec) Interface

Click > Electric Currents (ec),

Enter: t_Si in the Settings – Electric Currents – Thickness – Out-of-Plane-Thickness,

d edit window.

See Figure 5.29.

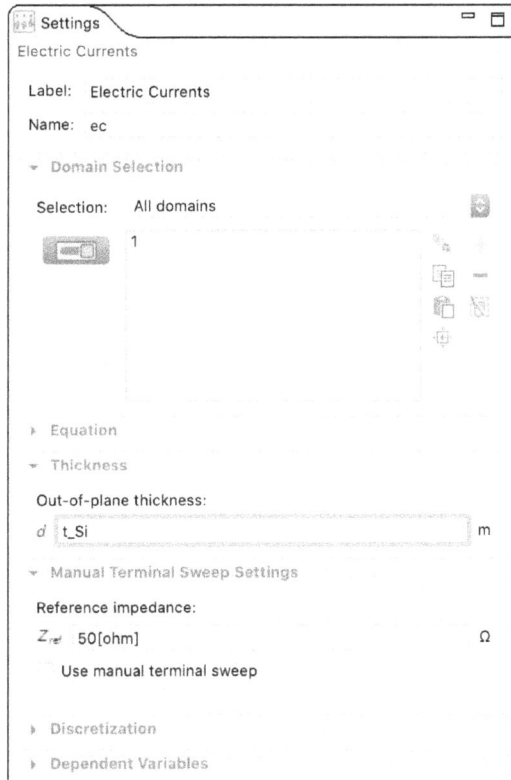

FIGURE 5.29 Settings electric currents.

Figure 5.29 shows the Settings Electric Currents.

Click > Model Builder – Component 1 – Electric Currents (ec) – Current Conservation 1.

Select: User defined from the pull-down menu in Settings – Current Conservation – Constitutive Relation Jc-E– Electrical conductivity.

Select > Full from the pull-down menu in Settings – Current Conservation – Constitutive Relation Jc-E – Electrical conductivity.

Enter the parameter values as indicated in Table 5.3 in the Settings – Current Conservation – Constitutive Relation Jc-E – Electrical conductivity edit window.

Be sure to enter the Silicon Conductivity Parameters in the four (4) upper-left positions of the matrix edit field, leave the other zeros (0) as they are.

TABLE 5.3 Silicon Conductivity Parameters.

Window #	Expression
11	s11
12	s12
21	s21
22	s22

Scroll-down, find the Settings – Current Conservation – Constitutive Relation D-E Relative permittivity pull-down menu.

Select: User defined from the pull-down menu in Settings – Current Conservation – Constitutive Relation D-E – Relative permittivity pull-down menu.

Enter:" epsilon_Si in the Settings – Current Conservation – Constitutive Relation D-E – Relative permittivity edit window.

NOTE *In this case, a relative permittivity of 1 is not appropriate. Since silicon has a different relative permittivity, then the modeler needs to enter that value as indicated (epsilon_Si).*

See Figure 5.30.

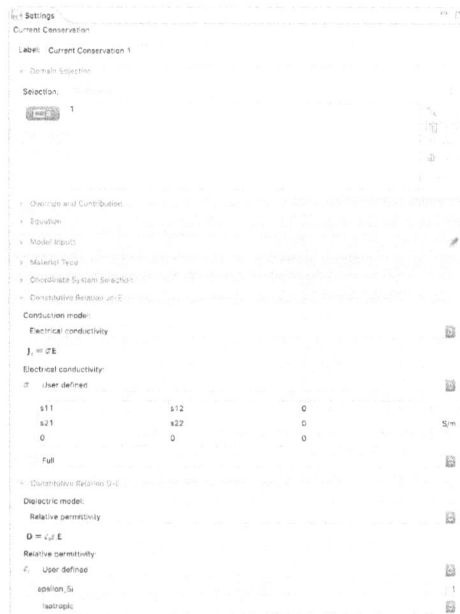

FIGURE 5.30 Settings – Current conservation edit windows.

Figure 5.30 shows the Settings – Current Conservation edit windows.

Right-Click > Model Builder > Component 1 (*comp1*) > Electric Currents.

Select > Electric Insulation from the Pop-up menu.

Click > Model Builder > Component 1 (*comp1*) > Electric Currents > Electric Insulation 2.

Click > Boundaries 2, 3, 6, 7 in the Graphics window.

See Figure 5.31.

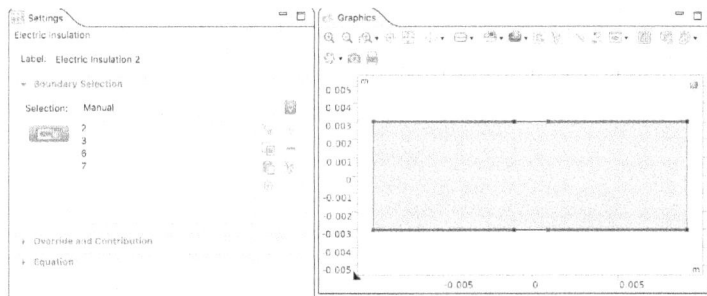

FIGURE 5.31 Model geometry electric insulation 2 boundaries.

Figure 5.31 shows the Model Geometry Electric Insulation 2 boundaries.

Right-Click > Model Builder > Component 1 (*comp1*) > Electric Currents.

Select > Ground from the Pop-up menu.

Click > Model Builder – Component 1 – Electric Currents – Ground 1.

Click > Boundary 1 in the Graphics window.

See Figure 5.32.

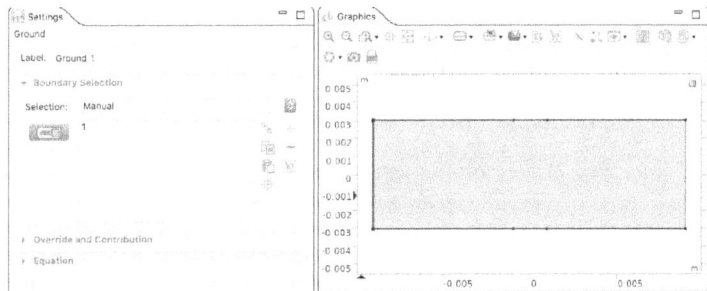

FIGURE 5.32 Model geometry ground 1 boundary.

Figure 5.32 shows the Model Geometry Ground 1 Boundary.

Right-Click > Model Builder > Component 1 (*comp1*) > Electric Currents.

Select > Electric Potential from the Pop-up menu.

Click > Model Builder > Component 1 (*comp1*) > Electric Currents > Electrical Potential 1.

Click > Boundary 8 in the Graphics window.

Enter: V0 in the Settings – Electric Potential – Electric Potential – Va edit window.

See Figure 5.33.

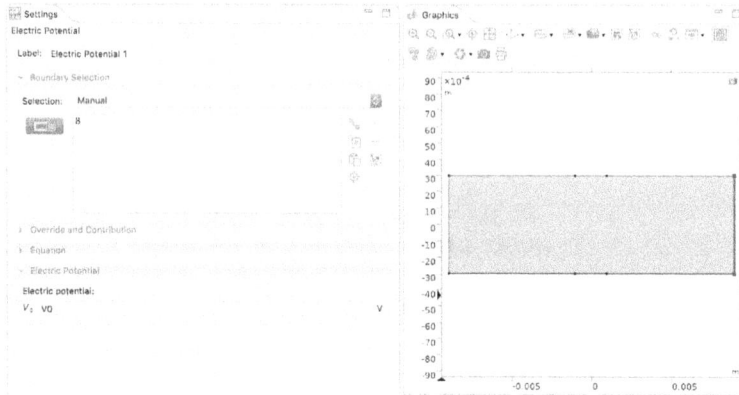

FIGURE 5.33 Model geometry electric potential 1.

Figure 5.33 shows the Model Geometry Electric Potential 1.

Right-Click > Model Builder > Component 1 (*comp1*) > Electric Currents.

Select > Floating Potential from the Pop-up menu.

Click > Model Builder > Component 1 (*comp1*) > Electric Currents – Floating Potential 1.

Click > Boundary 5 in the Graphics window.

NOTE

In this case, a Terminal current of zero (0) is appropriate. If a different value of current were to be drawn, then the modeler would need to enter that value as indicated.

See Figure 5.34.

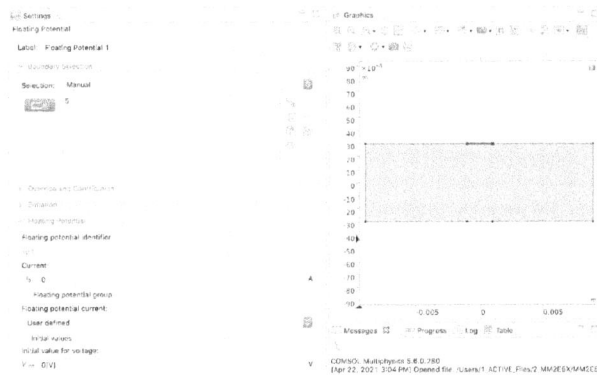

FIGURE 5.34 Model geometry floating potential 1 setting.

Figure 5.34 shows the Model Geometry Floating Potential 1 Setting.

Right-Click > Model Builder > Component 1 (*comp1*) > Electric Currents.

Select > Floating Potential from the Pop-up menu.

Click > Model Builder > Component 1 (*comp1*) > Electric Currents – Floating Potential 2.

Click > Boundary 4 in the Graphics window.

NOTE   *In this case, a Terminal current of zero (0) is appropriate. If a different value of current were to be drawn, then the modeler would need to enter that value as indicated.*

See Figure 5.35.

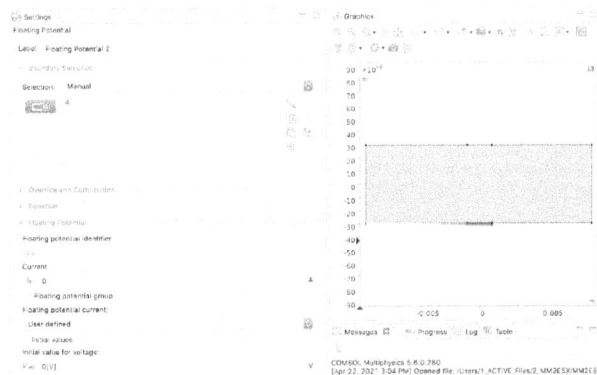

FIGURE 5.35 Model geometry floating potential 2 setting.

Figure 5.35 shows the Model Geometry Floating Potential 2 Setting.

Mesh 1

NOTE *The default settings for the Mesh Type are not appropriate for the solution of this problem and thus need to be modified.*

Click > Model Builder > Component 1 (*comp1*) > Mesh 1.

Click > Sequence type pull-down menu in Settings – Mesh – Mesh Settings – Sequence type.

Select > Physics-controlled mesh (as needed).

Click > Element size pull-down menu in Settings – Mesh – Mesh Settings – Element size.

Select: Extra fine.

Click > Build All button.

After meshing, the modeler should see a message in the message window about the number of elements (2212 elements) in the mesh.

See Figure 5.36.

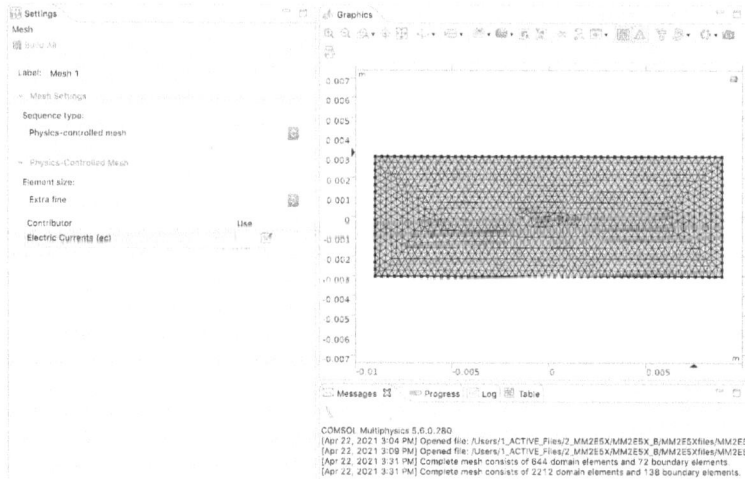

FIGURE 5.36 Desktop display – Graphics – Meshed domain.

Figure 5.36 shows the Desktop Display – Graphics – Meshed Domain.

Study 1

Right-Click > Model Builder – Study 1.

Select > Parametric Sweep from the Pop-up menu.

Click > Model Builder – Study 1 – Parametric Sweep.

Click > The Add button (Plus sign) in Settings – Parametric Sweep – Study Settings.

Select > Magnetic field (Bz) from the Pop-up menu.

Click > Range button in Settings – Parametric Sweep – Study Settings.

Enter > Start = 0, Step = 0.1, Stop = 2 in the Range Pop-up edit window.

Click > Replace button in the Range Pop-up edit window.

See Figure 5.37.

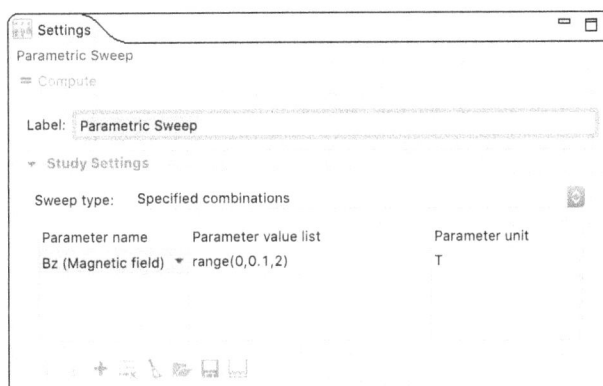

FIGURE 5.37 Settings – Parametric sweep – Study settings.

Figure 5.37 shows the Settings – Parametric Sweep – Study Settings.

Study 1

In Model Builder, Right-Click Study 1 > Select > Compute.

Computed results, using the default display settings, are shown in Figure 5.38.

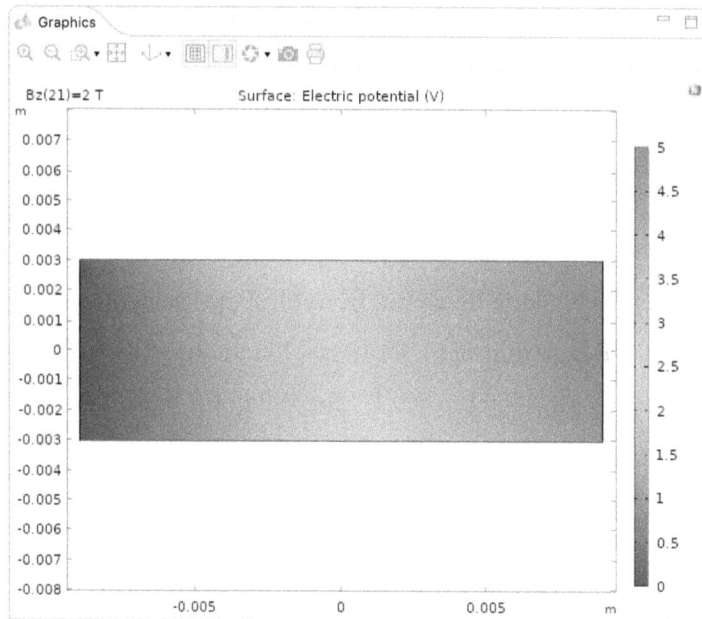

FIGURE 5.38 Hall effect model computed results, using the default display settings.

Figure 5.38 shows the Hall Effect Model Computed results, using the default display settings.

Results

In Model Builder,

Click > Results > Electric Potential (ec) twistie.

Right-Click > Electric Potential (ec).

Select: Contour from the pop-up menu.

Click > Contour 1 in Model Builder – Results – Electric Potential (ec).

Click > Settings – Contour – Coloring and Style – Color table.

Select > GrayScale from the pull-down menu.

Click > Plot, if needed.

See Figure 5.39.

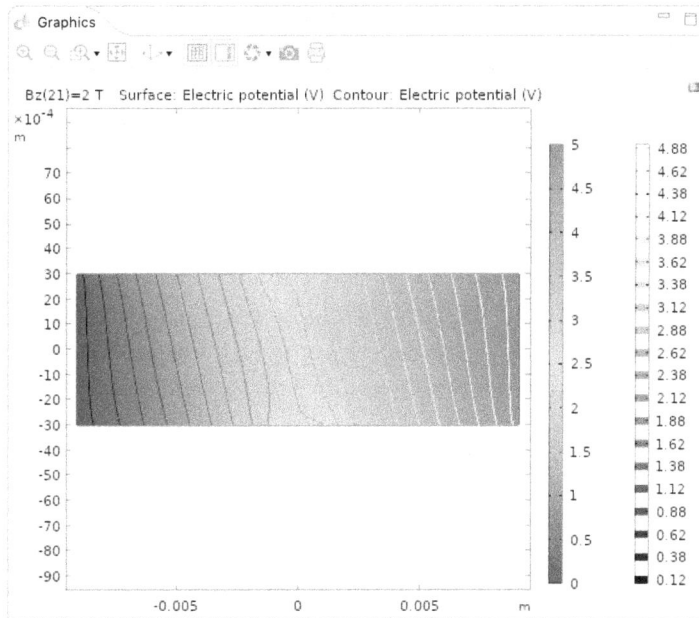

FIGURE 5.39 Hall effect model computed results, with contour plot.

Figure 5.39 shows the Hall Effect Model Computed results, with Contour Plot.

2D Hall Effect Model Animation

This animation shows how the Hall Voltage changes as the applied magnetic field changes.

In Model Builder,

Right-Click > Results – Export.

Select > Animation > Player.

In Settings Animation > Scene > Subject: Pull-down, Select: Electric Potential (ec) (as needed),

Right-Click > Animation 1: select > Play.

NOTE

The modeler is now able to observe the results of the modeling process dynamically. The above instructions allowed the modeler to build a movie with the default parameters. The modeler can now build other movies using different parameters that will be viewed in the Graphics window.

The next set of instructions allow the modeler to export the movie frames to a file.

In Model Builder,

Right-Click > Results – Export.

Select > Animation > File.

In Settings – Animation – Animation 2,

Select > Output type Movie from the pull-down menu (as needed).

Select > Format GIF from the pull-down menu (as needed).

NOTE *The detailed procedures for Animation in 5.x on the Macintosh and the PC are different. The Macintosh uses the GIF format. The PC uses the AVI format. If the modeler tries to use AVI on the Macintosh, an error results.*

In Settings – Animation – Animation 2 > Frames,

Click > Lock aspect ratio check box.

In Settings – Animation – Animation 2 > Advanced (Click > twistie (as needed)),

Click > Antialiasing check box (as needed).

Click > Settings – Export (located in the Settings toolbar).

Save the Animation as: MM2E5X_2D_HE_1_An.gif.

NOTE *The movie will be exported to a file.*

Click > Save for the completed MM2E5X_2D_HE_1.mph Hall Effect model.

2D Hall Effect Model Summary and Conclusions

The 2D Hall Effect model is a powerful tool that can be used to explore the Hall Effect in many different media (e.g. semiconductors, metals, semimetals, etc.). As has been shown earlier in this chapter, the Hall Effect is easily and simply modeled with a 2D Electric Currents (ec) Interface.

FIRST PRINCIPLES AS APPLIED TO 2D MODEL DEFINITION

First Principles Analysis derives from the fundamental laws of nature. In the case of models using this Classical Physics Analysis approach, the laws of conservation in physics require that what goes in (as mass, energy, charge, etc.) must come out (as mass, energy, charge, etc.) or must accumulate within the boundaries of the model.

The careful modeler must be knowledgeable of the implicit assumptions and default specifications that are normally incorporated into the COMSOL Multiphysics software model when a model is built using the default settings.

Consider, for example, the two 2D models developed in this chapter. In these models, it is implicitly assumed there are no thermally related changes (mechanical, electrical, etc.). It is also assumed the materials are homogeneous and isotropic, except as specifically indicated in the Hall Effect, and there are no thin electrical contact barriers at the electrical junctions. None of these assumptions are typically true in the general case. However, by making such assumptions, it is possible to easily build a 2D First Approximation Model.

NOTE *A First Approximation Model is one that captures all the essential features of the problem that needs to be solved, without dwelling excessively on small details. A good First Approximation Model will yield an answer that enables the modeler to determine if he needs to invest the time and the resources required to build a more highly detailed model.*

Also, the modeler needs to remember to name model parameters carefully as pointed out in Chapter 1.

REFERENCES

1. COMSOL Reference Manual (Version 5.6), p. 201

2. *https://en.wikipedia.org/wiki/Electrochemistry*

3. *https://en.wikipedia.org/wiki/Electropolishing*

4. *https://en.wikipedia.org/wiki/William_Gilbert_(Physician)*

5. https://en.wikipedia.org/wiki/Charles-Augustin_de_Coulomb

6. https://en.wikipedia.org/wiki/Joseph_Priestley

7. https://en.wikipedia.org/wiki/Georg_Ohm

8. https://en.wikipedia.org/wiki/Michael_Faraday

9. COMSOL Reference Manual (Version 5.6), pp. 1171-1198

10. COMSOL Reference Manual (Version 5.6), pp. 866-875

11. COMSOL Reference Manual (Version 5.6), pp. 1195

12. https://en.wikipedia.org/wiki/Electrical_conductor

13. *https://en.wikipedia.org/wiki/Georg_Ohm*

14. Kittel, Charles, "Introduction to Solid State Physics," ISBN 0-471-87474-4

15. Sze, S.M., "Semiconductor Devices, Physics and Technology," ISBN 0-471- 87424-8

16. Ziman, J.M., "Principles of the Theory of Solids," First Edition, p. 185

17. *https://en.wikipedia.org/wiki/Hall_effect*

18. *https://en.wikipedia.org/wiki/Edwin_Hall*

19. *https://en.wikipedia.org/wiki/Lorentz_force*

20. *https://en.wikipedia.org/wiki/Hall_effect*

21. COMSOL Reference Manual (Version 5.6), p. 954

22. COMSOL AC/DC Module Users Guide (Version 5.6), p. 135

SUGGESTED MODELING EXERCISES

1. Build, mesh, and solve the 2D Electropolishing Model as presented earlier in this chapter.

2. Build, mesh, and solve the 2D Hall Effect Model as presented earlier in this chapter.

3. Change the values of the materials parameters and then build, mesh, and solve the 2D Electropolishing Model as an example problem.

4. Change the values of the materials parameters and then build, mesh, and solve the 2D Hall Effect Model as an example problem.

5. Change the value of the proportionality constant and then build, mesh, and solve the 2D Electropolishing Model as an example problem.

6. Change the value of the Hall coefficient and then build, mesh, and solve the 2D Hall Effect Model as an example problem.

CHAPTER 6

2D AXISYMMETRIC MODELING USING COMSOL MULTIPHYSICS 5.x

In This Chapter

- Guidelines for 2D Axisymmetric Modeling in 5.x
    - 2D Axisymmetric Modeling Considerations
- 2D Axisymmetric Basic Models
    - 2D Axisymmetric Heat Conduction in a Cylinder Model
    - 2D Axisymmetric Transient Heat Transfer Model
- First Principles as Applied to 2D Axisymmetric Model Definition
- References
- Suggested Modeling Exercises

GUIDELINES FOR 2D AXISYMMETRIC MODELING IN 5.x

NOTE

In 5.x, there are fundamentally two types of 2D geometries: 2D and 2D Axisymmetric. Chapter 5 introduced basic 2D models (x, y). 2D models are assumed to be homogeneous and isotropic in the third (non-planar) dimension. In this chapter, the modeler is introduced to the development and analysis of 2D Axisymmetric models. 2D Axisymmetric models have two geometric dimensions (r, z). 2D Axisymmetric models are planar representations of cylindrical 3D objects that are rotationally homogeneous and isotropic.

In this chapter, introductory 2D Axisymmetric models will be presented. Such 2D Axisymmetric models are more geometrically complex than 1D or 2D models. 2D Axisymmetric models have proven to be very valuable to the science and engineering communities, both in the past and currently, as first-cut evaluations of potential physical behavior under the influence of external stimuli. The 2D Axisymmetric model responses and other such ancillary information are gathered and screened early in a project for initial evaluation and potentially for later use in building higher-dimensionality (3D) field-based (electrical, magnetic, etc.) models.

Since the models in this and subsequent chapters are more complex to build and more difficult to solve than most of the models presented thus far, it is important that the modeler have available the tools necessary to easily utilize the powerful capabilities of the 5.x software. In order to do that, the modeler should go to the main 5.x toolbar, Click > Options – Preferences – Show More Options. When the Preferences – Show More Options edit window is shown, Select > Equation View checkbox and Equation Sections checkbox. Click > OK {6.1}.

2D Axisymmetric Modeling Considerations

2D Axisymmetric Modeling can be less difficult in some cases than 1D or 2D modeling, having fewer implicit assumptions, and yet potentially, it can still be a very challenging type of model to build, depending on the underlying physics involved. The least difficult aspect of 2D Axisymmetric model creation arises from the fact that the geometry is relatively simple. (In a 2D Axisymmetric model, the modeler has only a single plane as the modeling space.) However, the physics in a 2D Axisymmetric model can range from relatively easy to extremely complex.

In compliance with the laws of physics, a 2D Axisymmetric model implicitly assumes that energy flow, materials properties, environment, and all other conditions and variables that are of interest are homogeneous, isotropic, and/or constant, unless otherwise specified, throughout the entire domain of interest both within the model and through the boundary conditions and in the environs of the model.

The modeler needs to bear the above-stated conditions in mind and insure that all of the modeling conditions and associated parameters (default settings) in each model created are properly considered, defined, verified, and/ or set to the appropriate values.

NOTE *It is always mandatory that the modeler be able to accurately anticipate the expected results of the model and accurately specify the manner in which those results will be presented. Never assume that any of the default values that are present when the model is created necessarily satisfy the needs or conditions of a particular model.*

Always verify that any parameters employed in the model are the correct value needed for that model. Calculated solutions that significantly deviate from the anticipated solution or from a comparison of values measured in an experimentally derived realistic model are probably indicative of one or more modeling errors either in the original model design, in the earlier model analysis, in the understanding of the underlying physics, or are simply due to human error.

2D Axisymmetric Coordinate System

In 2D Axisymmetric models, there are two geometric coordinates, space (r) and space (z), and the temporal coordinate, time (t). In a steady-state solution to a 2D Axisymmetric model, parameters can only vary as a function of position in space (r) and space (z) coordinates. Such a 2D Axisymmetric model represents the parametric condition of the model in a time-independent mode (quasi-static). In a transient solution model, parameters can vary both by position in space (r) and/or space (z) and in time (t).

The space coordinates (r) and (z) typically represent distance coordinates throughout which the model is to calculate the change of the specified observables (i.e. temperature, heat flow, pressure, voltage, current, etc.) over the range of coordinate values ($r_{min} <= r <= r_{max}$) and ($z_{min} <= z <= z_{max}$). The time coordinate (t) represents the range of temporal values ($t_{min} <= t <= t_{max}$) from the beginning of observation period (t_{min}) to the end of observation period (t_{max}).

Heat Transfer Theory

Two substantially different approaches to modeling heat transfer are presented in this chapter. These different approaches demonstrate the power of the COMSOL 2D Axisymmetric modeling software. The modeling examples in this chapter demonstrate heat transfer using a stationary technique in the first model and using a transient technique in the second model.

Heat transfer is an extremely important design consideration. It is one of the most widely needed and applied technologies employed in applied physics and engineering. Most modern products or processes require an understanding of heat transfer either during development or during the use of the product or process (automobiles, plate glass fabrication, plastic extrusion, plastic products, houses, ice cream, etc.).

Heat transfer concerns have existed since early in the beginnings of humanity. At this point, however, it is understood that the science of thermodynamics and consequently today's understanding of heat transfer began with the work of Nicolas Leonard Sadi Carnot, as published in his 1824 paper {6.2} "Reflections on the Motive Power of Fire." The actual term "thermodynamics" is attributed to William Thomson (Lord Kelvin) {6.3}. Later contributions to the understanding of heat, heat transfer, and thermodynamics, in general, were made by James Prescott Joule {6.4}, Ludwig Boltzmann {6.5}, James Clerk Maxwell {6.6}, Max Planck {6.7}, and many others. The physical understanding and engineering use of thermodynamics play a very important role in the technological aspects of machine and process design in modern applied science, engineering, and medicine.

The first example presented herein, Cylinder Conduction, explores the 2D Axisymmetric Stationary modeling of heat transfer and temperature profiling for a thermally conductive material, implemented through use of the COMSOL Heat Transfer Interface.

The second 2D Axisymmetric Transient Heat Transfer Model example in this chapter explores the modeling of heat transfer over time (transient thermal response).

Heat Conduction Theory

Heat conduction is a naturally occurring process in all materials. It is readily observed in all aspects of modern life (refrigerators, freezers, microwave ovens, thermal ovens, engines, etc.). The heat transfer process allows both linear and rotational work to be done in the generation of electricity and the movement of vehicles. The initial understanding of transient heat transfer was developed by Newton {6.8} and started with Newton's Law of Cooling {6.9}:

$$\frac{dQ}{dt} = h * A * (T_S - T_E) \tag{6.1}$$

Where $\frac{dQ}{dt}$ is the incremental energy lost in Joules per unit time (J/s).

A is the energy transmission surface area (m²).

h is the heat transfer coefficient (W/(m²*K)).

T_S is the surface temperature of the object losing heat (K).

T_E is the temperature of the environment gaining heat (K).

Subsequent work by Jean Baptiste Joseph Fourier {6.10}, based on Newton's Law of Cooling, developed the law for steady-state heat conduction (known as Fourier's Law {6.11}). Fourier's Law is expressed here in differential form as:

$$q = -k\nabla T \tag{6.2}$$

Where q is the heat flux in Watts per square meter (W/m²).

k is the thermal conductivity of the material (W/(m*K)).

∇T is the temperature gradient (K/m).

2D Axisymmetric Heat Conduction Modeling

The following numerical solution model (2D Axisymmetric Heat Conduction in a Cylinder Model) was originally developed by COMSOL as a tutorial model based on an example from the NAFEMS collection {6.12}. It was developed for distribution with the Multiphysics software as part of the Model Library. This model introduces two important basic concepts that apply to both applied physics and applied modeling: Axisymmetric Geometry (cylindrical) modeling and heat transfer modeling.

Heat transfer modeling is important in physical design and applied engineering problems. Typically, the modeler desires to understand heat generation during a process and either add heat or remove heat in order to achieve or maintain a desired temperature. Figure 6.1 shows a 3D rendition of the 2D Axisymmetric Cylinder Conduction geometry, as will be modeled herein. The dashed-line ellipses in Figure 6.1 indicate the 3D rotation that would need to occur to generate the 3D solid object from the 2D cross-section shown.

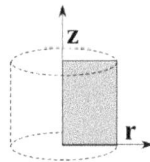

FIGURE 6.1 3D rendition of the 2D axisymmetric cylinder conduction model.

2D AXISYMMETRIC HEAT CONDUCTION IN A CYLINDER MODEL

NOTE

This model is derived from the COMSOL Cylinder Conduction Model. In this model, however, the selected thermally conductive solid is Niobium (Nb) {6.13, 6.14}. Niobium has a variety of uses, as an alloying element in steels, as an alloying element in titanium turbine blades, in superconductors, as an anti-corrosion coating, as an optical coating, and as an alloy in coinage.

2D AXISYMMETRIC BASIC MODELS

2D Axisymmetric Cylinder Conduction Model

Building the 2D Axisymmetric Cylinder Conduction Model

Startup 5.x, Click > Model Wizard button.

Select Space Dimension > 2D Axisymmetric.

Click > Twistie for Heat Transfer.

Click > Heat Transfer in Solids *(ht)*.

Click > Add.

Click > Study (Right Pointing Arrow button).

Select: General Studies > Stationary in the Select Study window.

Click > Done (Checked Box button).

Click > Save As.

Enter MM2E5X_2DA_HTC_1.mph.

See Figure 6.2.

FIGURE 6.2 Desktop display for the MM2E5X_2DA_HTC_1.mph model.

Figure 6.2 shows the Desktop Display for the MM2E5X_2DA_HTC_1.mph model.

NOTE *The axis of rotation (symmetry) is indicated in the Desktop Display – Graphics window by a non-solid (dots and dashes) line at r =0.*

Click > Global Definitions > Parameters 1.

In the Settings – Parameters – Parameters 1 edit window,

Enter the parameters shown in Table 6.1.

TABLE 6.1 Parameters Edit Window.

Name	Expression	Description
k_Nb	52.335[W/(m*K)]	Thermal conductivity Nb
rho_Nb	8.57e3[kg/m^3]	Density Nb
Cp_Nb	2.7e2[J/(kg*K)]	Heat capacity Nb
T_0	2.7315e2[K]	Boundary temperature
q_0	5e5[W/m^2]	Heat flux

See Figure 6.3.

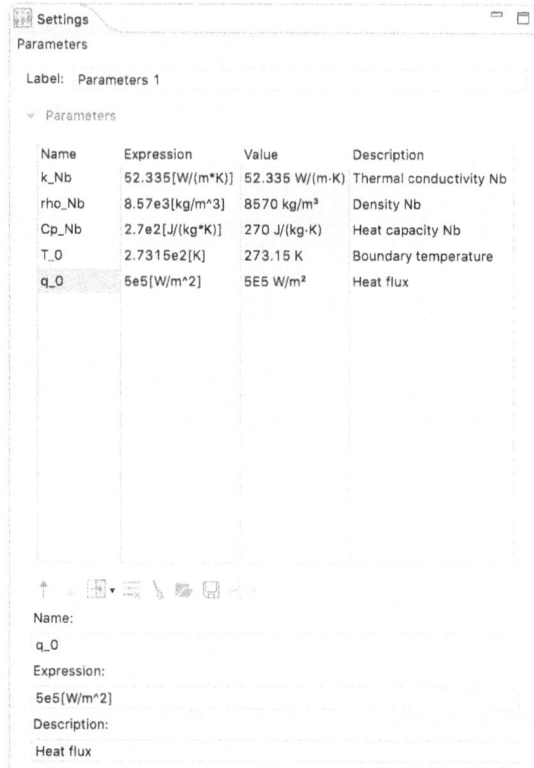

FIGURE 6.3 Settings – Parameters – Parameters 1 edit window filled.

Figure 6.3 shows the Settings – Parameters – Parameters 1 edit window filled.

Right-Click > Model Builder – Geometry 1.

Select > Rectangle from the Pop-up menu.

Enter > 8e-2 in the Settings – Rectangle – Size – Width entry window.

Enter > 1.4e-1 in the Settings – Rectangle – Size – Height entry window.

Enter > 2e-2 in the Settings – Rectangle – Position – r entry window.

Click > Build All.

See Figure 6.4.

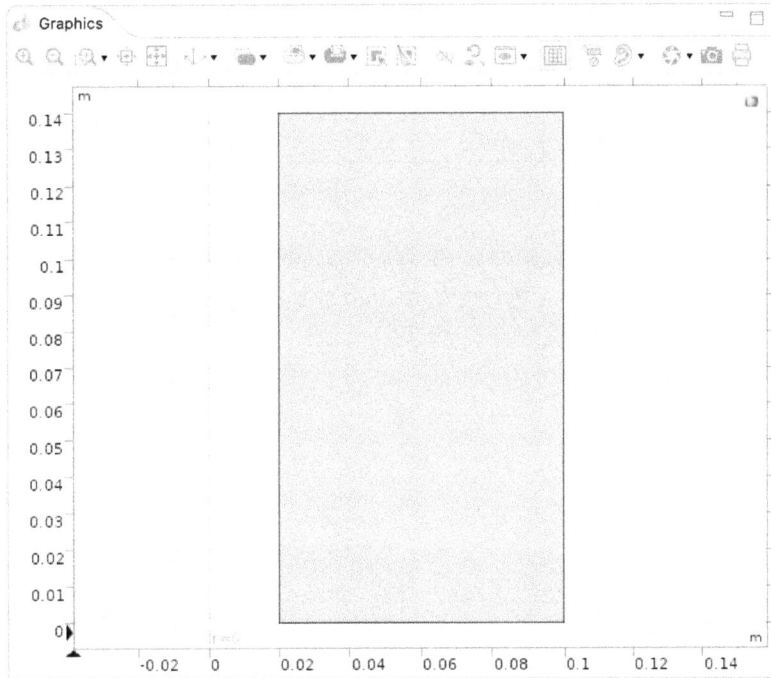

FIGURE 6.4 Settings – Rectangle entry windows and graphics window.

Figure 6.4 shows the Settings – Rectangle entry windows and Graphics window.

Right-Click > Model Builder – Geometry 1.

Select > Point from the Pop-up menu.

Enter > coordinates for point 1 using Table 6.2.

Click > Build Selected.

Right-Click > Model Builder – Geometry 1.

Select > Point from the Pop-up menu.

Enter > coordinates for point 2 using Table 6.2.

Click > Build Selected.

TABLE 6.2 Point Edit Windows.

Name	Location r	Location z
point 1	2e-2	4e-2
point 2	2e-2	1e-1

Click > Build All.

NOTE *The boundary points 1 & 2 are constraint points. They are added to the perimeter of the rectangle to indicate where the heated area ends and the thermally insulated area begins.*

See Figure 6.5.

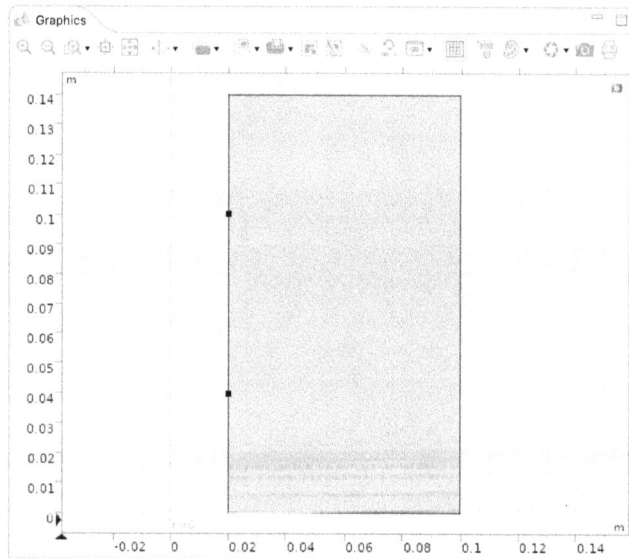

FIGURE 6.5 Model geometry in the graphics window.

Figure 6.5 shows the Model Geometry in the Graphics window.

Heat Transfer in Solids (ht) Interface

Click > Model Builder > Component 1 (*comp1*) > Twistie for the Heat Transfer in Solids (*ht*) (as needed).

Click > Model Builder > Component 1 (*comp1*) > Heat Transfer in Solids (*ht*) – Solid 1.

Verify that Domain 1 is Selected in Settings – Solid – Domain Selection.

Click > Settings –Solid – Heat Conduction – Thermal conductivity pull-down menu.

Select: User defined.

Enter: k_Nb in the Thermal conductivity edit window.

Click > Settings – Solid – Thermodynamics – Density pull-down menu.

Select: User defined.

Enter: rho_Nb in the Density edit window.

Click > Settings – Solid – Thermodynamics – Heat capacity at constant pressure pull-down menu.

Select: User defined.

Enter: Cp_Nb in the Heat capacity at constant pressure edit window.

See Figure 6.6.

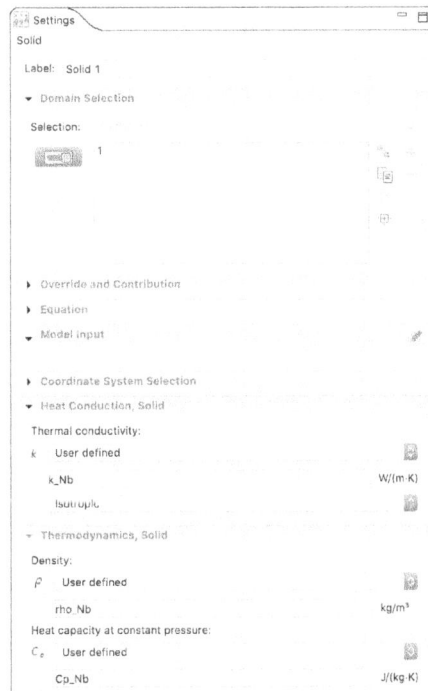

FIGURE 6.6 Settings – Solid edit window.

Figure 6.6 shows the Settings – Solid edit window.

Right-Click > Model Builder > Component 1 (*comp1*) > Heat Transfer in Solids (*ht*).

Select > Temperature from the Pop-up menu.

Shift-Click > Boundaries 2, 5, 6 in the Graphics window.

Enter > T_0 in the Settings – Temperature – Temperature edit window.

See Figure 6.7.

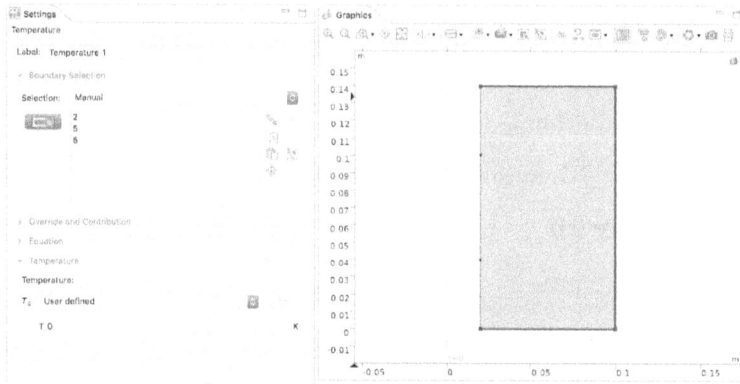

FIGURE 6.7 Settings – Temperature edit windows.

Figure 6.7 shows the Settings – Temperature edit windows.

Right-Click > Model Builder > Component 1 (*comp1*) > Heat Transfer in Solids (*ht*).

Select > Heat Flux from the Pop-up menu.

Click > Boundary 3 in the Graphics window.

Enter: q_0 in the Settings – Heat Flux – General inward heat flux edit window.

See Figure 6.8.

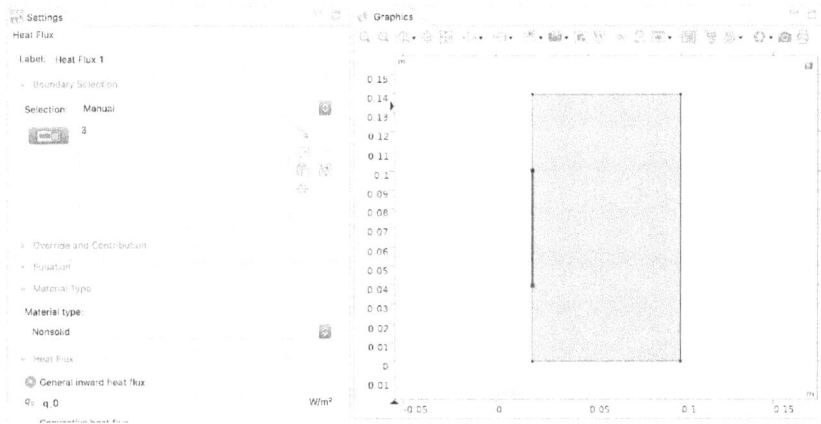

FIGURE 6.8 Settings – Heat flux edit windows.

Figure 6.8 shows the Settings – Heat Flux edit windows.

Mesh 1

Right-Click > Model Builder > Component 1 (*comp1*) > Mesh 1.

Select > Free Triangular from the pop-up menu.

Click > Settings – Build All button.

After meshing, the modeler should see a message in the message window about the number of elements (360 domain and 50 boumdary elements) in the mesh.

NOTE
The Free Triangular Mesh Type is chosen because it provides an adequate number of mesh elements and is relatively easy to solve.

See Figure 6.9.

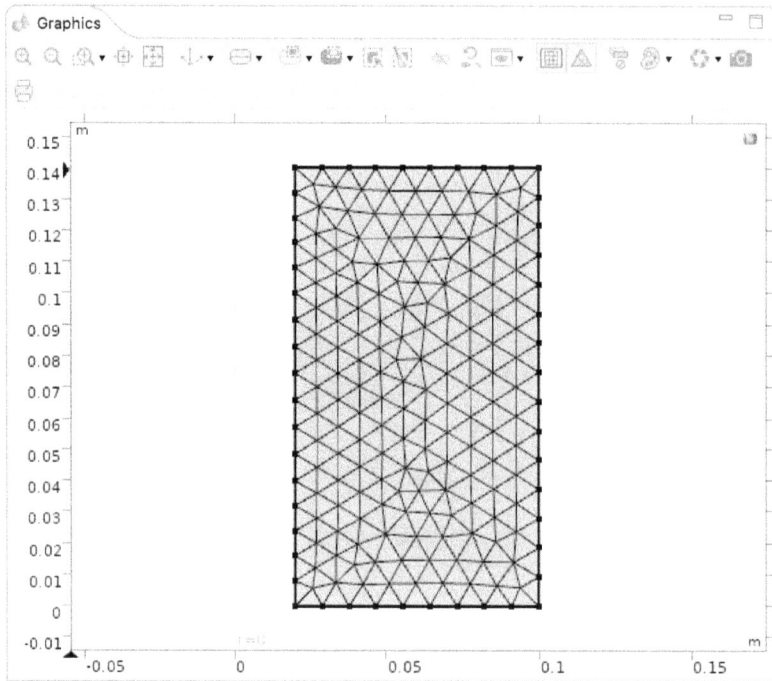

FIGURE 6.9 Desktop display – Graphics – Meshed domain.

Figure 6.9 shows the Desktop Display – Graphics – Meshed Domain.

Study 1

NOTE *The default settings for the Stationary Study Type are appropriate for the solution of this problem and thus do not need to be modified.*

In Model Builder,

Right-Click > Study 1 >Select > Compute.

Computed results, using the default display settings, are shown in Figure 6.10.

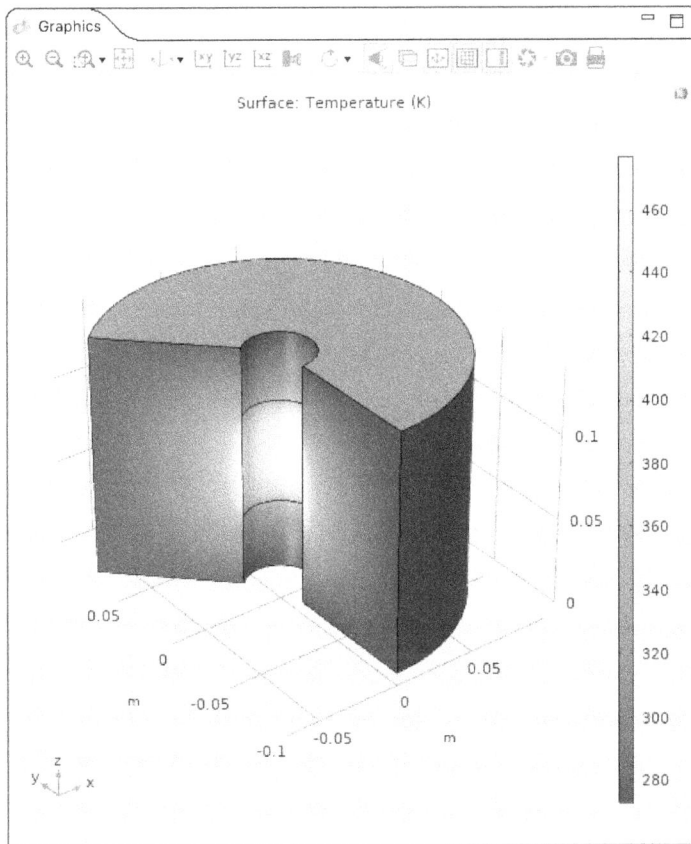

FIGURE 6.10 Computed results, Using the default display settings.

Figure 6.10 shows the computed results, using the default display settings.

Results

In Model Builder,

Click > Results > Isothermal Contours.

See Figure 6.11.

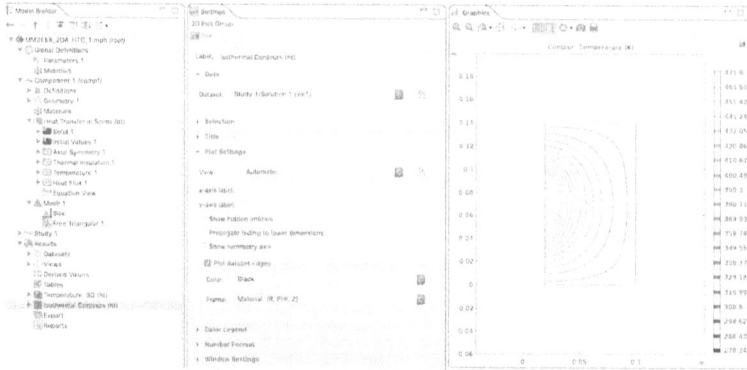

FIGURE 6.11 Desktop display – Default contour plot.

Figure 6.11 shows the Desktop Display – Default Contour Plot.

The modeler should be sure to save the model before closing the COMSOL Desktop.

2D Axisymmetric Heat Conduction in a Cylinder Model Summary and Conclusions

The 2D Axisymmetric Heat Conduction in a Cylinder Model is a powerful tool that can be used to explore heat transfer techniques in many different conductive media (e.g. metals, semimetals, alloys, graphene films, graphite films, etc.). The contour plot easily shows the non-linearity of the temperature distribution. As has been shown earlier in this chapter, the 2D Axisymmetric Heat Conduction in a Cylinder Model is easily and simply modeled with the Heat Transfer Interface.

2D Axisymmetric Transient Heat Transfer Model Considerations

In 1855, Adolf Eugen Fick {6.15} published his now fundamental and famous laws {6.16} for the diffusion (mass transport) of a first item (e.g. a gas, a liquid, etc.) through a second item (e.g. another gas, liquid, etc.).

Fick's First Law is written:

$$J = -D\nabla\Phi \qquad (6.3)$$

Where J = Diffusion Flux in moles per square meter per second [mol/(m²*s)].

D = Diffusion Coefficient (diffusivity) in square meters per second [m²/s].

Φ = Concentration in moles per cubic meter [mol/m³].

∇ = Gradient operator.

NOTE *It should be noted that the minus sign in front of the diffusivity indicates that the item that is flowing (gas, heat, money, etc.) flows from the higher concentration to the lower concentration.*

See Figure 6.12

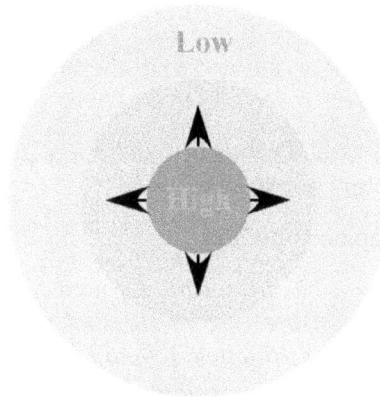

FIGURE 6.12 Graphical rendition of Fick's First Law.

Fick's Second Law calculates the details of how diffusion (mass transport) causes the diffusing item's concentration to change as a function of time (under transient conditions).

Fick's Second Law is written:

$$\frac{\partial \Phi}{\partial t} = D\nabla^2 \Phi \qquad (6.4)$$

Where D = Diffusion Coefficient (diffusivity) in square meters per second [m²/s].

Φ = Concentration vector in moles per cubic meter [mol/m³].

∇^2 = Laplace operator.

If modified slightly and applied to the flow of heat, where:

$$D = \alpha = \frac{k}{\rho C_p} \qquad (6.5)$$

Where α = Thermal Diffusion Coefficient (thermal diffusivity) in square meters per second [m²/s].

k = Thermal conductivity in Watts per meter-degree Kelvin [W/(m*K)].

ρ = Material density in kilograms per cubic meter [kg/m³].

C_p = Specific heat in Joules per kilograms-degree Kelvin [J/(kg*K)].

Then, Fick's Second Law becomes the Heat Equation:

$$\frac{\partial u}{\partial t} = \alpha \nabla^2 u \qquad (6.6)$$

Where u = u(x, y, z, t), a function of the three special parameters (x, y, z) and time (t).

α = Thermal Diffusion Coefficient (thermal diffusivity) in square meters per second [m²/s].

∇^2 = Laplace operator.

NOTE

The heat equation has been broadly used to solve many different diffusion problems {6.17}. In the case of thermal diffusion, the special parameters are the geometric coordinates. Examples of other uses of the heat equation are easily found in physics, finance, chemistry, and many other application areas.

2D Axisymmetric Transient Heat Transfer Model

NOTE

This is the model of a sphere of Niobium. The sphere of Niobium has an initial temperature of 273.15 K (0 degC) and is suspended (immersed) in a medium that applies a constant temperature (1000 degC) to the exterior surface of that sphere, starting at time zero.

Building the 2D Axisymmetric Transient Heat Transfer Model

Startup 5.x, Click > Model Wizard button.

Select Space Dimension > 2D Axisymmetric.

Click > Twistie for Heat Transfer in the Add Physics window.

Click > Heat Transfer – Heat Transfer in Solids *(ht)*.

Click > Add.

Click > Study (Right Pointing Arrow button).

Select: General Studies > Time Dependent in the Select Study window.

Click > Done (Checked Box button).

Click > Save As.

Enter MM2E5X_2DA_HTTD_1.mph.

See Figure 6.13.

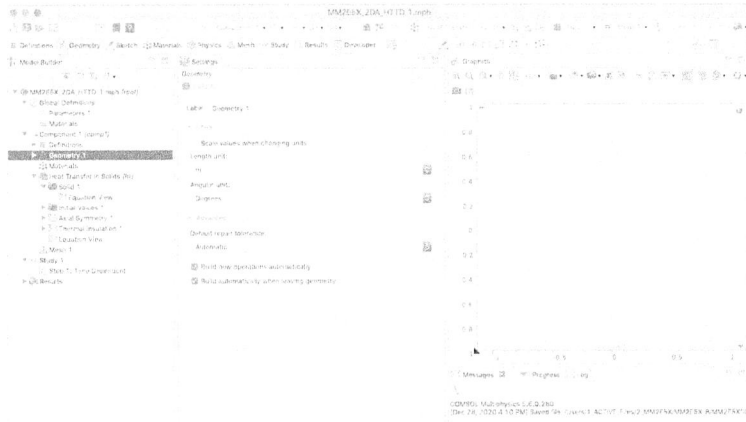

FIGURE 6.13 Desktop display for the MM2E5X_2DA_HTTD_1.mph model.

Figure 6.13 shows the Desktop Display for the MM2E5X_2DA_HTTD_1.mph model.

Click > Global Definitions > Parameters 1.

In the Settings – Parameters – Parameters 1 edit window,

Enter the parameters shown in Table 6.3.

TABLE 6.3 Parameters 1 Edit Window.

Name	Expression	Description
k_Nb	52.335[W/(m*K)]	Thermal conductivity Nb
rho_Nb	8.57e3[kg/m^3]	Density Nb
Cp_Nb	2.7e2[J/(kg*K)]	Heat capacity Nb
T_0	2.7315e2[K]	Initial temperature
T_appl	1.27315e3[K]	Boundary temperature

See Figure 6.14.

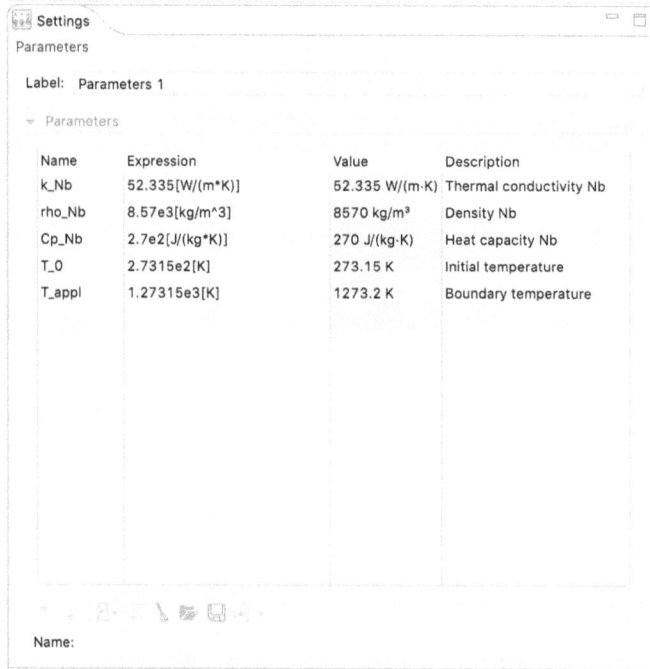

FIGURE 6.14 Settings parameters – Parameters 1 edit window filled.

Figure 6.14 shows the Settings Parameters – Parameters 1 edit window filled.

Create 2D Axisymmetric Sphere

NOTE   *In the next few steps, a circle will be created and then overlaid by a rectangle. Once the Boolean difference is taken between the two objects, the geometric remainder is the 2D Axisymmetric cross-section of a sphere.*

In Model Builder – Component 1 (*comp1*),

Right-Click > Model Builder – Geometry 1.

Select > Circle from the Pop-up menu.

Enter > 0.4 in the Settings – Circle – Size – Radius entry window.

Click > Build All.

See Figure 6.15.

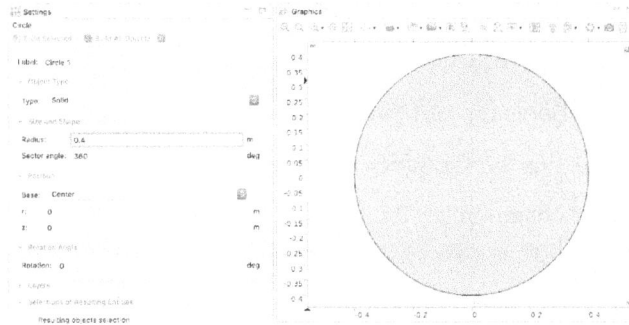

FIGURE 6.15 Settings – Circle entry windows.

Figure 6.15 shows the Settings – Circle entry windows.

Right-Click > Model Builder – Geometry 1.

Select > Rectangle from the Pop-up menu.

Enter > 0.4 in the Settings – Rectangle – Size – Width entry window.

Enter > 0.8 in the Settings – Rectangle – Size – Height entry window.

Click > Settings – Rectangle – Position – Base.

Select > Corner from the Pop-up menu (as needed).

Enter > –0.4 in the Settings – Rectangle – Position – r entry window.

Enter > –0.4 in the Settings – Rectangle – Position – z entry window.

Click > Build All.

See Figure 6.16.

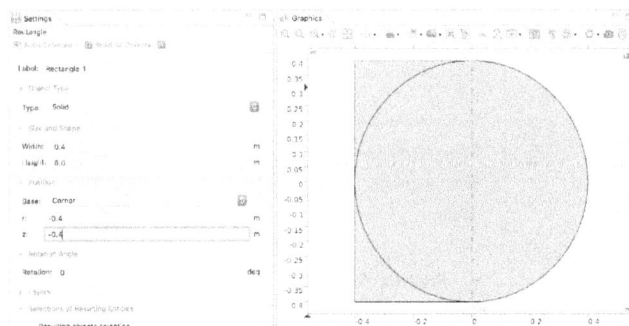

FIGURE 6.16 Settings – Rectangle entry windows.

Figure 6.16 shows the Settings – Rectangle entry windows.

Right-Click > Model Builder – Geometry 1.

Select > Booleans and Partitions > Difference from the Pop-up menu.

Click > c1 in the Graphics window.

Click > Activate Selection for the Settings – Difference – Objects to subtract window (bottom window).

Click > r1 in the Graphics window.

Click > Repair tolerance pop-up menu,

Select: Relative.

See Figure 6.17.

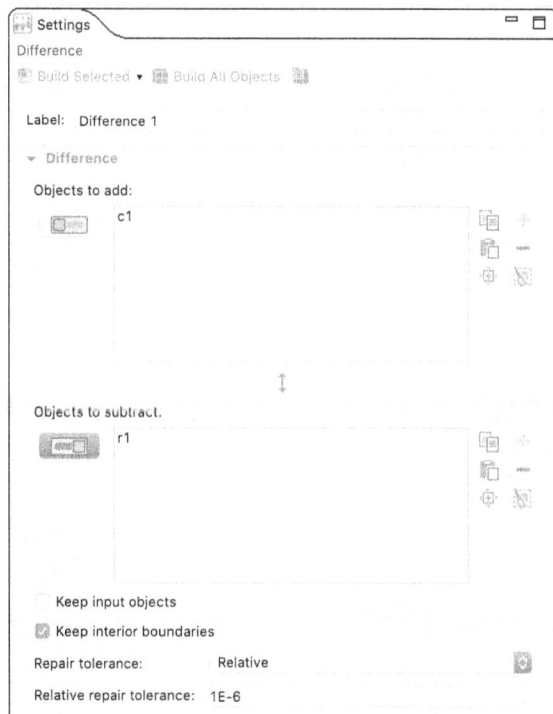

FIGURE 6.17 Settings – Difference entry windows.

Figure 6.17 shows the Settings – Difference entry windows.

Click > Build All.

See Figure 6.18.

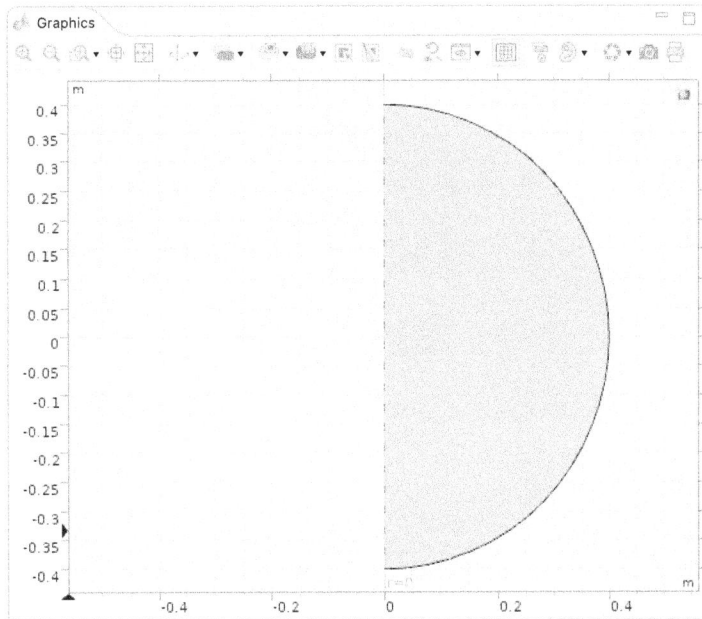

FIGURE 6.18 Graphics window – Axisymmetric sphere.

Figure 6.18 shows the Graphics Window – Axisymmetric Sphere.

Heat Transfer in Solids (ht) Interface

Click> Twistie for the Model Builder > Component 1 (*comp1*) > Heat Transfer in Solids (ht) (as needed).

Click> Model Builder > Component 1 (*comp1*) > Heat Transfer in Solids (ht) > Solid 1.

Click > Settings Solid - Solid 1 > Heat Conduction, Solid > Thermal conductivity k pull-down menu.

Select > User defined.

Enter: k_Nb in the Thermal conductivity edit window.

Click > Settings Solid - Solid 1 > Thermodynamics, Solid > Density ρ pull-down menu.

Select > User defined.

Enter: rho_Nb in the Density edit window.

Click > Settings Solid - Solid 1 > Thermodynamics, Solid > Heat capacity at constant pressure pull-down menu.

Select > User defined.

Enter: Cp_Nb in the Heat capacity at constant pressure edit window.

See Figure 6.19.

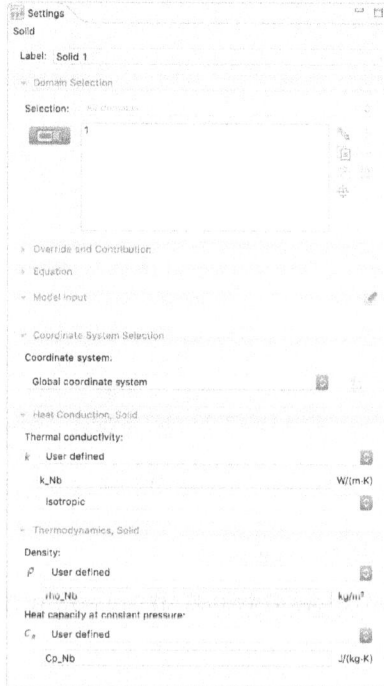

FIGURE 6.19 Settings solid - Solid 1 edit windows.

Figure 6.19 shows the Settings Solid - Solid 1 edit windows.

Click > Model Builder > Component 1 (*comp1*) > Heat Transfer in Solids (ht) > Initial Values 1.

Enter > T_0 in Settings – Initial Values – Initial Values – Temperature edit window.

See Figure 6.20.

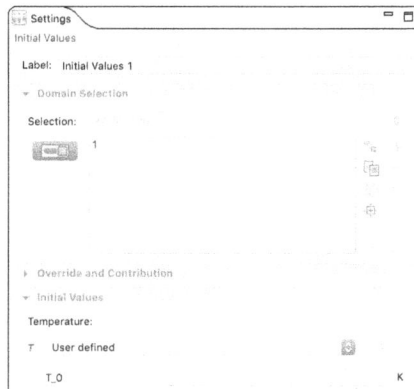

FIGURE 6.20 Settings – Initial values – Initial values 1 – Temperature wdit window.

Figure 6.20 shows the Settings – Initial Values – Initial Values 1 – Temperature edit window.

Right-Click > Model Builder > Component 1 (*comp1*) > Heat Transfer in Solids (ht).

Select > Temperature from the Pop-up menu.

Click > Model Builder > Component 1 (*comp1*) > Heat Transfer in Solids (ht) > Temperature 1.

Select > Boundaries 2 & 3 in the Graphics window.

Enter > T_appl in the Settings > Temperature > Temperature 1 > Temperature: edit window.

See Figure 6.21.

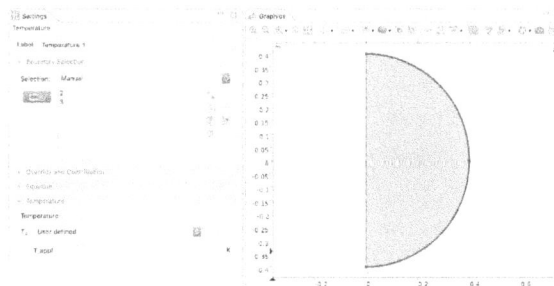

FIGURE 6.21 Settings – Temperature and graphics windows.

Figure 6.21 shows the Settings – Temperature and Graphics windows.

Mesh 1

————
NOTE *The Free Triangular Mesh is easy to use and converges well and so it is chosen to use with this model.*

Right-Click > Model Builder > Component 1 (*comp1*) > Mesh 1.

Select > Free Triangular.

Click > Build All button.

After meshing, the modeler should see a message in the message window about the number of elements (265 domain elements and 39 boundary elements) in the mesh.

See Figure 6.22.

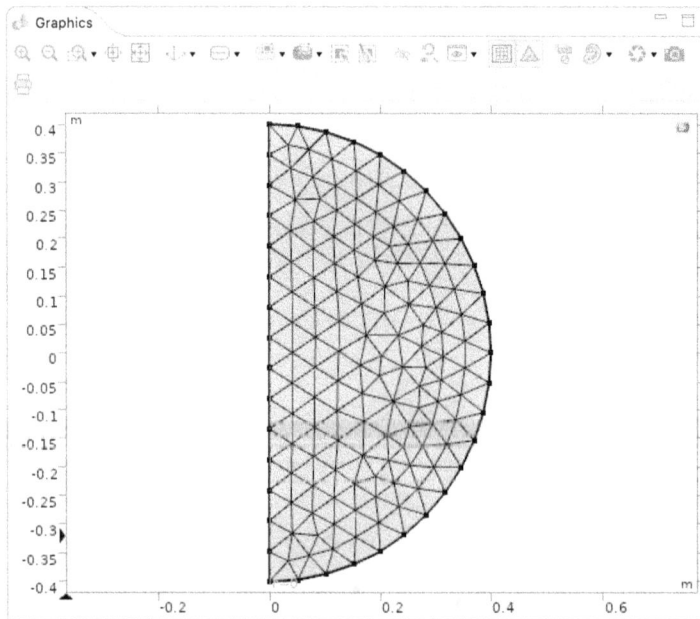

FIGURE 6.22 Desktop display – Graphics – Meshed domain.

Figure 6.22 shows the Desktop Display – Graphics – Meshed Domain.

Study 1

Click > Model Builder – Study 1 – Twistie (as needed).

Click > Step 1 > Time Dependent.

Click > Range button in Settings > Time Dependent > Study Settings > Output Times.

Enter > Start = 0, Step = 2, Stop = 300 in the Range Pop-up edit window.

Click > Replace button in the Range Pop-up edit window.

See Figure 6.23.

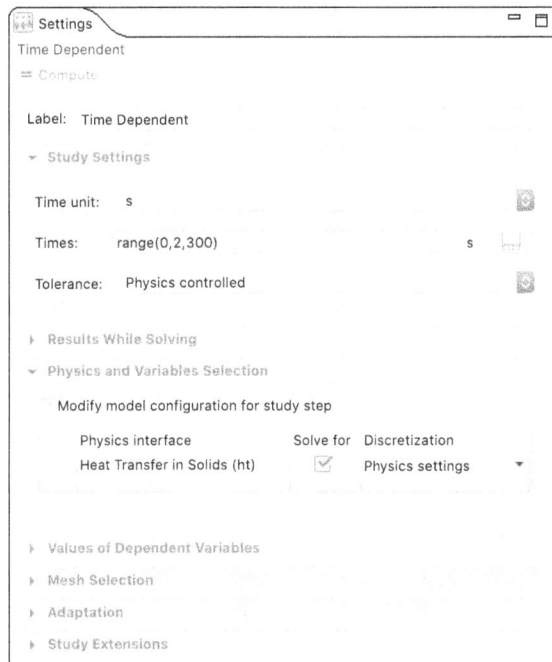

FIGURE 6.23 Settings – Time dependent – Study settings.

Figure 6.23 shows the Settings – Time Dependent – Study Settings.

In Model Builder,

Right-Click Study 1 > Select > Compute.

Click > Zoom Extents.

Computed results, using the default display settings, plus Zoom Extents, are shown in Figure 6.24.

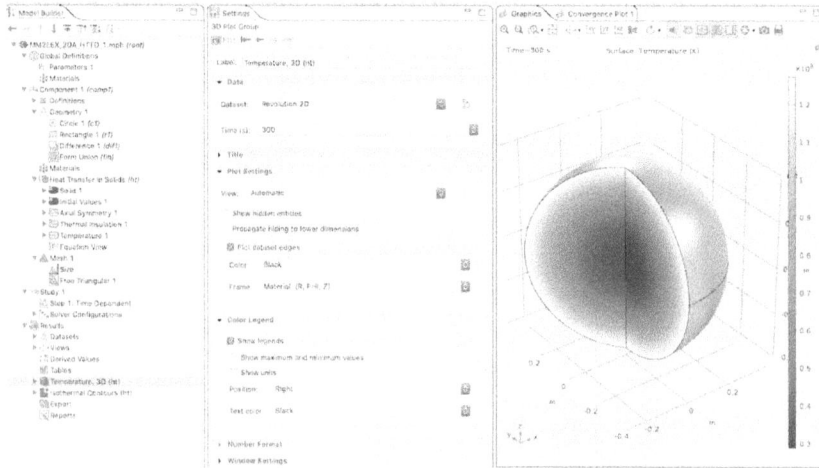

FIGURE 6.24 2D Axisymmetric transient heat transfer model results, using the default display settings, plus zoom extents.

Figure 6.24 shows the 2D Axisymmetric Transient Heat Transfer Model results, using the default display settings, plus Zoom Extents.

2D Results

NOTE   *This step converts the units of the plot from Kelvin to Celsius.*

In Model Builder,

Click > Results > Temperature, 3D (ht) > twistie.

Click > Results > Temperature, 3D (ht) > Surface.

Click > Settings – Surface – Expression – Unit.

Select > degC from the pull-down menu.

Click > Plot.

See Figure 6.25.

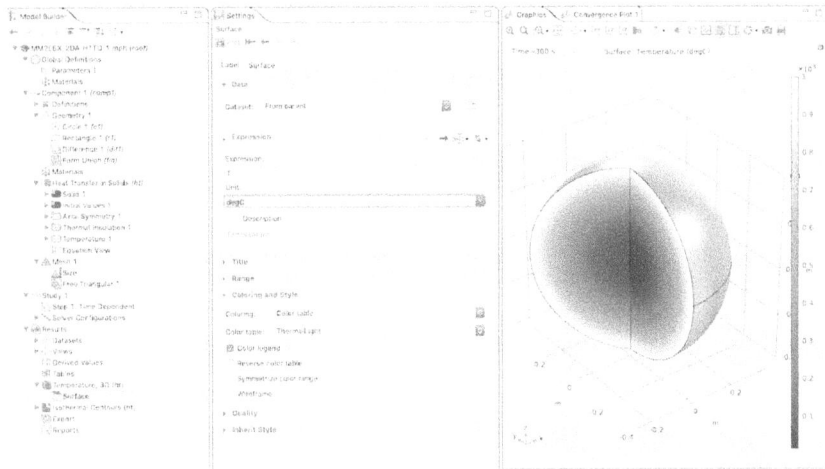

FIGURE 6.25 2D Axisymmetric transient heat transfer model results in degrees celsius.

Figure 6.25 shows the 2D Axisymmetric Transient Heat Transfer Model results in degrees Celsius.

NOTE *Since this model began as a 2D Axisymmetric model, it should be noted that the default plot in 5.x is a 3D model generated from the 2D solution by creating a revolution data set {6.18}. This saves all the physical and computational effort that would be required to build and calculate the 3D version of this model.*

3D Results

Revolution 2D Axisymmetric Transient Heat Transfer Model (3D) Animation

This animation shows how the Revolution 2D Axisymmetric Transient Heat Transfer Model (3D) temperature changes as time changes.

In Model Builder,

Right-Click > Results – Export.

Select > Animation > Player.

In Settings Animation > Scene > Subject: Pull-down, Select: Temperature, 3D (ht) (as needed),

Right-Click > Animation 1: select > Play.

The modeler is now able to observe the results of the modeling process dynamically. The above instructions allowed the modeler to build a movie with the default parameters. The modeler can now build other movies using different parameters that will be viewed in the Graphics window.

The next set of instructions allow the modeler to export the movie frames to a file.

This animation shows how the temperature changes as the time changes.

In Model Builder,

Right-Click > Results – Export.

Select > Animation > File.

In Settings Animation – Scene,

Select > Temperature, 3D (ht) from the pull-down menu (as needed).

In Settings – Animation – Output,

Select > Output type Movie from the pull-down menu.

Select > Format GIF from the pull-down menu.

Enter > 10 in the Frames per second edit window (as needed).

The detailed procedures for Animation in 5.x on the Macintosh and the PC are different. The Macintosh uses the GIF format. The PC uses the AVI format. If the modeler tries to use AVI on the Macintosh, an error results.

In Settings – Animation – Frames,

Click > Lock aspect ratio check box.

In Settings – Animation – Click > Advanced twistie (as needed),

Click > Antialiasing check box (as needed).

Click > Settings – Export (in the Settings Animation toolbar)..

Enter: MM2E5X_2DA_HTTD_1_An.gif as the file name and Click > Save.

The movie has been exported and Saved.

Click > Save the completed Model MM2E5X_2DA_HTTD_1.mph Revolution 2D Axisymmetric Transient Heat Transfer Model (3D) model.

Revolution 2D Axisymmetric Transient Heat Transfer Model (3D) Model Summary and Conclusions

The 2D Axisymmetric Transient Heat Transfer Model is a powerful tool that can be used to explore the heat transfer in many different media (e.g. semiconductors, metals, semimetals, insulators, etc.). It can also easily be converted to a 3D as has been shown earlier in this chapter.

FIRST PRINCIPLES AS APPLIED TO 2D AXISYMMETRIC MODEL DEFINITION

First Principles Analysis derives from the fundamental laws of nature. In the case of models using this Classical Physics Analysis approach, the laws of conservation in physics require that what goes in (as mass, energy, charge, etc.) must come out (as mass, energy, charge, etc.) or must accumulate within the boundaries of the model.

The careful modeler must be knowledgeable of the implicit assumptions and default specifications that are normally incorporated into the COMSOL Multiphysics software model when a model is built using the default settings.

Consider, for example, the two 2D models developed in this chapter. In these models, it is implicitly assumed that there are no thermally related changes (mechanical, electrical, etc.). It is also assumed that the materials are homogeneous and isotropic, except as specifically indicated and that there are no thin insulating contact barriers at the thermal junctions. None of these assumptions are typically true in the general case. However, by making such assumptions, it is possible to easily build a 2D Axisymmetric and/or a 3D (2D Revolution) First Approximation Model.

NOTE

A First Approximation Model is one that captures all the essential features of the problem that needs to be solved, without dwelling excessively on all of the small details. A good First Approximation Model will yield an answer that enables the modeler to determine if he needs to invest the time and the resources required to build a more highly detailed model.

Also, the modeler needs to remember to name model parameters carefully as pointed out in Chapter 1.

REFERENCES

1. COMSOL Reference Manual, p. 201

2. *https://en.wikipedia.org/wiki/Nicolas_Léonard_Sadi_Carnot*

3. *https://en.wikipedia.org/wiki/William_Thomson,_1st_Baron_Kelvin*

4. *https://en.wikipedia.org/wiki/James_Prescott_Joule*

5. *https://en.wikipedia.org/wiki/Ludwig_Boltzmann*

6. *https://en.wikipedia.org/wiki/James_Clerk_Maxwell*

7. *https://en.wikipedia.org/wiki/Max_Planck*

8. *https://en.wikipedia.org/wiki/Isaac_Newton*

9. *https://en.wikipedia.org/wiki/Convective_heat_transfer#Newton%27s_law_of_cooling*

10. *https://en.wikipedia.org/wiki/Joseph_Fourier*

11. *https://en.wikipedia.org/wiki/Thermal_conduction#Fourier%27s_law*

12. A.D. Cameron, et al., *NAFEMS Benchmark Tests for Thermal Analysis (Summary)*, NAFEMS Ltd., 1986

13. *https://en.wikipedia.org/wiki/Niobium*

14. Kittel, Charles, "Introduction to Solid State Physics", ISBN 0-471-87474-4, pp. 324

15. *https://en.wikipedia.org/wiki/Adolf_Eugen_Fick*

16. *https://en.wikipedia.org/wiki/Fick%27s_laws_of_diffusion*

17. *https://en.wikipedia.org/wiki/Heat_equation*

18. COMSOL Reference Manual, p. 1559

SUGGESTED MODELING EXERCISES

1. Build, mesh, and solve the 2D Axisymmetric Cylinder Conduction Model as presented earlier in this chapter.

2. Build, mesh, and solve the 2D Axisymmetric Transient Heat Transfer Model as presented earlier in this chapter.

3. Change the values of the materials parameters and then build, mesh, and solve the 2D Axisymmetric Cylinder Conduction Model as an example problem.

4. Change the values of the materials parameters and then build, mesh, and solve the 2D Axisymmetric Transient Heat Transfer Model as an example problem.

5. Change the value of the heat flux and then build, mesh, and solve the 2D Axisymmetric Cylinder Conduction Model as an example problem.

6. Change the value of the radius to reduce the size of the sphere and then build, mesh, and solve the 2D Axisymmetric Transient Heat Transfer Model as an example problem.

2D SIMPLE AND ADVANCED MIXED MODE MODELING USING COMSOL MULTIPHYSICS 5.x

In This Chapter

- Guidelines for 2D Simple Mixed Mode Modeling in 5.x
  - 2D Simple Mixed Mode Modeling Considerations
- 2D Simple Mixed Mode Models
  - 2D Electrical Impedance Sensor Model
  - 2D Metal Layer on a Dielectric Block Model
- 2D Advanced Mixed Mode Model
  - 2D Metal Layer on a Diamond Substrate Model
- First Principles as Applied to 2D Simple and Advanced Mixed Mode Model Definition
- References
- Suggested Modeling Exercises

GUIDELINES FOR 2D SIMPLE MIXED MODE MODELING IN 5.x

NOTE

In this chapter, 2 simple and 1 advanced mixed mode 2D models will be presented. Such 2D models are typically more conceptually complex than the models that were presented in earlier chapters of this text. 2D simple mixed mode models have proven to be very valuable to the science and engineering communities, both in the past and currently, as first-cut evaluations of potential systemic physical behavior under the influence of mixed external stimuli. The 2D mixed mode model responses and other such ancillary information can be gathered and screened early in a project for a first-cut evaluation. That initial information can potentially be used later as guidance in building higher-dimensionality (3D) field-based (electrical, magnetic, etc.) models.

Since the models in this and subsequent chapters are more complex and more difficult to solve than the models presented thus far, it is important that the modeler have available the tools necessary to most easily utilize the powerful capabilities of the 5.x software. In order to do that, if you have not done this previously, the modeler should go to the main 5.x toolbar, Click > Options – Preferences – Show More Options. When the Preferences – Show More Options edit window is shown, Select > Equation view checkbox and Equation Section checkbox. Click > OK {7.1}.

2D Simple Mixed Mode Modeling Considerations

2D Models can in some cases be less difficult than some 1D or 3D models, having fewer implicit assumptions, and yet potentially, 2D models can still be a very challenging type of model to build, depending on the underlying physics. The least difficult aspect of 2D model creation arises from the fact that the geometry is relatively simple. However, the physics in a 2D simple mixed mode model can range from relatively simple to extremely complex.

In compliance with the laws of physics, a 2D model implicitly assumes that energy flow, materials properties, environment, and all other conditions and variables that are of interest are homogeneous, isotropic and/or constant, unless otherwise specified, throughout the entire domain of interest both within the model and through the boundary conditions and in the environs of the model.

The modeler needs to bear the above-stated conditions in mind and ensure that all of the modeling conditions and associated parameters (default settings) in each model created are properly considered, defined, verified and/or set to the appropriate values.

NOTE *It is always mandatory that the modeler be able to accurately anticipate the expected results of the model and accurately specify the manner in which those results will be presented. Never assume that any of the default values* *that are present when the model is created necessarily satisfy the needs or conditions of a particular model.*

Always verify that any parameters employed in the model are the correct value needed for that model. Calculated solutions that significantly deviate from the anticipated solution or from a comparison of values measured in an experimentally derived realistic model are probably indicative of one or more modeling errors either in the original model design, in the earlier model analysis, in the understanding of the underlying physics or are simply due to human error.

2D Coordinate System

In a 2D model, if parameters can only vary as a function of the position (x) and (y) coordinates, then such a 2D model represents the parametric condition of the model in a time-independent mode (stationary). In a time-dependent study or frequency domain study model, parameters can vary both with a position in (x) and/or (y) and with time (t).

See Figure 7.1.

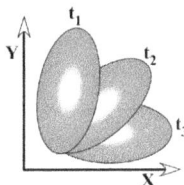

FIGURE 7.1 2D Coordinate system, plus time.

Figure 7.1 shows the 2D Coordinate System, plus Time.

NOTE

In all 2D models, there is a specified third geometric dimension with a default value of 1[m]. The properties of the model being constructed are isotropically mapped throughout that depth. That depth is known as the Out-of-Plane Thickness. That Out-of-Plane Thickness should be changed as needed by the modeler to ensure that the resulting calculations reflect the parameters of the physical system being modeled. The Out-of-Plane Thickness edit window can be found by Clicking on the 2D Physics Interface being employed.

See Figure 7.2.

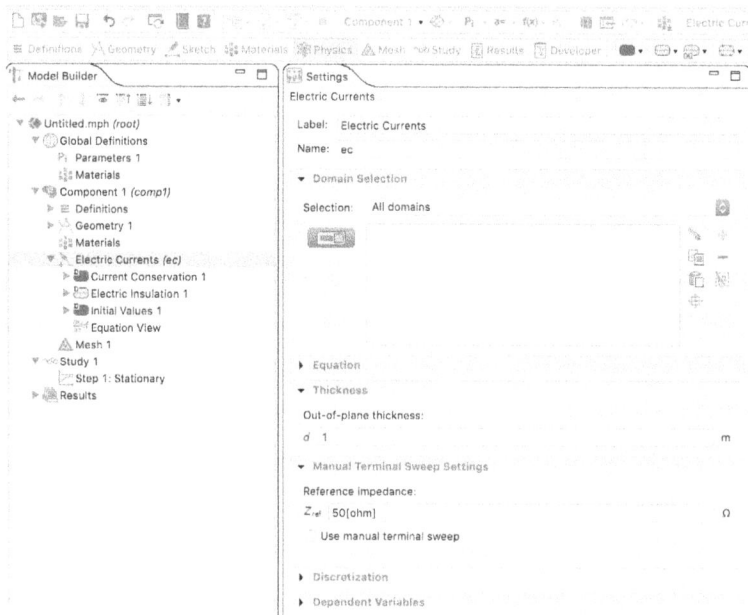

FIGURE 7.2 Out-of-plane thickness edit window.

Figure 7.2 shows the Out-of-Plane Thickness edit window.

In this chapter, two new primary concepts are introduced to the modeler: the Frequency Domain study {7.2} and the Highly Conductive Layer {7.3}. The Frequency Domain study corresponds to a frequency sweep. That is, the model solution is essentially a collection of (quasi-static) solutions, where the input signals are sinusoidal waves, except that one or more of the dependent variables $(f(x, y, t))$ changes with time. The space coordinates (x) and (y) typically represent a distance coordinate throughout which the model is to calculate the change of the specified observables (i.e. temperature, heat flow, pressure, voltage, current, etc.) over the range of values $(x_{min} <= x <= x_{max})$ and $(y_{min} <= y <= y_{max})$. The time coordinate (t) represents the range of values $(t_{min} <= t <= t_{max})$ from the beginning of the observation period (t_{min}) to the end of the observation period (t_{max}).

Electrical Impedance Theory

The concept of electrical impedance {7.4}, as used in Alternating Current (AC) theory {7.5}, is an expansion on the basic concept of resistance as illustrated by Ohm's Law {7.6}, in Direct Current theory.

NOTE   *Ohm's Law was discovered by Georg Ohm and as published in 1827, is:*

$$I = \frac{V}{R} \tag{7.1}$$

Where I = Current in Amperes [A].

V = Voltage (Electromotive Force) in Volts [V].

R = Resistance in Ohms [ohm].

In AC theory, both voltage (V) and current (I) alternate periodically as a function of time. Typically, the alternating behavior (frequency (f)) of the voltage and current are separately represented as either a single sinusoidal wave or as a sum of several sinusoidal waves.

The analysis of complex waveforms is typically handled by Fourier Analysis {7.7}. That topic will not be presented herein. However, the reader is encouraged to expand his modeling horizons by exploring waveform analysis further.

In this case, however, for clarity, the exploration of the concept of impedance will be confined to single-frequency analysis. The concept of impedance was developed and named by Oliver Heaviside {7.8} in 1886. Arthur E. Kennelly {7.9} reformulated impedance in the currently used complex number formulation in 1893.

The first factor that needs to be considered, when expanding modeling calculations from the DC realm (frequency equals zero (f = 0)) to the AC realm (frequency greater than zero (f > 0)), is that the resistance (R) maps into the impedance (Z), as follows {7.10}:

$$Z = R + j\left(\omega L - \frac{1}{\omega C}\right) = R + jX = \left(R^2 + X^2\right)^{1/2} e^{j\tan^{-1}(X/R)} \tag{7.2}$$

Where: Z = Complex Impedance [ohm].

R = Resistance in Ohms [ohm].

j = $(-1)^{1/2}$ (imaginary unit).

ω = 2πf = angular frequency {7.11}.

X = Reactance [ohm] {7.12}.

L = Inductance [henry].

C = Capacitance [farad].

The relative vector phase relationship of an AC voltage applied to a simple series circuit containing resistance, inductance and capacitance is shown in Figure 7.3.

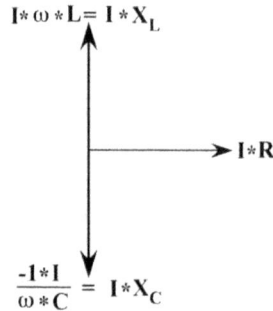

FIGURE 7.3 AC Voltage resistive/reactive vector phase diagram.

Figure 7.3 shows the AC voltage resistive/reactive vector phase diagram.

A second factor that needs to be considered by the modeler, when modeling in the AC realm, is the skin depth (δ) {7.13}. In any material, as a function of the complex permittivity {7.14}, electromagnetic waves (AC) will be attenuated (dissipated, turned into heat, etc.) and shifted in phase as a function of the distance (depth) traveled in that material.

Consider, for example, for a transverse electromagnetic wave propagating in the Z direction, the voltage relationship would be expressed as follows:

$$E_x = E_0 * e^{-kz} = E_0 * e^{-\alpha z} * e^{-j*\beta z} \tag{7.3}$$

Where: E_x = Transverse Electromagnetic wave propagating in the Z Direction.

E_0 = Scalar Voltage Amplitude.

k = Complex Propagation constant.

$j = (-1)^{1/2}$.

e = Base of Natural Logarithms.

α = Attenuation Constant.

β = Wave Solution Constant.

and where α is:

$$\alpha = \omega * \left(\frac{\mu\varepsilon}{2} \left(1 + \left(1 + \left(\frac{\sigma}{\omega\varepsilon} \right)^2 \right)^{\frac{1}{2}} \right) \right)^{\frac{1}{2}} \tag{7.4}$$

and where

ε = permittivity.

μ = permeability.

ω = angular frequency.

σ = conductivity.

For a good conductor, where 1<< σ/ωε, the 1s in the above equation can be ignored and then α becomes:

$$\alpha = \sqrt{\frac{\omega\mu\sigma}{2}} \tag{7.5}$$

The skin depth (δ) is the point at which the amplitude of the signal of interest decreases to E_0* e^{-1}. Therefore δ is:

$$\delta = \frac{1}{\alpha} \tag{7.6}$$

The first model presented in this chapter, 2D Electric Impedance Sensor Model (MM2E5X_2D_EIS_1.mph), explores the sensing of two (2) small volume differential conductivity regions in a block of material that has a bulk conductivity of $1e^{-3}$ S/m and a relative permittivity of 12. The model is implemented using the AC/DC Electric Currents Physics Interface and is solved using a Frequency Domain Study.

NOTE *In the process of building and solving the MM2E5X_2D_EIS_1.mph model, it will become obvious to the modeler that the power of the Electric Impedance Sensing technique lies in the non-invasive nature of this tool. In order to use this tool, the modeler does not need to know in advance what is inside the "black box" and does not need to destroy (cut open) the "black box" to determine the presence of areas of differential electronic properties and their physical locations.*

2D SIMPLE MIXED MODE MODELS

2D Electric Impedance Sensor Model

Building the 2D Electric Impedance Sensor Model

Startup 5.x, Click > Model Wizard button.

Select Space Dimension > 2D.

Click > Twistie for AC/DC.

Click > Twistie for Electric Fields and Currents

Click > Electric Currents (ec).

Click > Add.

Click > Study (Right Pointing Arrow button).

Select > General Studies > Frequency Domain in the Select Study window.

Click > Done (Checked Box button).

Click > Save As.

Enter MM2E5X_2D_EIS_1.mph.

Click > Save.

See Figure 7.4.

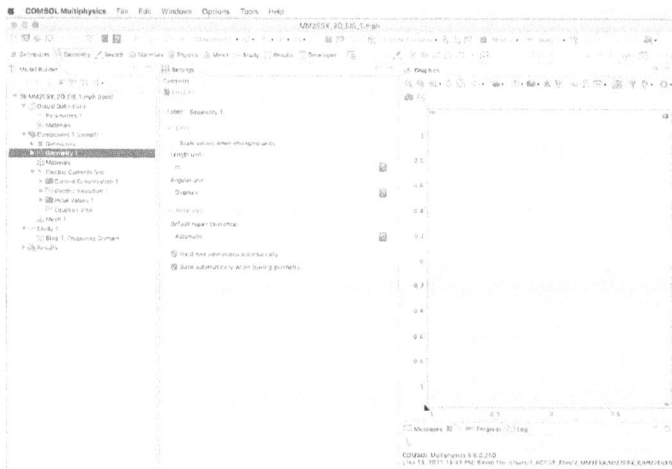

FIGURE 7.4 Desktop display for the MM2E5X_2D_EIS_1.mph model.

Figure 7.4 shows the Desktop Display for the MM2E5X_2D_EIS_1.mph model.

Click > Global Definitions > Parameters 1.

In the Settings – Parameters – Parameters 1 edit window,

Enter the parameters shown in Table 7.1.

TABLE 7.1 Parameters Edit Window.

Name	Expression	Description
sig_bulk	1[mS/m]	Conductivity bulk
eps_r_bulk	12	Permittivity relative bulk
y_0	−0.3[m]	Cavity 0 center y
x_0	0.3[m]	Cavity 0 center x
r_0	0.053[m]	Cavity 0 radius
y_1	−0.3[m]	Cavity 1 center y
x_1	−0.3[m]	Cavity 1 center x
r_1	0.047[m]	Cavity 1 radius
freq_0	1[MHz]	Frequency of measurement
x0	1[m]	Parametric range variable

See Figure 7.5.

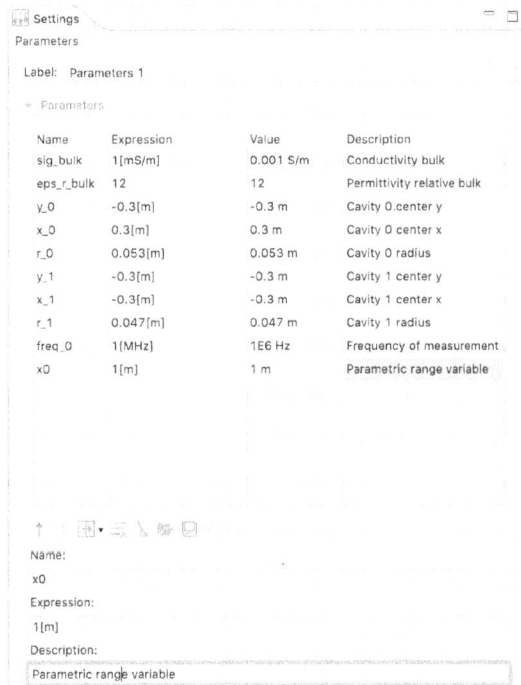

FIGURE 7.5 Settings – Parameters – Parameters edit window filled.

Figure 7.5 shows the Settings – Parameters – Parameters edit window filled.

Right-Click > Global Definitions.

Select > Variables from the Pop-up menu.

In the Settings – Variables – Variables 1 edit window,

Enter the variables shown in Table 7.2.

TABLE 7.2 Variables Edit Window Table.

Name	Expression	Description
sigma_1	sig_bulk*(((x-x_0-x0)^2+(y-y_0)^2)>r_0^2)	Conductivity cavity 0 local
eps_r_1	1+(eps_r_bulk−1)*(((x-x_0-x0)^2+(y-y_0)^2)>r_0^2)	Permittivity cavity 0 local
sigma_2	sig_bulk*(((x-x_1-x0)^2+(y-y_1)^2)>r_1^2)	Conductivity cavity 1 local
eps_r_2	1+(eps_r_bulk−1)*(((x-x_1-x0)^2+(y-y_1)^2)>r_1^2)	Permittivity cavity 1 local
sigma_local	(sigma_1+sigma_2)/2	Conductivity total local
eps_r_local	(eps_r_1+eps_r_2)/2	Permittivity total local

See Figure 7.6.

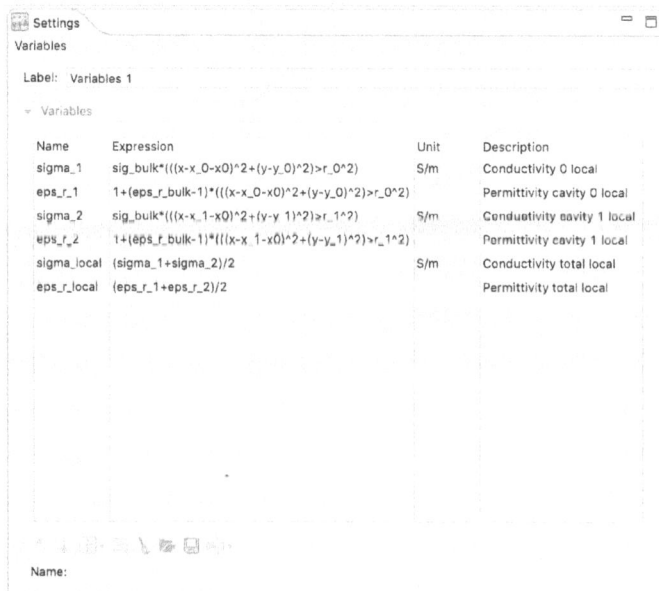

FIGURE 7.6 Settings – Variables – Variables 1 edit window filled.

Figure 7.6 shows the Settings – Variables – Variables 1 edit window filled.

Geometry

Right-Click > Model Builder > Component 1 (*comp1*) > Geometry 1.

Select > Rectangle from the Pop-up menu.

Enter > 1 in the Settings – Rectangle – Size – Width entry window.

Enter > 0.5 in the Settings – Rectangle – Size – Height entry window.

Enter > –0.5 in the Settings – Rectangle – Position – x entry window.

Enter > –0.5 in the Settings – Rectangle – Position – y entry window.

Click > Build All.

Right-Click > Model Builder > Component 1 (*comp1*) > Geometry 1.

Select > Point from the Pop-up menu.

Enter > coordinates for point 1 using Table 7.3.

Click > Build Selected.

Right-Click > Model Builder > Component 1 (*comp1*) > Geometry 1.

Select > Point from the Pop-up menu.

Enter > coordinates for point 2 using Table 7.3.

Click > Build All Objects.

TABLE 7.3 Point Edit Windows.

Name	Location x	Location y
point 1	–0.01	0
point 2	0.01	0

NOTE *The boundary points 1 & 2 are constraint points. They are added to the perimeter of the rectangle to indicate where the two sides of the electrode area end and where each of the two electrically insulated areas begin.*

See Figure 7.7.

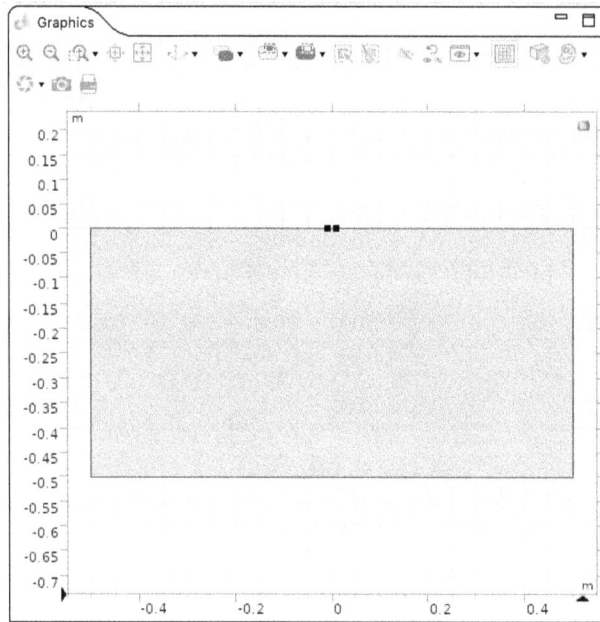

FIGURE 7.7 Model geometry in the graphics window.

Figure 7.7 shows the Model Geometry in the Graphics window.

Electric Currents (ec) Interface

Click > Model Builder > Component 1(*comp1*) > Twistie for the Electric Currents (ec) (as needed).

Click > Model Builder > Component 1(*comp1*) > Electric Currents (ec).

Verify that Domain 1 is Selected in Settings > Electric Currents > Domains > Selection.

Click > Settings > Electric Currents > Twistie for Equation.

Select > Frequency Domain from the Equation form pull-down menu.

Select > User defined from the Frequency pull-down menu.

Enter > freq_0 in the Frequency edit window.

NOTE   *The modeler should note that the default setting of the Out-of-Plane Thickness (1[m]) is correct for this model and should not be changed.*

Click> Model Builder > Component 1 (*comp1*) > Electric Currents (ec) > Current Conservation 1.

Click > Settings – Current Conservation > Constitutive Relation Jc-E > Electrical conductivity pull-down menu.

Select: User defined from the Electrical conductivity pull-down menu.

Enter: sigma_local in the Electrical conductivity edit window.

Click > Settings – Current Conservation > Constitutive Relations D-E > Relative Permittivity.

Select: User defined from the Relative permittivity pull-down menu.

Enter: eps_r_local in the Electrical conductivity edit window.

See Figure 7.8.

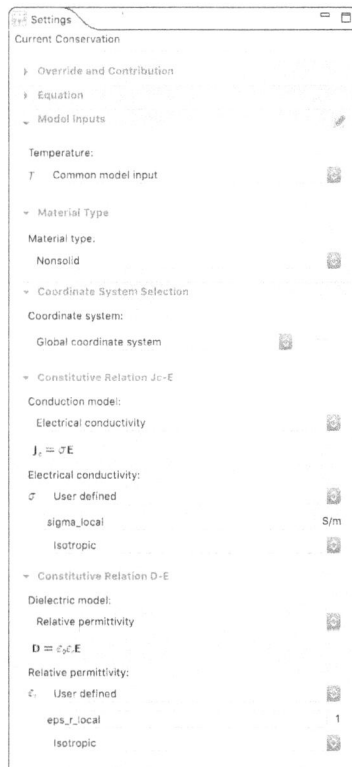

FIGURE 7.8 Settings – Current conservation edit windows.

Figure 7.8 shows the Settings – Current Conservation edit windows.

Right-Click > Model Builder > Component 1 (*comp1*) > Electric Currents (ec).

Select > Terminal from the pop-up menu.

Click > Boundary 4 only in the Graphics window.

Enter > 1 in the Settings – Terminal – Current edit window.

See Figure 7.9.

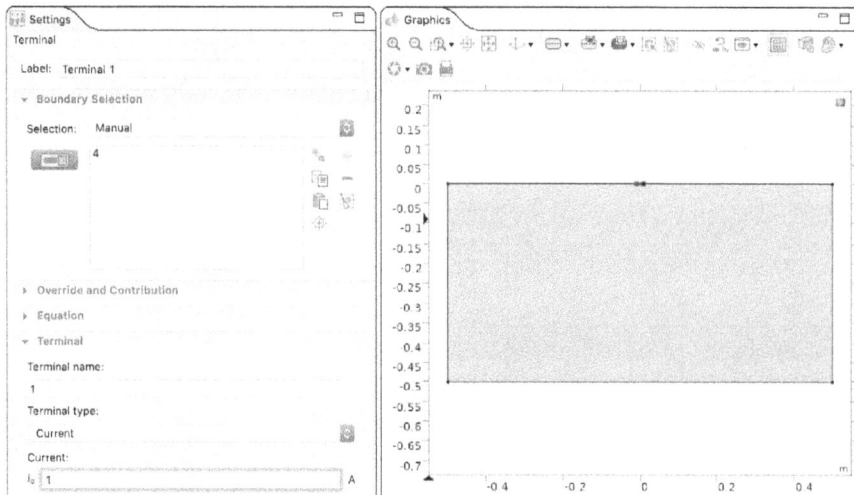

FIGURE 7.9 Settings – Terminal edit windows.

Figure 7.9 shows the Settings – Terminal edit windows.

Right-Click > Model Builder > Component 1 (*comp1*) > Electric Currents (ec).

Select > Ground from the Pop-up menu.

Click > Boundaries 1, 2, 6 in the Graphics window.

See Figure 7.10.

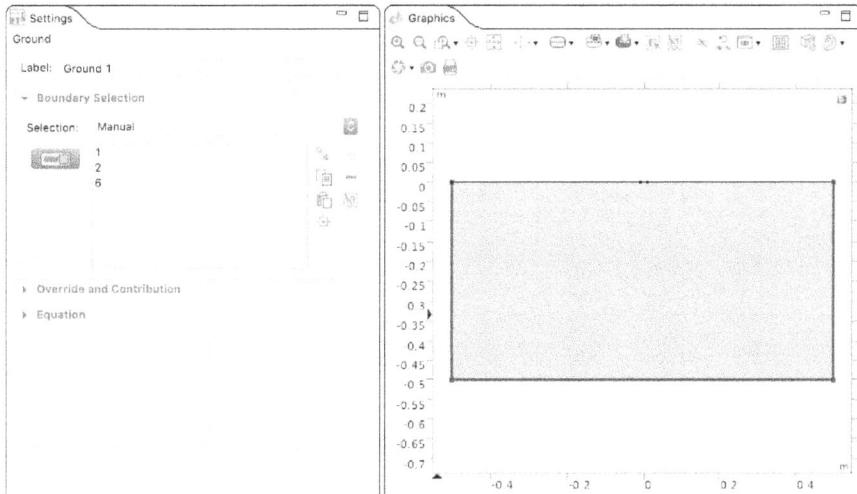

FIGURE 7.10 Settings – Ground edit windows.

Figure 7.10 shows the Settings – Ground edit windows.

Mesh 1

Right-Click > Model Builder – Component 1 – Mesh 1.

Select > Free Triangular from the Pop-up menu.

NOTE *The Free Triangular Mesh Type is chosen because, once adjusted, it will provide an adequate number of mesh elements and is relatively easy to solve.*

Click > Size.

Click > Settings – Size – Twistie for Element Size Parameters.

Enter > 0.01 in the Maximum element size edit window.

Click > Build All.

After meshing, the modeler should see a message in the message window about the number of elements (12474 domain elements and 300 boundary elements) in the mesh.

See Figure 7.11.

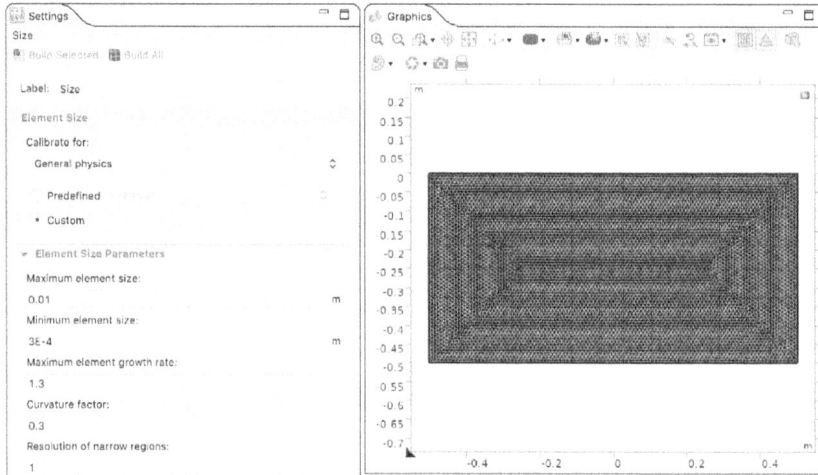

FIGURE 7.11 Settings – Size – Graphics – Meshed domain.

Figure 7.11 shows the Settings – Size – Graphics – Meshed Domain.

Study 1

NOTE
The default settings for the Frequency Domain Study Type are not appropriate for the solution of this problem and thus need to be modified.

Click > Model Builder > Study 1 > Step 1: Frequency Domain.

Enter > 1[MHz] in Settings – Frequency Domain > Study Settings > Frequencies edit window.

See Figure 7.12a.

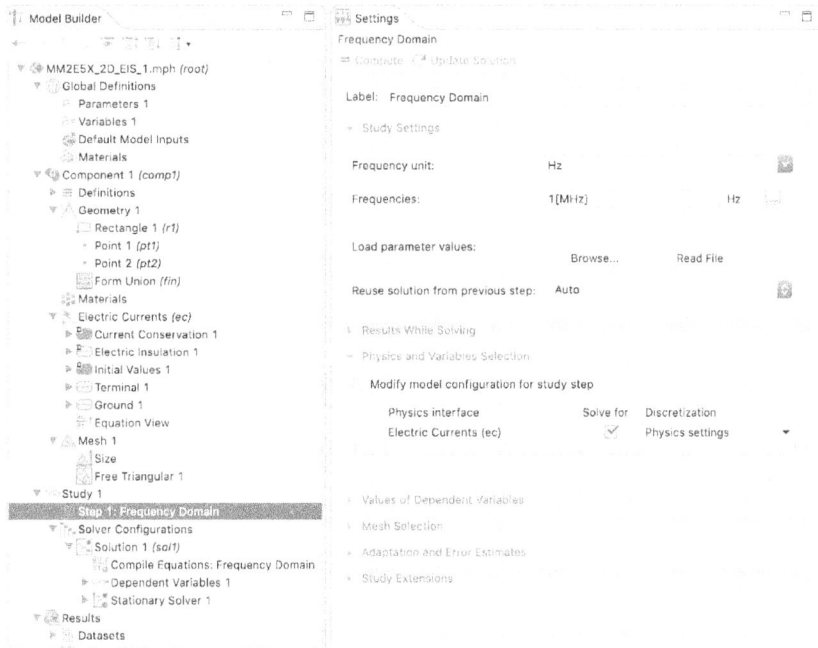

FIGURE 7.12a Settings – Frequency domain > Study settings > Frequencies edit window.

Figure 7.12a Settings – Frequency Domain > Study Settings > Frequencies edit window.

Right-Click > Model Builder > Study 1.

Select > Show Default Solver.

Click > Model Builder > Study 1> Solver Configurations twistie (as needed).

Click > Model Builder > Study 1> Solver Configurations > Solution 1> Stationary Solver 1 twistie.

Click > Model Builder > Study 1> Solver Configurations > Solution 1> Stationary Solver 1 twistie.

Click > Model Builder > Study 1> Solver Configurations > Solution 1> Stationary Solver 1 > Parametric 1.

Click > Settings – Parametric > General > Defined by study step pop-up menu.

Select: User defined in the pop-up menu.

Enter: x0 in the Settings – Parametric – General – Parameter names edit window.

Enter: range(−0.5,0.01,0.5) in the Parameter values list.

See Figure 7.12b.

FIGURE 7.12b Settings – Parametric – General – Parameter names edit windows.

Figure 7.12b Settings – Parametric – General – Parameter names edit windows.

In Model Builder, Right-Click Study 1 > Select > Compute.

Computed results, using the default display settings, are shown in Figure 7.13.

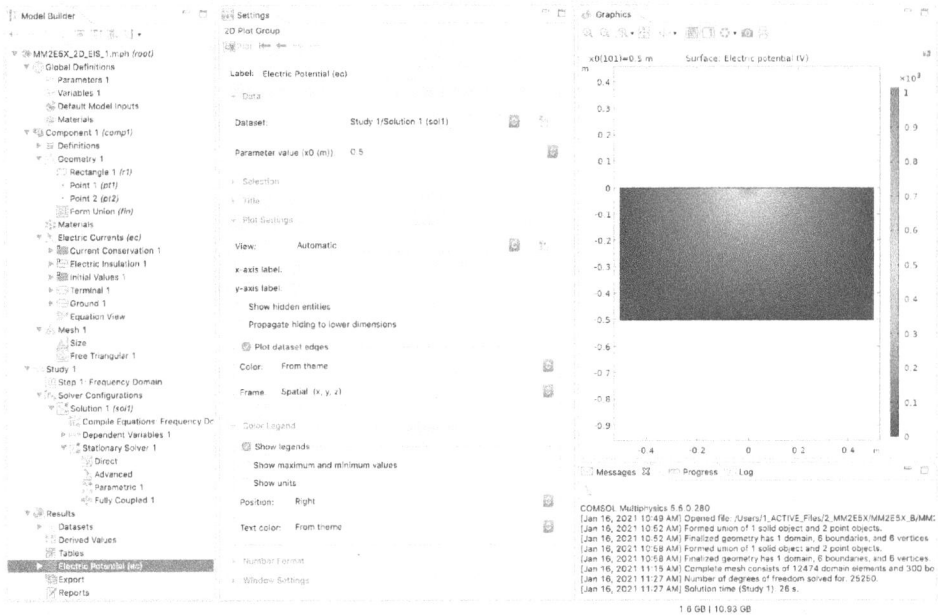

FIGURE 7.13 Computed results, using the default display settings.

Figure 7.13 shows the computed results, using the default display settings.

NOTE

The modeler should note that the computed results, using the default display settings, do not readily show the location and size of the calculated voids. In order to easily observe those results, it is necessary to adjust the display settings parameters to increase the solution contrast in the display plot.

Results

In Model Builder,

Click > Results > Electric Potential (ec).

Right-Click > Results > Electric Potential (ec),

Select > Duplicate from the Pop-up menu.

Click > Results > Electric Potential (ec) 1.

Select > x0 = 0 from the Pop-up menu, Settings-2D Plot Group > Data> Parameter value (x0 (m)):

See Figure 7.14a.

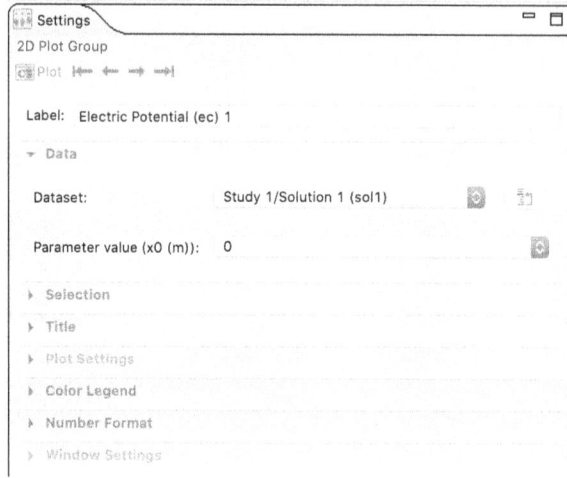

FIGURE 7.14a Desktop display – 2D Plot group modified display parameters.

Figure 7.14a shows the Desktop Display – 2D Plot Group modified display parameters.

Click > Results > Electric Potential (ec) 1 twistie.

Click > Results > Electric Potential (ec) 1 > Surface 1.

Enter > 20*log10(root.comp1.ec.normJ) in the Settings – Surface – Expression edit window.

The equation 20*log10 (variable) transforms the variable data from a linear plot into a logarithmic plot. The variable data is converted into decibels {7.15}. The net effect is to improve the plot contrast so that the calculated voids are displayed properly.

NOTE *Similarly, the minimum, maximum, and color table parameters are chosen to improve the plot display. The modeler should try different values and observe the quality and contrast of the resulting plots.*

The variable name {7.16} beginning with (root.) establishes the exact path within the COMSOL Multiphysics model. In this case, the path goes as follows: root (Model Builder) > comp1 (Component 1) > ec (Electric Currents) > normJ (Normal Current).

Click > Plot.

Click > Settings – Surface – Range twistie.

Select > Settings – Surface – Range – Manual color range check box.

Enter > Minimum = –15 and Maximum = 15.

Click > Plot (if needed).

See Figure 7.14b.

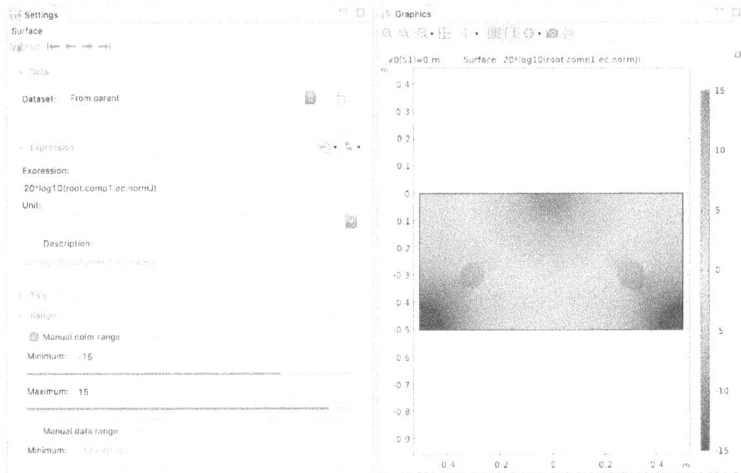

FIGURE 7.14b Desktop display – Computed solution with modified display parameters.

Figure 7.14b shows the Desktop Display – Computed Solution with modified display parameters.

The modeler should note that the calculated voids are now displayed clearly.

More Results

The computed solution can also be displayed as a function of the impedance.

Right-Click > Model Builder – Results.

Select > 1D Plot Group from the Pop-up menu.

Right-Click > Model Builder – Results – 1D Plot Group 3.

Select > Global from the Pop-up menu.

Click > Settings – Global – y-Axis Data – Replace Expression button.

Select > Component 1 > Electric Currents > Terminals > ec.Z11 - Impedance from the Pop-up menu.

Click > Plot.

See Figure 7.15.

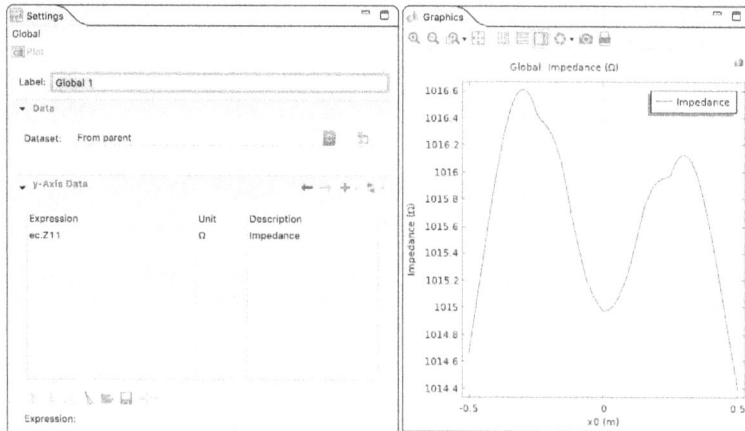

FIGURE 7.15 1D Impedance plot for the two void 2D EIS solution.

Figure 7.15 shows the 1D Impedance Plot for the two void 2D EIS Solution.

The phase angle of the impedance can also be displayed.

Right-Click > Model Builder – Results.

Select > 1D Plot Group from the Pop-up menu.

Right-Click > Model Builder – Results – 1D Plot Group 4.

Select > Global from the Pop-up menu.

Click > Settings – Global – y-Axis Data – Replace Expression.

Select > Component 1 > Electric Currents > Terminals > ec.Z11 - Impedance from the Pop-up menu.

Enter: $(180*arg(ec.Z11))/pi$ in the Expression edit window.

NOTE *The function arg {7.17} yields the phase angle of the function enclosed within the parentheses.*

Enter > Phase Angle (degrees) in the Description edit window.

Click > Title twistie, Select > Title type: None from the Pop-up menu.

Click > Legends twistie, Select: Manual from the Legends Pop-up menu.

Enter > Phase Angle (degrees) in the Legends edit window.

Click > Plot.

See Figure 7.16.

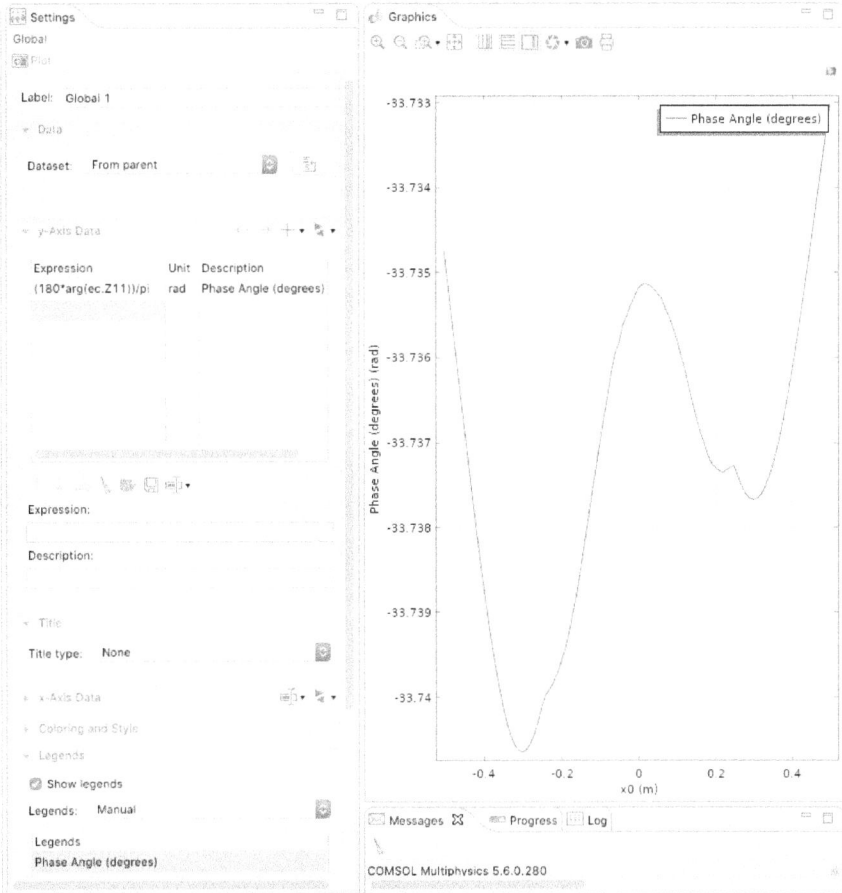

FIGURE 7.16 Phase angle of the 1D impedance plot for the two void 2D EIS solution.

Figure 7.16 shows the Phase Angle of the 1D Impedance Plot for the two void 2D EIS Solution.

2D Electric Impedance Sensor Model Summary and Conclusions

The 2D Electric Impedance Sensor Model is a powerful tool that can be used to develop the design of an Electric Impedance Sensor Machine to nondestructively explore materials and search for inclusions, voids, or any other differences in the electrical properties of imbedded regions of a block of material. With this model, the modeler can easily vary all of the machines parameters and optimize the design before the first prototype is physically built. This technique works best when the contrast in the properties of the materials is largest. The technique can be used either statically or dynamically and has broad application support in industry.

Save > MM2E5X_2D_EIS_1.mph and close the COMSOL Desktop.

2D Metal Layer on a Dielectric Block Model Considerations

The modeler will recall from Chapter 6, that in 1855, Adolf Eugen Fick {7.18} published his now fundamental and famous laws {7.19} for the diffusion (mass transport) of a first item (e.g., a gas, a liquid, etc.) through a second item (e.g., another gas, liquid, etc.).

Fick's First Law is written:

$$J = -D\nabla\Phi \qquad (7.7)$$

Where J = Diffusion Flux in moles per square meter per second [mol/(m²*s)].

D = Diffusion Coefficient (diffusivity) in square meters per second [m²/s].

Φ = Concentration in moles per cubic meter [mol/m³].

∇ = Gradient operator.

NOTE *It should be noted that the minus sign in front of the diffusivity indicates that the item that is flowing (gas, heat, money, etc.). It flows from the higher concentration to the lower concentration.*

See Figure 7.17

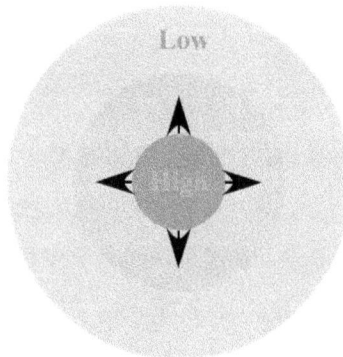

FIGURE 7.17 Graphical rendition of Fick's First Law.

Figure 7.17 shows a graphical rendition of Fick's First Law.

Fick's Second Law calculates the details of how diffusion (mass transport) causes the diffusing item's concentration to change as a function of time (under transient conditions).

Fick's Second Law is written:

$$\frac{\partial \Phi}{\partial t} = D\nabla^2 \Phi \tag{7.8}$$

Where D = Diffusion Coefficient (diffusivity) in square meters per second [m²/s].

Φ = Concentration vector in moles per cubic meter [mol/m³].

∇^2 = Laplace operator.

If modified slightly, where:

$$D = \alpha = \frac{k}{\rho C_p} \tag{7.9}$$

Where α = Thermal Diffusion Coefficient (thermal diffusivity) in square meters per second [m²/s].

k = Thermal conductivity in Watts per meter-degree Kelvin [W/(m*K)].

ρ = Material density in kilograms per cubic meter [kg/m³].

C_p = Specific heat in Joules per kilograms-degree Kelvin [J/(kg*K)].

Then, Fick's Second Law becomes the Heat Equation:

$$\frac{\partial u}{\partial t} = \alpha\nabla^2 u \tag{7.10}$$

Where u = u(x, y, z, t), a function of the three spatial parameters (x, y, z) and time (t).

α = Thermal Diffusion Coefficient (thermal diffusivity) in square meters per second [m²/s].

∇^2 = Laplace operator.

The heat equation has been broadly used to solve many different diffusion problems {7.20}. In the case of thermal diffusion, the spatial parameters are the geometric coordinates. Examples of other uses of the heat equation are easily found in physics, finance, chemistry, and many other application areas.

NOTE

2D Metal Layer on a Dielectric Block Model

NOTE

This is a 2D model of heat conduction through a silica glass block that has a physically thin (0.0001m), high thermal conductivity metallic layer (copper) on the top of the block. The silica glass block has an initial temperature of 300K. At time zero, the left side of the block is fixed at a temperature of 300K (T_L) and the right side of the block is fixed at 600K (T_H). The model calculates the temperature distribution within the block as time progresses and heat flows from right to left in the block. This model is derived from COMSOL Model 452.

See Figure 7.18.

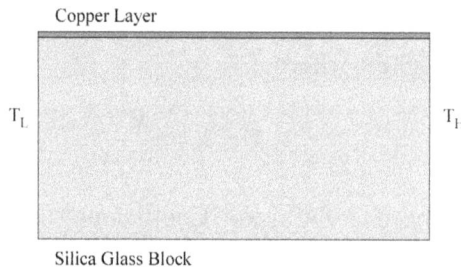

Copper Layer

T_L T_H

Silica Glass Block

FIGURE 7.18 Silica glass block with a copper layer.

Figure 7.18 shows a Silica Glass Block with a Copper Layer.

Building the 2D Metal Layer on a Dielectric Block Model

Startup 5.x, Click > Model Wizard button.

Select Space Dimension > 2D.

Click > Twistie for Heat Transfer in the Select Physics window.

Click > Heat Transfer – Heat Transfer in Solids (ht).

Click > Add button.

Click > Study (Right Pointing Arrow button).

Select > General Studies -Time Dependent in the Select Study window.

Click > Done (Checked Box button).

Click > Save As.

Enter MM2E5X_2D_CpSG_1.mph.

See Figure 7.19.

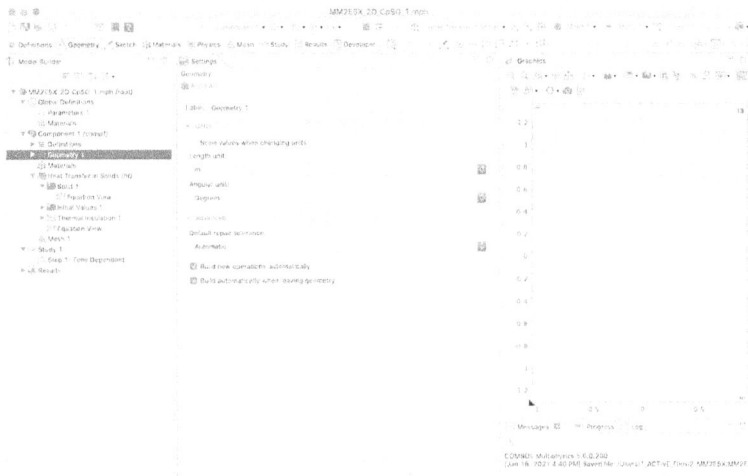

FIGURE 7.19 Desktop display for the MM2E5X_2D_CpSG_1.mph model.

Figure 7.19 shows the Desktop Display for the MM2E5X_2D_CpSG_1.
mph model.

Geometry 1

NOTE

In the first study implemented in this model, the copper layer will be approximated by the Highly Conductive Layer function in the Heat Transfer in Solids Interface. In the second study implemented in this model, the copper layer will be a physical copper layer. The modeler can then directly compare the results of the calculations done for both studies, on nominally the same problem, and then the modeler can determine the trade-offs for modeling by either route.

In Model Builder – Component 1 (*comp1*),

Right-Click > Model Builder – Geometry 1.

Select > Rectangle from the Pop-up menu.

Enter > 0.02[m] in the Settings – Rectangle – Width entry window.

Enter > 0.01[m] in the Settings – Rectangle – Height entry window.

Click > Build All.

See Figure 7.20.

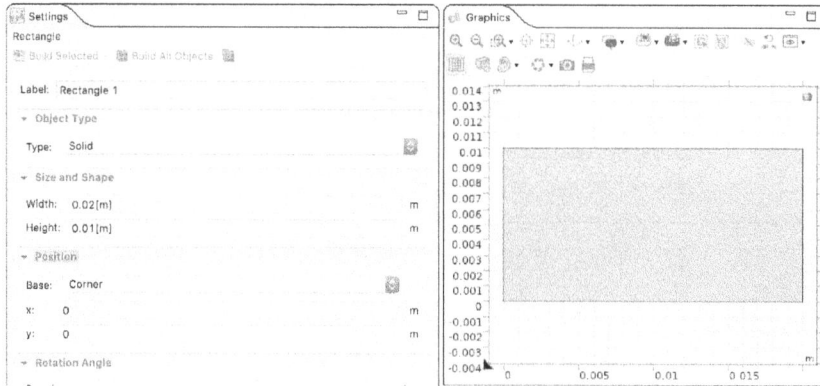

FIGURE 7.20 Settings – Rectangle entry windows.

Figure 7.20 shows the Settings – Rectangle entry windows.

Materials

Right-Click> Model Builder > Component 1 (*comp1*) > Materials,

Select > Add Material from Library in the Pop-up menu.

Click > Add Material > Click > Built-in materials library twistie.

Select > Built-in Silica glass.

Right-Click > Silica glass.

Select > Add to Component 1 from the Pop-up menu.

Select > Built-in Copper.

Right-Click > Copper.

Select > Add to Component 1.

Click > Close on the Add Material Tab.

In the Model Builder,

Click > Copper.

Click > Settings – Material > Geometric Entity Selection > Geometric entity level.

Select > Boundary from the pull-down menu.

Click > Boundary 3 in the Graphics window.

See Figure 7.21.

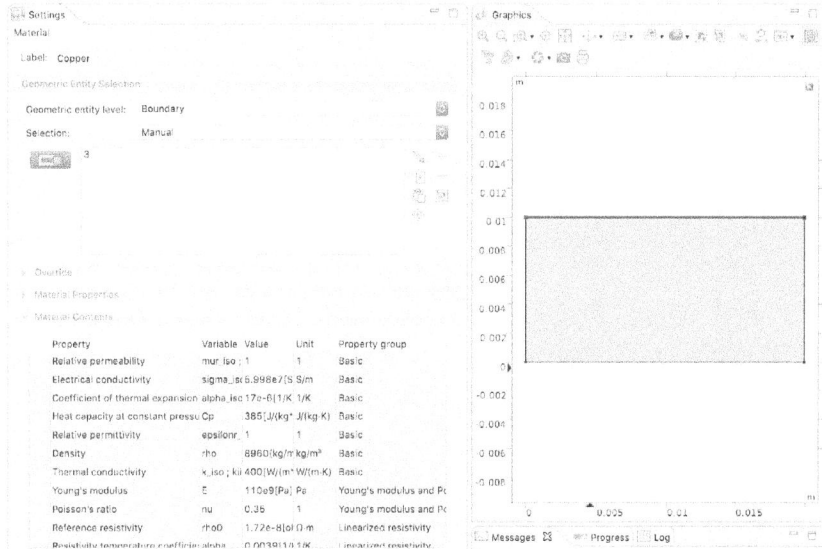

FIGURE 7.21 Settings – Material > Geometric entity selection > Geometric entity level edit window.

Figure 7.21 shows the Settings – Material > Geometric Entity Selection > Geometric entity level edit window.

Heat Transfer in Solids (ht) Interface

Right-Click> Model Builder > Component 1 (*comp1*) > Heat Transfer in Solids (ht).

Select > Temperature from the Pop-up menu.

Click > Boundary 1 in the Graphics window.

Enter > 300[K] in the Settings – Temperature – Temperature – Temperature edit window.

See Figure 7.22.

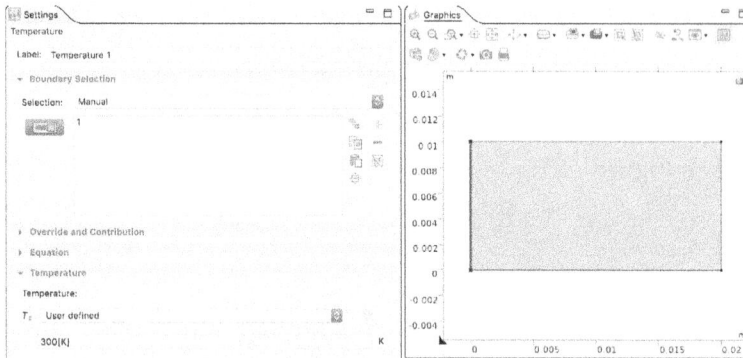

FIGURE 7.22 Settings – Temperature – Temperature – Temperature edit window.

Figure 7.22 shows the Settings – Temperature – Temperature – Temperature edit window.

Right-Click> Model Builder > Component 1 (*comp1*) > Heat Transfer in Solids (ht).

Select > Temperature from the Pop-up menu.

Click > Model Builder > Component 1 (*comp1*) > Heat Transfer in Solids (ht) > Temperature 2.

Click > Boundary 4 in the Graphics window.

Enter > 600[K] in the Settings – Temperature > Temperature > Temperature edit window.

See Figure 7.23.

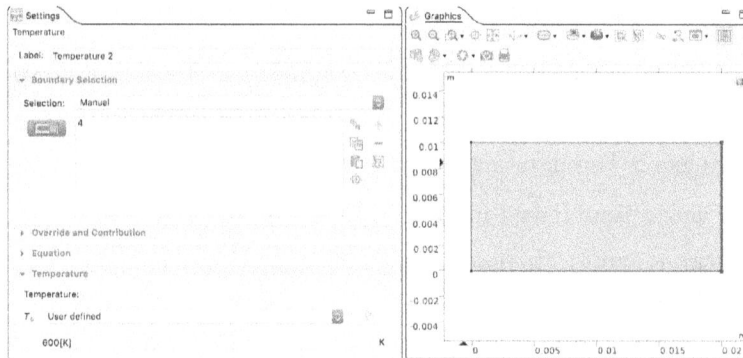

FIGURE 7.23 Settings – Temperature – Temperature – Temperature edit window.

Figure 7.23 shows the Settings – Temperature – Temperature – Temperature edit window.

Right-Click> Model Builder > Component 1 (*comp1*) > Heat Transfer in Solids (ht).

Select > Thin Structures > Thin Layer from the Pop-up menu.

Click > Boundary 3 in the Graphics window.

Enter > 1e-4[m] in the Settings – Thin Layer > Layer Model > Layer type: pull-down menu,

Select: Thermally thin approximation.

See Figure 7.24a.

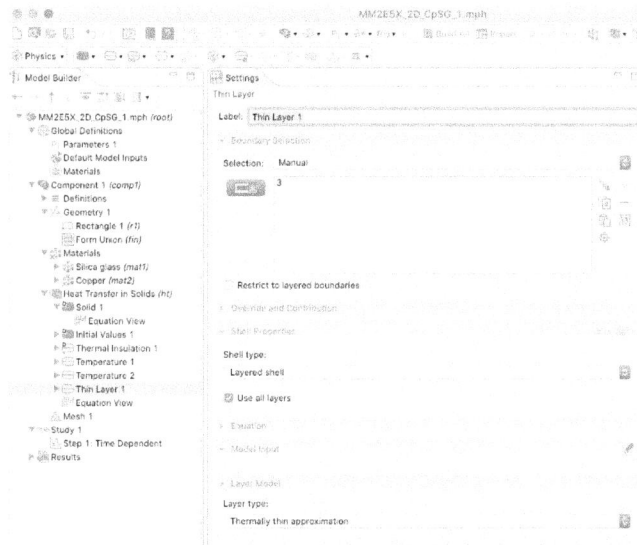

FIGURE 7.24a Settings – Thin layer > Thin layer > Layer thickness parameters window.

Figure 7.24a shows the Settings – Thin Layer > Thin Layer > Layer Thickness Parameters Window.

Click> Model Builder > Component 1 (*comp1*) > Materials – Copper (mat2)

In the Settings Material window, in the Material Contents - Thickness edit window, in the Value column, Enter: 1e-4[m].

See Figure 7.24b.

FIGURE 7.24b Settings – Thin layer > Thin layer > Layer thickness edit window.

Figure 7.24b shows the Settings – Thin Layer > Thin Layer > Layer Thickness Edit Window.

Click > Model Builder > Component 1 (*comp1*) > Heat Transfer in Solids (ht) > Initial Values 1.

Enter > 300[K] in the Settings – Initial Values > Initial Values > Temperature edit window.

See Figure 7.25.

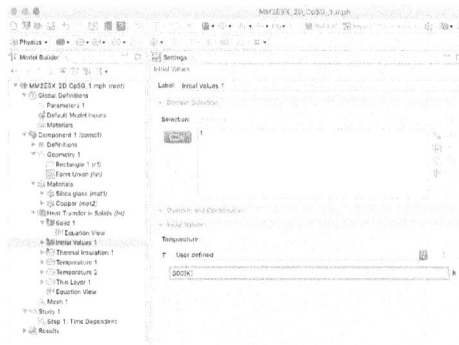

FIGURE 7.25 Settings – Initial values > Initial values > Temperature edit window.

Figure 7.25 shows the Settings – Initial Values > Initial Values > Temperature edit window.

Mesh 1

NOTE *The Free Triangular Mesh is easy to use and converges well and so it is chosen to use with this model.*

Right-Click > Model Builder > Component 1 (*comp1*) > Mesh 1.

Select > Free Triangular.

Click > Build All button.

After meshing, the modeler should see a message in the message window about the number of elements (310 domain elements and 46 boundary elements) in the mesh.

See Figure 7.26.

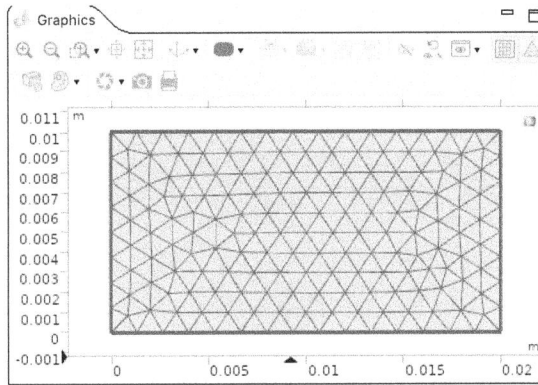

FIGURE 7.26 Desktop display – Graphics – Meshed domain.

Figure 7.26 shows the Desktop Display – Graphics – Meshed Domain.

Study 1

Click > Model Builder – Study 1– Twistie (as needed).

Click > Study 1 – Step 1: Time Dependent.

Click > Range button in Settings – Time Dependent – Study Settings – Output Times.

Enter > Start = 0, Step = 5, Stop = 60 in the Range Pop-up edit window.

Click > Replace button in the Range Pop-up edit window.

See Figure 7.27.

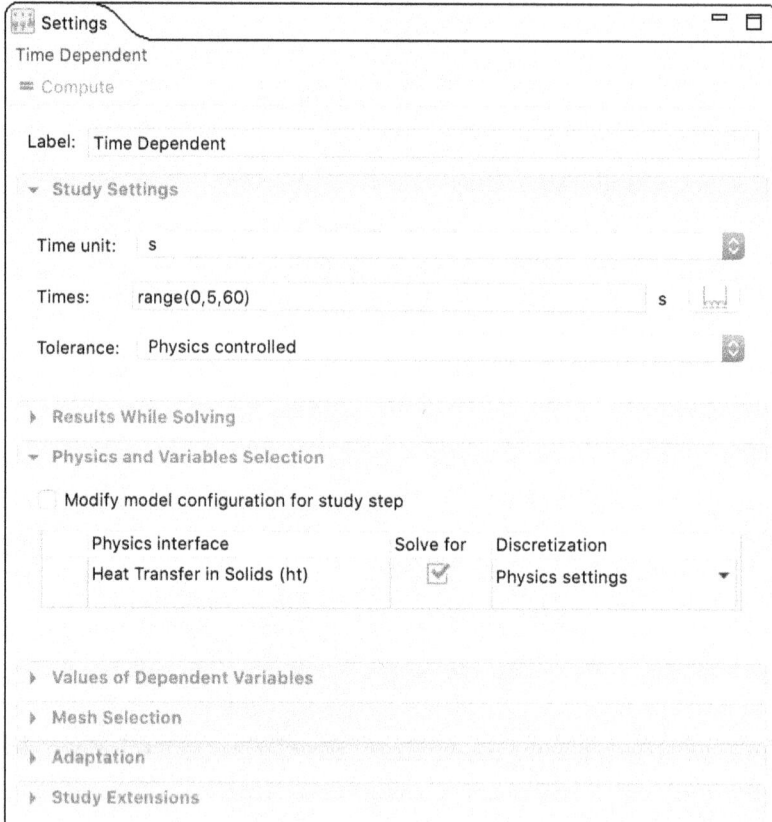

FIGURE 7.27 Settings – Time dependent – Study settings.

Figure 7.27 shows the Settings – Time Dependent – Study Settings.

In Model Builder,

Right-Click > Study 1 >Select > Compute.

Results

Click > Zoom Extents.

Computed results, using the default display settings, plus Zoom Extents, are shown in Figure 7.28.

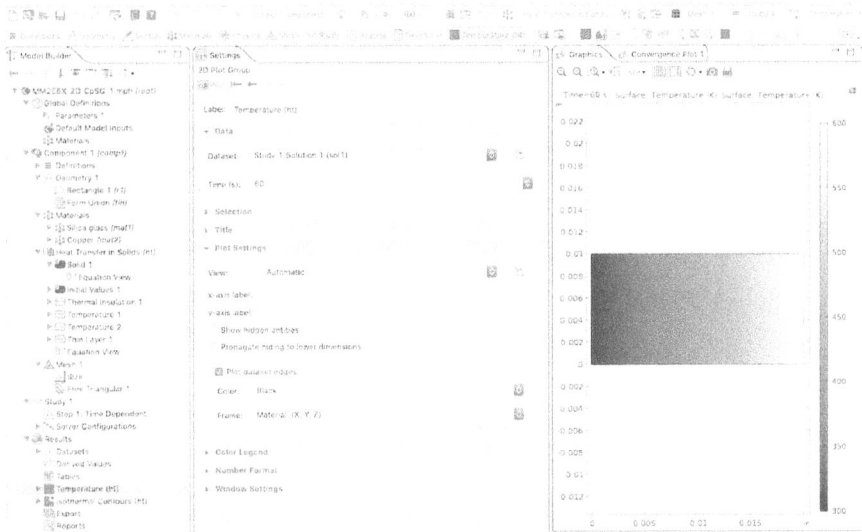

FIGURE 7.28 2D Thin layer on a dielectric block model results, using the default display settings, plus zoom extents.

Figure 7.28 shows the 2D Thin Layer on a Dielectric Block Model results, using the default display settings, plus Zoom Extents.

Click > Save As.

Enter MM2E5X_2D_CpSG_1.mph.

Click > Save As.

Enter MM2E5X_2D_CpSG_2.mph.

Building the 2nd 2D Metal Layer on a Dielectric Block Model

NOTE

In the first study implemented in this model, the copper layer was approximated by the Thin Conductive Layer function in the Heat Transfer in Solids Interface. In this, the second study implemented in this model, the copper layer will be a geometrical copper layer. The modeler will now be able to directly compare the results of both the first model and the second model calculations, on nominally the same problem. The modeler can now determine the relative trade-offs required when he chooses to model by one approach or the other.

Startup 5.x,

File > Open > MM2E5X_2D_CpSG_2.mph.

Right-Click > Model Builder – MM2E5X_2D_CpSG_2.mph (root).

Select > Add Component > 2D from the Pop-up menus.

Right-Click > Component 2 (*comp2*).

Select > Add Physics from the Pop-up menu.

Select > Heat Transfer in Solids (ht) in the Add Physics window.

Select > Add to Component 2.

Close > Add Physics window.

Right-Click > Model Builder – MM2E5X_2D_CpSG_2.mph (root).

Select > Add Study from the Pop-up menu.

Select > General studies > Time Dependent in the Add Study window.

Select > Add Study.

Close > Add Study window.

Geometry 2

In Model Builder > Component 2 (*comp2*),

Right-Click > Model Builder – Geometry 2.

Select > Rectangle from the Pop-up menu.

Enter > 0.02[m] in the Settings – Rectangle – Width entry window.

Enter > 0.01[m] in the Settings – Rectangle – Height entry window.

Click > Build All.

See Figure 7.29.

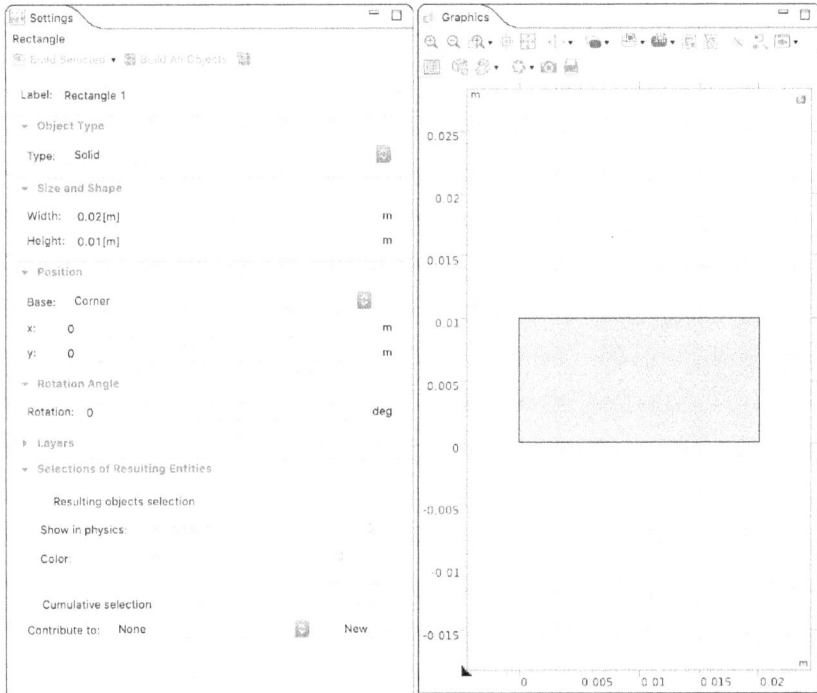

FIGURE 7.29 Settings – Rectangle entry windows.

Figure 7.29 shows the Settings – Rectangle Entry Windows.

In Model Builder > Component 2 (*comp2*),

Right-Click > Model Builder – Geometry 2.

Select > Rectangle from the Pop-up menu.

Enter > 0.02[m] in the Settings – Rectangle – Width entry window.

Enter > 1e-4[m] in the Settings – Rectangle – Height entry window.

Enter > 0.01[m] in the Settings – Rectangle – Position y entry window.

Click > Build All.

See Figure 7.30.

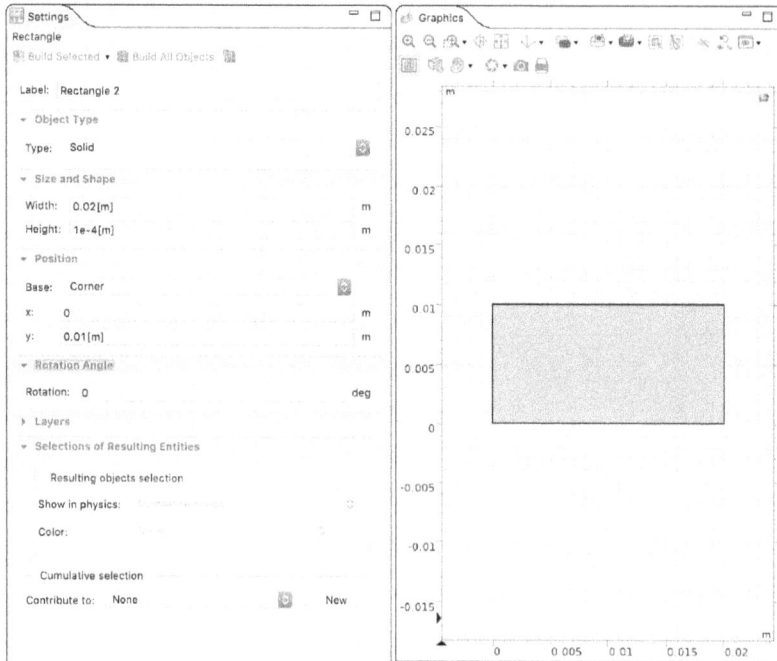

FIGURE 7.30 Settings – Rectangle entry windows.

Figure 7.30 shows the Settings – Rectangle Entry Windows.

Materials

Right-Click > Model Builder > Component 2 (*comp2*) > Materials.

Select > Add Material from Library in the Pop-up menu.

Click > Add Material > Twistie of the Built-in materials library.

Select > Built-in Silica glass.

Right-Click > Silica glass.

Select > Add to Component 2.

Right-Click > Add Material > Built-in Copper.

Select > Add to Component 2.

Click > Add Material close button (hollow X)

Click > Model Builder – Component 2 – Materials – Silica Glass.

Click > Settings > Material > Geometric Entity Selection,

Click > 2 in the edit window,

Click > Minus sign to remove the number 2.

See Figure 7.31a.

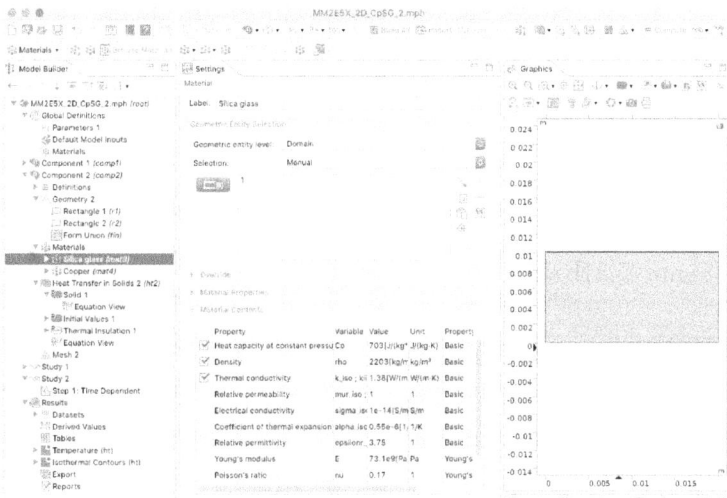

FIGURE 7.31a Silica glass block and settings window.

Figure 7.31a shows the Silica Glass Block and Settings Window.

Click > Model Builder – Component 2 – Materials – Copper.

Click > Settings > Material > Geometric Entity Selection,

Enter: 2 in the edit window (using the Clipboard),

See Figure 7.31b.

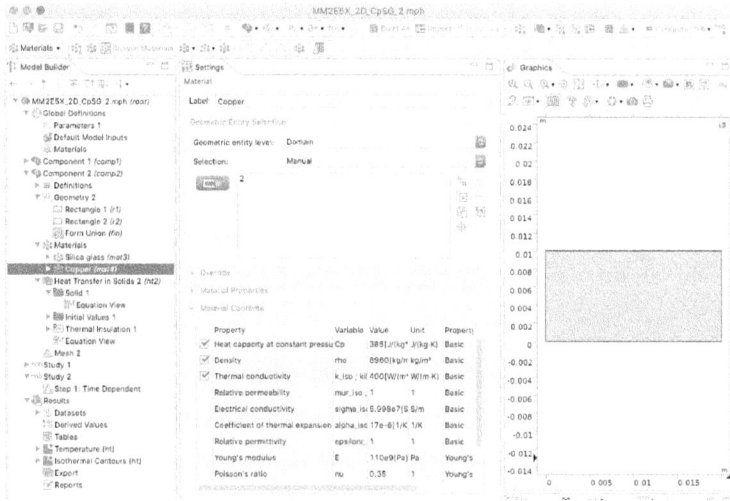

FIGURE 7.31b Copper layer and settings window.

Figure 7.31b shows the Copper Layer and Settings Window.

Heat Transfer 2 (ht2) Interface

Right-Click> Model Builder > Component 2 (*comp2*) > Heat Transfer in Solids 2 (ht2).

Select > Temperature from the Pop-up menu.

Click > Model Builder > Component 2 (*comp2*) > Heat Transfer in Solids 2 (ht2) > Temperature 1

Enter: Boundaries 1 and 3 using the Clipboard.

Expand the upper Left Corner in the Graphics window.

Enter > 300[K] in the Settings – Temperature – Temperature – Temperature edit window.

See Figure 7.32.

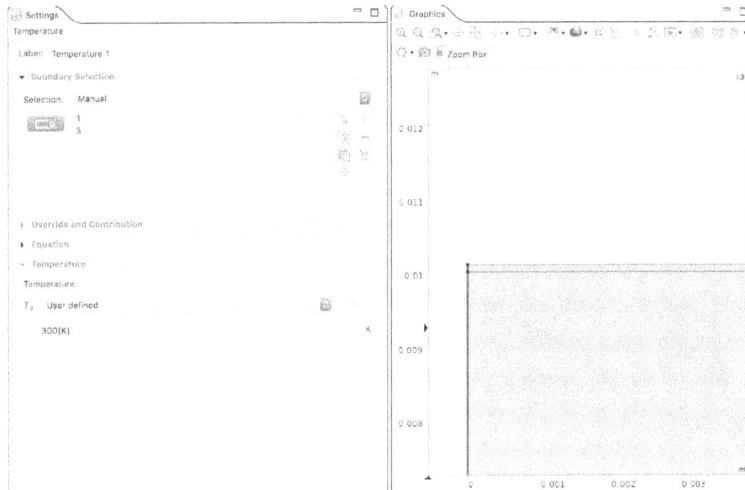

FIGURE 7.32 Settings – Temperature – Temperature – Temperature edit window.

Figure 7.32 shows the Settings – Temperature – Temperature – Temperature edit window.

Click > Zoom Extents.

Right-Click> Model Builder > Component 2 (*comp2*) > Heat Transfer in Solids 2 (ht2).

Select > Temperature from the Pop-up menu.

Click > Model Builder > Component 2 (*comp2*) > Heat Transfer in Solids 2 (ht2) > Temperature 2

Enter: Boundaries 6 and 7 using the Clipboard.

Expand the upper Right Corner in the Graphics window.

Enter > 600[K] in the Settings – Temperature – Temperature – Temperature edit window.

See Figure 7.33.

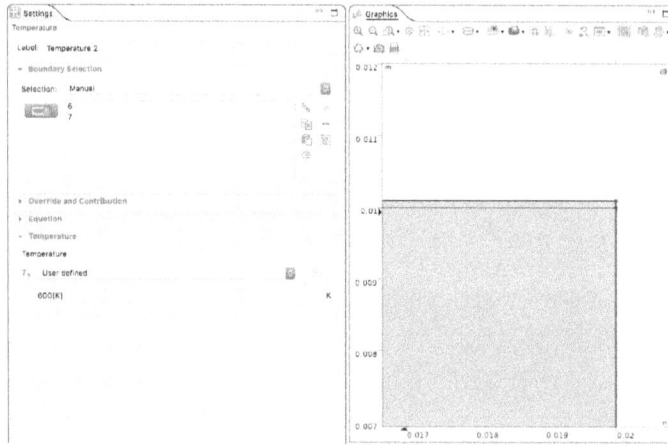

FIGURE 7.33 Settings – Temperature – Temperature – Temperature edit window.

Figure 7.33 shows the Settings – Temperature – Temperature – Temperature edit window.

Click > Zoom Extents.

Click > Model Builder > Heat Transfer in Solids 2 (ht2) – Initial Values 1.

Enter > 300[K] in the Settings – Initial Values – Initial Values – Temperature edit window.

See Figure 7.34.

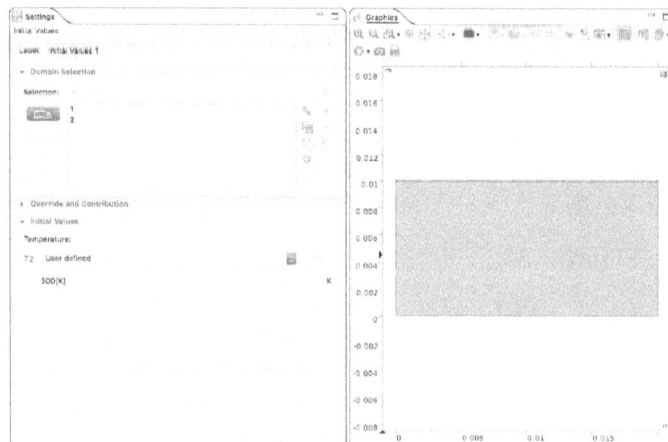

FIGURE 7.34 Settings – Initial values – Initial values – Temperature edit window.

Figure 7.34 shows the Settings – Initial Values – Initial Values – Temperature edit window.

Mesh 2

NOTE *The Free Triangular Mesh is easy to use and converges well and so it is chosen to use with this model.*

Right-Click > Model Builder > Mesh 2.

Select > Free Triangular.

Click > Build All button.

After meshing, the modeler should see a message in the message window about the number of elements (2079 domain elements and 331 boundary elements) in the mesh.

See Figure 7.35.

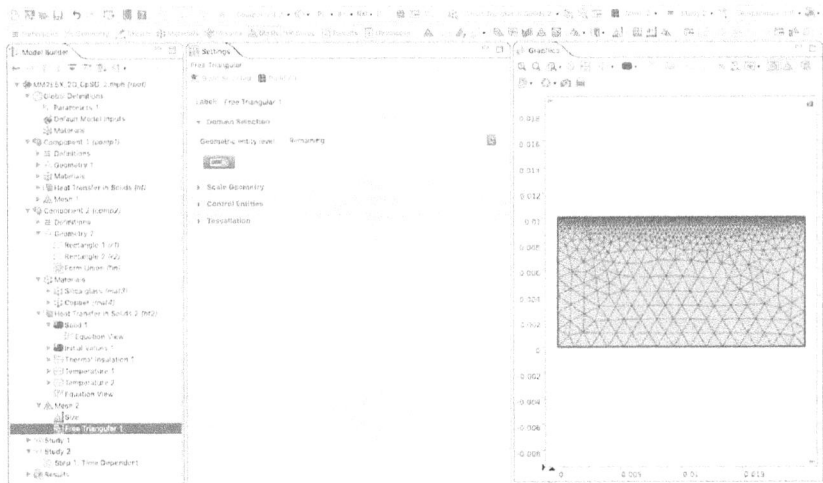

FIGURE 7.35 Desktop display – Graphics – Meshed domain.

Figure 7.35 shows the Desktop Display – Graphics – Meshed Domain.

Study 2

Click > Model Builder – Study 2 – Twistie (as needed).

Click > Study 2 – Step 1: Time Dependent.

Click > Range button in Settings – Time Dependent – Study Settings – Output Times.

Enter > Start = 0, Step = 5, Stop = 60 in the Range Pop-up edit window.

Click > Replace button in the Range Pop-up edit window.

See Figure 7.36.

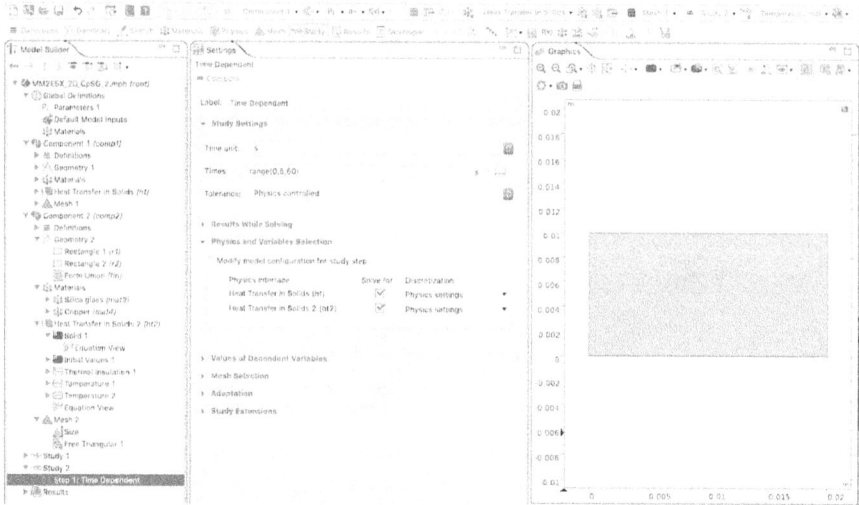

FIGURE 7.36 Settings – Time dependent – Study settings.

Figure 7.36 shows the Settings – Time Dependent – Study Settings.

In Model Builder,

Right-Click Study 2 > Select > Compute.

Results

Click > Temperature (ht2).

Computed results, using the default display settings, plus Zoom Extents, are shown in Figure 7.37.

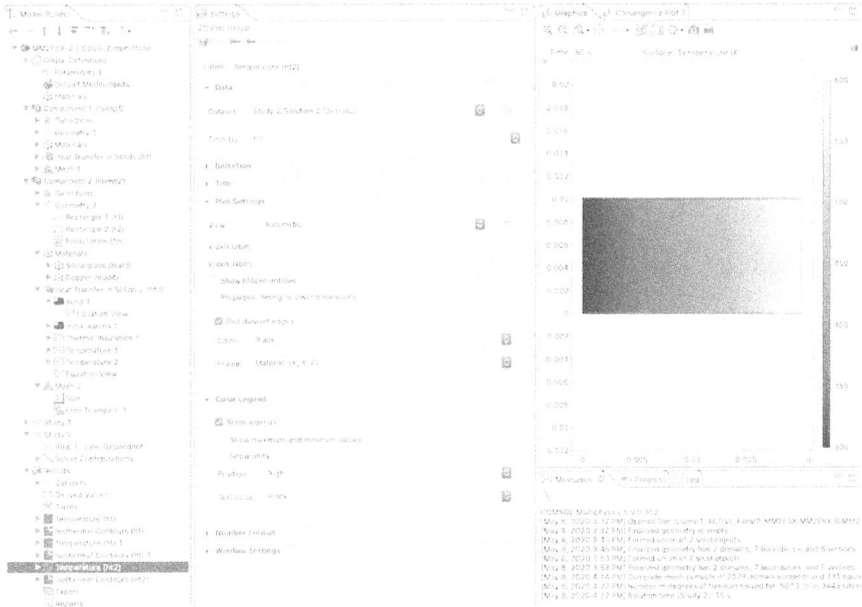

FIGURE 7.37 2D Metal layer on a dielectric block model results, using the default display settings.

Figure 7.37 shows the 2D Metal Layer on a Dielectric Block Model results, using the default display settings.

Results Comparison of Study 1 and Study 2

Right-Click > Temperature (ht) 1,

Select > Plot in > New Window.

Click > Settings – 2D Plot Group – Windows Settings Twistie.

Click > Plot window: pull-down menu,

Select: Plot 1.

Click > Windows title check box.

Enter > Thin Layer Plot.

Click > Settings – 2D Plot Group > Plot button.

NOTE *To see the Plots from both models at the same time, use the following instructions.*

Right-Click > Temperature (ht2),

Select > Plot in > New Window.

Click > Settings – 2D Plot Group – Windows Settings Twistie.

Click > Plot window: pull-down menu,

Select: Plot 2.

Click > Windows title check box.

Enter > Copper Layer Plot.

Click > Settings – 2D Plot Group > Plot button.

Click and Drag the upper edge of the Copper Layer Plot window to the boundary between the Graphics and Messages windows (very carefully) and then Click.

Drag the border between the two plot windows up until the two windows are the same size.

Click > Zoom Extents for each plot window.

See Figure 7.38.

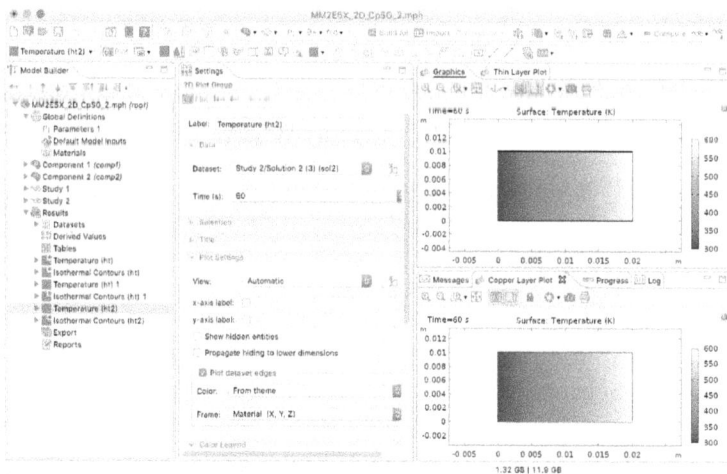

FIGURE 7.38 Simultaneous comparison of the thin layer plot and the copper layer plot solutions.

Figure 7.38 shows the Simultaneous Comparison of the Thin Layer Plot and the Copper Layer Plot Solutions.

Be sure and Save the model before closing the desktop.

NOTE *The modeler can readily observe that the Highly Conductive Layer Model and the Copper Layer Model result in almost exactly the same solution. To a First Approximation, they are essentially the same solution.*

2D Advanced Model

Building a 2D Metal Layer on a Diamond Block Model

NOTE *In the first study implemented above, a copper layer was approximated by the Thin Conductive Layer function in the Heat Transfer in Solids Interface. In this, Advanced Model, the third study implemented in this chapter, the copper layer will be a geometrical copper layer. The modeler will now be able to compare the results of the earlier models, to the results of this Advanced model calculation, on nominally the same modeling problem, employing diamond as the substrate. The modeler can now determine the relative trade-offs required when he chooses different materials to model one methodology or another.*

Startup 5.x,

In the New window, Click > Model Wizard.

In the Model Wizard window, Click > 2D.

In the Select Physics window, Click > the Heat Transfer twistie.

Select: Heat Transfer in Solids (ht).

Click > Add.

Click > Study (right pointing arrow button).

Select: General studies > Time Dependent in the Select Study window.

Click > Done.

Click > Save As.

Enter: MM2E5X_2D_CpSG_Dia_3.mph.

See Figure 7.39.

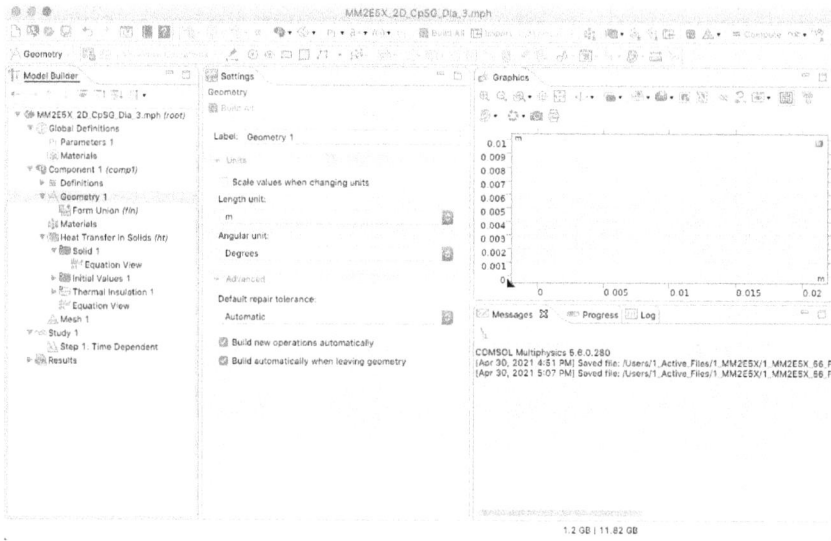

FIGURE 7.39 Copper layer on diamond block COMSOL desktop.

Figure 7.39 shows the Copper Layer on Diamond Block COMSOL Desktop.

Geometry 1

In Model Builder > Component 1 (*comp1*),

Right-Click > Model Builder – Geometry 1.

Select > Rectangle from the Pop-up menu.

Enter > 0.02[m] in the Settings – Rectangle – Width entry window.

Enter > 0.01[m] in the Settings – Rectangle – Height entry window.

Click > Build Selected.

See Figure 7.40.

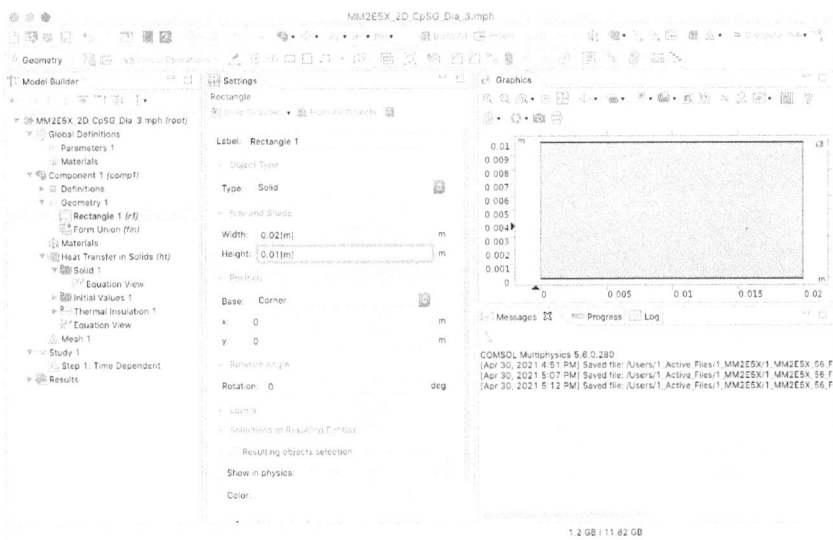

FIGURE 7.40 Block.

Figure 7.40 shows the Block.

In Model Builder > Component 1 (*comp1*),

Right-Click > Model Builder – Geometry 1.

Select > Rectangle from the Pop-up menu.

Enter > 0.02[m] in the Settings – Rectangle – Width entry window.

Enter > 1e-4[m] in the Settings – Rectangle – Height entry window.

Enter > 0.01[m] in the Settings – Rectangle – Position y entry window.

Click > Build All.

See Figure 7.41.

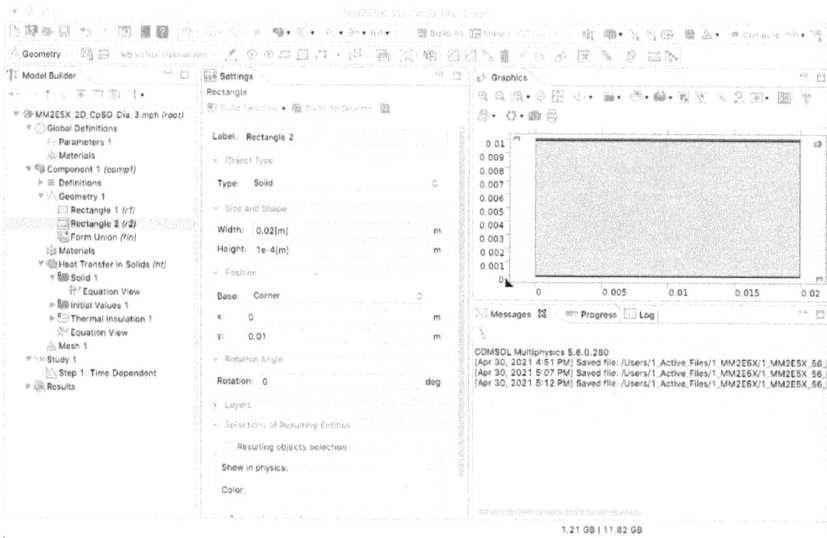

FIGURE 7.41 Copper layer.

Figure 7.41 shows the Copper Layer.

Materials

Right-Click > Model Builder > Component 1 (*comp1*) > Materials.

Select: Add Material from Library in the Pop-up menu.

Click > Add Material > Twistie of the Built-in materials library.

Select > Built-in Copper.

Click > Add to Component.

Click > Add Material close button (hollow X)

In Settings Material Copper > Geometric Entity Selection

Click > 2, remove 2 using the Minus (remove sign).

Right-Click > Model Builder > Component 1 (*comp1*) > Materials.

Select: Blank Material from the pop-up menu.

In Settings Material Label:,

Enter: Diamond.

In Settings Material Diamond > Geometric Entity Selection

Enter: 2, using the Clipboard.

In the Settings Material Diamond window,

Click > Material Contents twistie (as needed),

Enter in the following values in the Value column

Thermal conductivity > 1,000[W/(m*K)] {7.21}.

Density > 3,500[kg/m^3] {7.22}.

Heat capacity at constant pressure > 509.1[J/(kg*K)] {7.23}.

See Figure 7.42a.

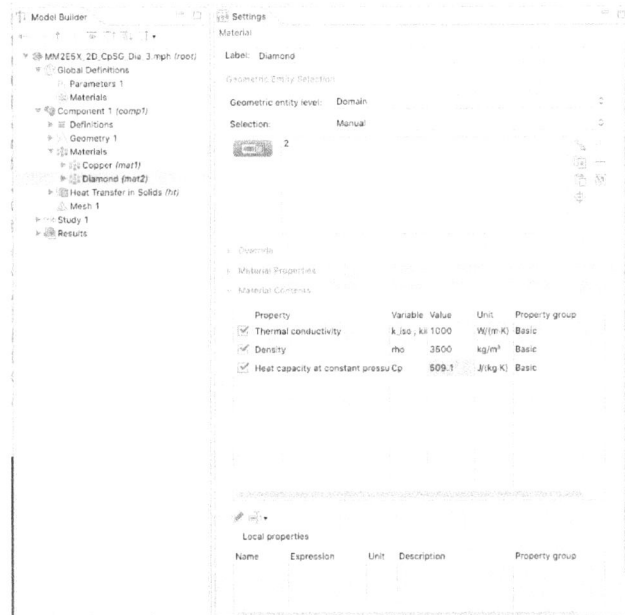

FIGURE 7.42a Diamond block and settings window.

Figure 7.42a shows the Diamond Block and Settings Window.

Click > Model Builder – Component 1 – Materials – Copper.

See Figure 7.42b.

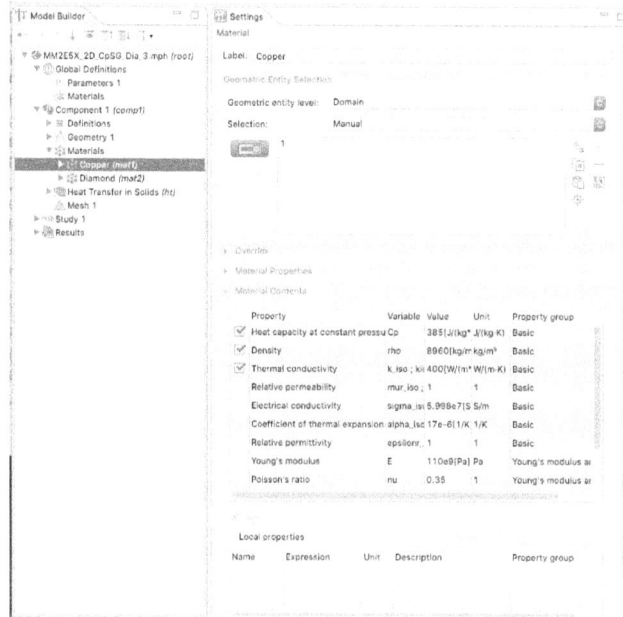

FIGURE 7.42b Copper layer and settings window.

Figure 7.42b shows the Copper Layer and Settings Window.

Heat Transfer in Solids (ht) Interface

Right-Click> Model Builder > Component 1 (*comp1*) > Heat Transfer in Solids 1 (ht1).

Select > Temperature from the Pop-up menu.

Click > Model Builder > Component 1 (*comp1*) > Heat Transfer in Solids (ht) > Temperature 1

Enter: Boundaries 1 and 3 using the Clipboard.

Expand the upper Left Corner in the Graphics window.

Enter > 300[K] in the Settings – Temperature – Temperature – Temperature edit window.

See Figure 7.43.

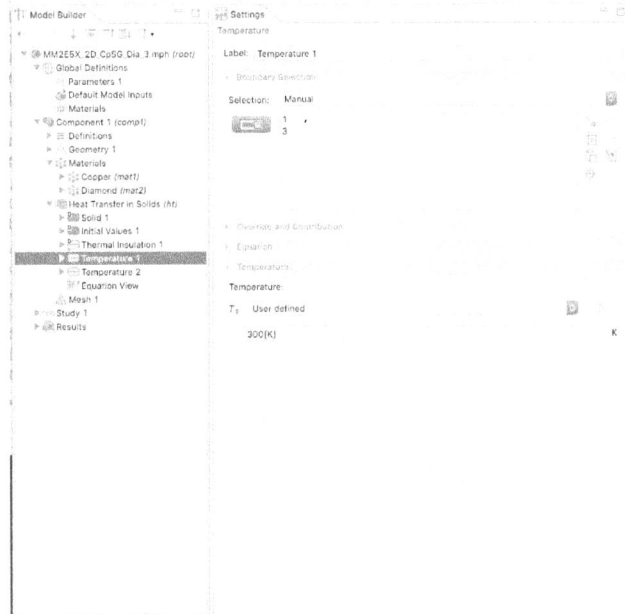

FIGURE 7.43 Settings – Temperature – Temperature – Temperature edit window.

Figure 7.43 shows the Settings – Temperature – Temperature – Temperature edit window.

Click > Zoom Extents.

Right-Click> Model Builder > Component 1 (*comp1*) > Heat Transfer in Solids (ht).

Select > Temperature from the Pop-up menu.

Click > Model Builder > Component (*comp1*) > Heat Transfer in Solids (ht) > Temperature 2

Enter: Boundaries 6 and 7 using the Clipboard.

Expand the upper Right Corner in the Graphics window.

Enter > 600[K] in the Settings – Temperature – Temperature – Temperature edit window.

See Figure 7.44.

FIGURE 7.44 Settings – Temperature – Temperature – Temperature edit window.

Figure 7.44 shows the Settings – Temperature – Temperature – Temperature edit window.

Click > Zoom Extents.

Click > Model Builder > Heat Transfer in Solids (ht) – Initial Values 1.

Enter > 300[K] in the Settings – Initial Values – Initial Values – Temperature edit window.

See Figure 7.45.

FIGURE 7.45 Settings – Initial values – Initial values – Temperature edit window.

Figure 7.45 shows the Settings – Initial Values – Initial Values – Temperature edit window.

Mesh 1

NOTE *The Free Triangular Mesh is easy to use and converges well and so it is chosen to use with this model.*

Right-Click > Model Builder > Mesh 1.

Select > Free Triangular.

Click > Build All button.

After meshing, the modeler should see a message in the message window about the number of elements (2079 domain elements and 331 boundary elements) in the mesh.

See Figure 7.46.

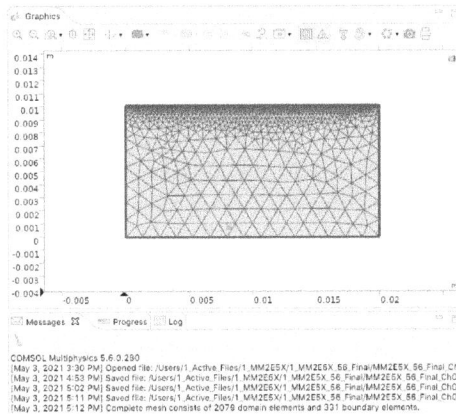

FIGURE 7.46 Desktop display – Graphics – Meshed domain.

Figure 7.46 shows the Desktop Display – Graphics – Meshed Domain.

Study 1

Click > Model Builder – Study 1 – Twistie (as needed).

Click > Study 1 – Step 1: Time Dependent.

Click > Range button in Settings – Time Dependent – Study Settings – Output Times.

Enter > Start = 0, Step = 5, Stop = 60 in the Range Pop-up edit window.

Click > Replace button in the Range Pop-up edit window.

See Figure 7.47.

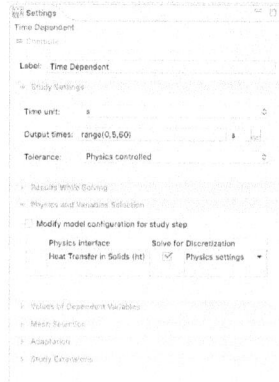

FIGURE 7.47 Settings – Time dependent – Study settings.

Figure 7.47 shows the Settings – Time Dependent – Study Settings.

In Model Builder,

Right-Click Study 1 > Select > Compute.

Results

Click > Temperature (ht).

Computed results, using the default display settings, plus Zoom Extents, are shown in Figure 7.48.

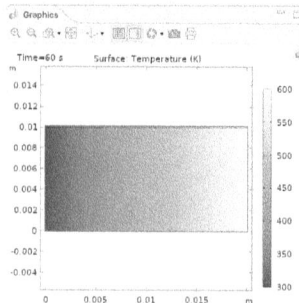

FIGURE 7.48 2D Metal layer on a diamond block model results, using the default display settings.

Figure 7.48 shows the 2D Metal Layer on a Diamond Block Model results, using the default display settings.

2D Metal Layer on a Dielectric Block Models (2) Summary and Conclusions

The 2D Metal Layer on a Dielectric Block Model is a powerful tool that can be used to develop the design of numerous different models for heat transfer problems. With this model, the modeler can easily vary all of the machine's parameters and optimize the design to obtain the best heat transfer before the first prototype is physically built. This technique works best when the contrast in the properties of the materials is largest. The technique can be used either statically or dynamically and has broad application support in industry.

2D Metal Layer on a Diamond Block Model Summary and Conclusions

The 2D Metal Layer on a Diamond Block Model is a powerful tool that can be used to develop and compare the design of numerous different models, using different materials for heat transfer problems. With this model, the modeler can easily vary all of the machine's parameters and optimize the design to obtain the best heat transfer before the first prototype is physically built. This technique works best when the contrast in the properties of the materials is largest. The technique can be used either statically or dynamically and has broad application support in industry.

FIRST PRINCIPLES AS APPLIED TO 2D SIMPLE MIXED MODE MODEL DEFINITION

First Principles Analysis derives from the fundamental laws of nature. In the case of models using this Classical Physics Analysis approach, the laws of conservation in physics require that what goes in (as mass, energy, charge, etc.) must come out (as mass, energy, charge, etc.) or must accumulate within the boundaries of the model.

The careful modeler must be knowledgeable of the implicit assumptions and default specifications that are normally incorporated into the COMSOL Multiphysics software model when a model is built using the default settings.

Consider, for example, the two 2D models developed in this chapter. In these models, it is implicitly assumed that there are no thermally related changes (mechanical, electrical, etc.). It is also assumed that the materials are homogeneous and isotropic, except as specifically indicated, and that there are no thin insulating contact barriers at the thermal junctions. None of these assumptions are typically true in the general case. However, by making such assumptions, it is possible to easily build a 2D Simple Mixed Mode First Approximation Model.

NOTE

A First Approximation Model is one that captures all the essential features of the problem that needs to be solved, without dwelling excessively on all of the small details. A good First Approximation Model will yield an answer that enables the modeler to determine if he needs to invest the time and the resources required to build a more highly detailed model.

Also, the modeler needs to remember to name model parameters carefully as pointed out in Chapter 1.

REFERENCES

1. COMSOL Reference Manual, p. 254

2. COMSOL Reference Manual, p. 901

3. Heat Transfer Module Users Guide, p. 217

4. *https://en.wikipedia.org/wiki/Electrical_impedance*

5. *https://en.wikipedia.org/wiki/Alternating_current*

6. *https://en.wikipedia.org/wiki/Ohm%27s_law*

7. *https://en.wikipedia.org/wiki/Fourier_analysis*

8. *https://en.wikipedia.org/wiki/Oliver_Heaviside*

9. *https://en.wikipedia.org/wiki/Arthur_E._Kennelly*

10. Scott, W.T., The Physics of Electricity and Magnetism, John Wiley and Sons, Second Edition, 1966, Chapter 9.2-9.4, pp 461-485

11. *https://en.wikipedia.org/wiki/Angular_frequency*

12. *https://en.wikipedia.org/wiki/Electrical_reactance*

13. *https://en.wikipedia.org/wiki/Skin_effect*

14. *https://en.wikipedia.org/wiki/Permittivity*

15. *https://en.wikipedia.org/wiki/Decibel*

16. COMSOL Reference Manual, p. 300

17. COMSOL Reference Manual, p. 320

18. *https://en.wikipedia.org/wiki/Adolf_Eugen_Fick*

19. *https://en.wikipedia.org/wiki/Fick%27s_laws_of_diffusion*

20. *https://en.wikipedia.org/wiki/Heat_equation*

21. *https://en.wikipedia.org/wiki/Thermal_conductivity*

22. *https://en.wikipedia.org/wiki/Density*

23. *https://en.wikipedia.org/wiki/Table_of_specific_heat_capacities*

SUGGESTED MODELING EXERCISES

1. Build, mesh, and solve the 2D Electric Impedance Sensor Model as presented earlier in this chapter.

2. Build, mesh, and solve the 2D Metal Layer on a Dielectric Block Model as presented earlier in this chapter.

3. Change the values of the materials parameters and then build, mesh, and solve the 2D Electric Impedance Sensor Model as an example problem.

4. Change the values of the materials parameters and then build, mesh, and solve the 2D Metal Layer on a Dielectric Block Model as an example problem.

5. Change the value of the boundary temperatures and then build, mesh, and solve the 2D Metal Layer on a Dielectric Block Model as an example problem.

6. Change the value of the geometries of the layer and the block and then solve the 2D Metal Layer on a Dielectric Block Model as an example problem.

CHAPTER 8

2D COMPLEX MIXED MODE MODELING USING COMSOL MULTIPHYSICS 5.x

In This Chapter

- Guidelines for 2D Complex Mixed Mode Modeling in 5.x
  - 2D Complex Mixed Mode Modeling Considerations
- 2D Complex Mixed Mode Models
  - 2D Finding the Impedance of a
  - Two (2) Wire, Parallel-Wire, Transmission Line
  - 2D Finding the Impedance of a Concentric,
  - Two (2) Wire, Transmission Line (Coaxial Cable)
  - 2D Axisymmetric Transient Modeling of a Coaxial Cable
- First Principles as Applied to 2D Complex Mixed Mode Model Definition
- References
- Suggested Modeling Exercises

GUIDELINES FOR 2D COMPLEX MIXED MODE MODELING IN 5.x

NOTE *In this chapter, complex mixed-mode 2D models will be presented. Such 2D models are typically more conceptually complex than the models that were presented in earlier chapters of this text. 2D complex mixed-mode models have proven to be very valuable to the science and engineering communities, both in the past and currently, as first-cut evaluations of potential systemic physical behavior under the influence of mixed external stimuli. The 2D mixed-mode model responses and other such ancillary information can be gathered and screened early in a project for a first-cut evaluation. That initial information can potentially be used later as guidance in building higher-dimensionality (3D) field-based (electrical, magnetic, etc.) models, as needed.*

Since the models in this and subsequent chapters are more complex and more difficult to solve than the models presented thus far, it is important that the modeler have available the tools necessary to most easily utilize the powerful capabilities of the 5.x software

In order to do that, if you have not done this previously, the modeler should go to the main 5.x toolbar, Click > Options – Preferences – Show More Options. When the Preferences – Show More Options edit window is shown, Select > Equation view checkbox and Equation Section checkbox. Click > OK.

2D Complex Mixed Mode Modeling Considerations

2D Models can in some cases be less difficult than some 1D or 3D models, having fewer implicit assumptions. Potentially, 2D models can still be a very challenging type of model to build, depending on the underlying physics. The least difficult aspect of 2D model creation arises from the fact that the geometry is relatively simple. However, the physics in a 2D complex mixed-mode model can range from relatively simple to extremely complex. This is especially true of the RF Module models featured in this chapter.

In compliance with the laws of physics, a 2D model implicitly assumes that energy flow, materials properties, environment, and all other conditions and variables that are of interest are homogeneous, isotropic, and/or constant, unless otherwise specified. This condition is assumed to be true throughout the entire domain of interest both within the model and through the boundary conditions and in the environs of the model.

The model builder needs to bear the above-stated conditions in mind and carefully ensure that all of the modeling conditions and associated parameters (default settings) in each model created are properly considered, defined, verified, and/or set to the appropriate values.

It is always mandatory that the modeler be able to accurately anticipate the expected results of the model and accurately specify the manner in which those results will be presented. Never assume that any of the default values that are present when the model is created necessarily satisfy the needs or conditions of a particular model.

NOTE   *Always verify that any parameters employed in the model are the correct value needed for that model. Calculated solutions that significantly deviate from the anticipated solution or from a comparison of values measured in an experimentally derived realistic model are probably indicative of one or more modeling errors either in the original model design, in the earlier model analysis, in the understanding of the underlying physics, or are simply due to human error.*

2D Coordinate System

In a 2D model, if parameters can only vary as a function of the position (x) and (y) coordinates, then such a 2D model represents the parametric condition of the model in a time-independent mode (stationary). In a time-dependent study or frequency domain study model, parameters can vary both with position in (x) and/or (y) and with time (t).

See Figure 8.1.

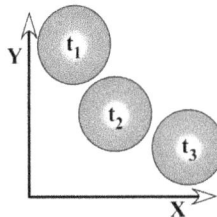

FIGURE 8.1 2D Coordinate system, plus time.

Figure 8.1 shows the 2D coordinate system, plus time.

In this chapter, three fundamental concepts are introduced to the modeler: the impedance of a simple, unshielded, air-dielectric, two-wire transmission line {8.2}, the characteristic impedance of a coaxial cable {8.3} and the transient response of a coaxial cable {8.4}.

Radio Frequency Waves

The electromagnetic transmission methods that are utilized in the currently presented material are an outgrowth of the discoveries made during the underlying scientific work done in the study of the fundamentals of electromagnetic theory {8.6} by James Clerk Maxwell (1865){8.7} and functional application of Radio Waves by Gugliemo Marconi (1897) {8.8}.

NOTE *There is a very large literature on Radio Frequency Engineering {8.9} and on Radio Frequency Systems {8.10}. Exploration of that literature is left to the modeler.*

2D COMPLEX MIXED MODE MODELS USING THE RF MODULE

Finding the Impedance of a Two (2) Wire, Parallel-Wire, Air-Dielectric, Transmission Line

Building a Two (2) Wire, Parallel-Wire, Air-Dielectric, Transmission Line Model

NOTE *In this complex mixed-mode 2D model, the basics of a parallel 2-wire, air-dielectric, flat (not twisted), the transmission line is presented. In order to simplify this First Approximation Model, it is assumed that the wires and media are each perfectly flat and that both extend to Infinity (eliminates the end effects). Once the calculated solution has been obtained, that solution is then compared to the analytic solution for the same problem. This comparison demonstrates the proximity of the solutions obtained by the two different paths.*

This 2D model (derived from COMSOL Model 12403) is somewhat more conceptually complex than the models that were previously presented in the earlier chapters of this text. The physics of a parallel-wire transmission line dictates that any applied signal must propagate in the TEM (Transverse Electromagnetic mode). That statement means that E_z and B_z both vanish uniquely (i.e. all the Electric (E) and Magnetic (B) fields lie in the X-Y Plane).

Figure 8.1a shows a cross-section of a two-wire transmission line, embedded in a circular air domain.

See Figure 8.1a.

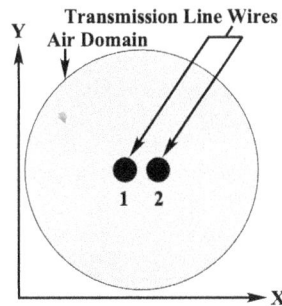

FIGURE 8.1a Two-wire 2D transmission line and domain.

Figure 8.1a shows Two-Wire 2D Transmission Line and Domain.

It is assumed herein, for the purposes of this model, that the air surrounding the wires is a perfect insulator (no leakage current) and that the air is surrounded by a perfect magnetic conductor (PMC) (no flux leakage). Figure 8.1b shows how a Line- Integral can be used to calculate the current flowing through wire 2.

See Figure 8.1b.

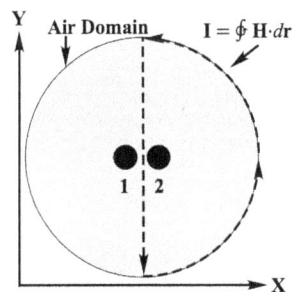

FIGURE 8.1b Wire 2 current flow line integral.

Figure 8.1b shows Wire 2 Current Flow Line Integral.

It is also assumed herein, for the purposes of this model, that the wires are perfect conductors (no internal potential drop). Figure 8.1c shows how a Line- Integral can be used to calculate the potential difference between wire 1 and wire 2.

See Figure 8.1c.

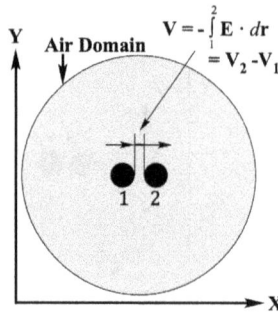

$$V = -\int_1^2 \mathbf{E} \cdot d\mathbf{r}$$
$$= V_2 - V_1$$

FIGURE 8.1c Wire 1 - Wire 2 potential difference line integral.

Figure 8.1c shows Wire 1 - Wire 2 Potential Difference Line Integral.

Startup 5.x, Click > Model Wizard button.

Select Space Dimension > 2D.

Click > Twistie for Radio Frequency.

Select: > Electromagnetic Waves, Frequency Domain (emw).

Click > Add.

Click > Study (Right Pointing Arrow button).

Select > Preset Studies for Selected Physics interfaces > Mode Analysis.

Click > Done.

Save > MM2E5X_2D_TWTL_RF_1.mph

See Figure 8.2.

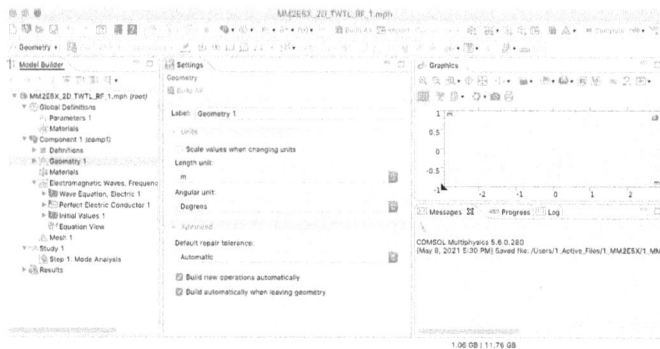

FIGURE 8.2 MM2E5X_2D_TWTL_RF_1 Desktop.

Figure 8.2 shows MM2E5X_2D_TWTL_RF_1 Desktop.

Global Definitions

Parameters 1

In the Model Builder window, under Global Definitions,

Click > Parameters 1.

In the Settings Parameters window,

Enter the Parameters contained in Table 8.1.

NOTE *Numerical parameter values are followed by [units] except when the numerical value is dimensionless. Then the numerical parameter is followed by [1].*

TABLE 8.1 Parameters.

Name	Expression	Value	Description
r_a	1.0[mm]	0.001[m]	Wire, radius
r_d	4.0[mm]	0.004[m]	Center-to-center spacing
r_air	5*r_d	0.02[m]	Air-domain radius
Z0_analytic	(Z0_const/pi)* log(r_d/r_ a+ sqrt((r_d/r_a)^2-1))	247.44[Ohm]	Characteristic impedance, analytic

NOTE *In this model, a predefined COMSOL constant (Z0_const) for the characteristic impedance of vacuum (free space) {8.11} is employed.*

See Figure 8.3.

FIGURE 8.3 Parameters entry windows.

Figure 8.3 shows the Parameters entry windows.

Geometry 1

NOTE *A sequence of geometrical figures will now be selected to create the geometric elements of this 2D Model.*

Circle 1 (c1)

Right-Click > Geometry 1,

Select: Circle from the pop-up menu.

In the Settings Circle window > tab Size and Shape,

Enter: r_air in the Radius text field.

Click > Build Selected in the window tool bar.

See Figure 8.4.

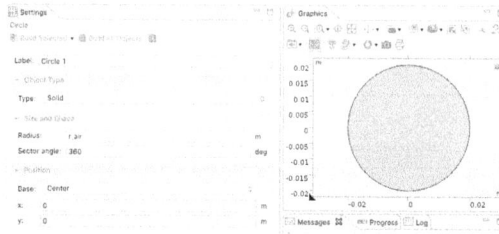

FIGURE 8.4 Circle 1 and air domain parameters.

Figure 8.4 shows Circle 1 and Air Domain Parameters.

Circle 2 (c2)

Right-Click > Geometry 1,

Select: Circle from the pop-up menu.

In the Settings Circle window > Size and Shape tab,

Enter: r_a in the Radius text field.

In the Settings Circle window > Position tab,

Enter: r_d in the x text field.

Click > Build Selected in the window tool bar.

See Figure 8.5.

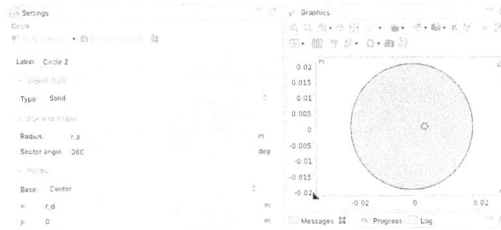

FIGURE 8.5 Circle 1 and circle 2.

Figure 8.5 shows Circle 1 and Circle 2.

NOTE *The second wire of the Two-Wire Transmission Line (TWTL) will be created by using the mirroring operation.*

Mirror 1 (mir1)

Click > Geometry 1.

Right-Click > Geometry 1,

Select: Transforms > Mirror from the pop-up menu.

In the Settings Mirror window, in the Input tab,

Use the Clipboard to enter c2 in the Input objects: window.

Select the Keep input objects check box, by clicking.

See Figure 8.6.

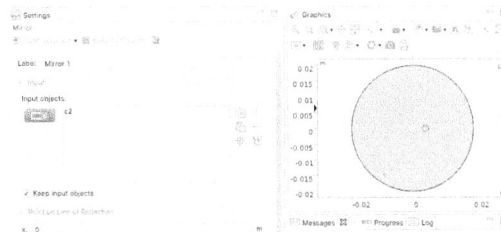

FIGURE 8.6 Settings mirror window.

Figure 8.6 shows the Settings Mirror window.

Click > Build Selected

See Figure 8.7.

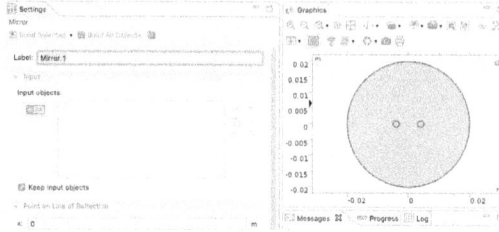

FIGURE 8.7 Settings mirror window after build.

Figure 8.7 shows the Settings Mirror window After Build.

Difference 1 (dif1)

Click > Geometry 1.

Right-Click > Geometry 1,

Select: Booleans and Partitions > Difference.

Use the Clipboard to enter c1 in the Objects to add: window.

In the Settings Difference window,

Click > Difference > Objects to subtract: Activator button.

Use the Clipboard to enter c2 and mir1 in the Objects to subtract: window.

Click > Build Selected

See Figure 8.8.

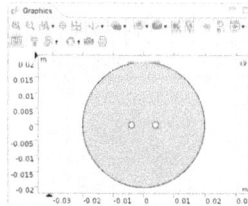

FIGURE 8.8 Difference graphics window after build.

Figure 8.8 shows the Difference Graphics window After Build.

NOTE   *The integration line will next be created for the Electric Field line integral to determine the potential difference.*

Line Segment 1 (ls1)

Click > Geometry 1.

Right-Click > Geometry 1,

Select: Line Segment.

In Settings Line Segment > Starting Point,

Click > Specify: >

Select: Coordinates from the pull-down menu.

In Settings Line Segment > Starting Point > x: text field,

Enter: -r_d+r_a

In the Settings Line Segment window > Endpoint,

Click > Specify: >

Select: Coordinates from the pull-down menu.

In Settings Line Segment > Endpoint > x: text field,

Enter: r_d-r_a

Click > Build Selected.

See Figure 8.9.

FIGURE 8.9 Voltage line integral parameters settings window.

Figure 8.9 shows the Voltage Line Integral Parameters Settings Window.

NOTE
The integration line will next be created for the Magnetic Field line integral to determine the current flow.

Line Segment 2 (ls2)

Click > Geometry 1.

Right-Click > Geometry 1,

Select: Line Segment.

In Settings Line Segment > Starting Point,

Click > Specify: > Choose: Coordinates from the pull-down menu.

In Settings Line Segment > Starting Point > y: text field,

Enter: -r_air

In the Settings Line Segment window > Endpoint,

Click > Specify: >

Select: Coordinates from the pull-down menu.

In Settings Line Segment > Endpoint > y: text field,

Enter: r_air

Click > Build Selected

See Figure 8.10.

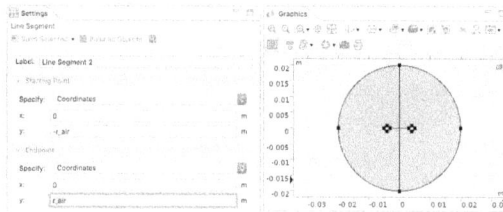

FIGURE 8.10 Voltage and current line integral build.

Figure 8.10 shows the Voltage and Current Line Integral Build.

DEFINITIONS

NOTE

This section of the model adds a variable for the Characteristic Impedance computed as the Voltage (between the wires) divided by the Current (through the wires) (i.e. Z0 = V/I). It also defines two integration coupling operators that enable the computation of the Voltage(V) and the Current (I).

Integration 1 (intop1)

Click > Component 1 (comp1) > Definitions.

Right-Click > Definitions > Nonlocal Couplings,

Select: Integration.

In the Settings Integration window > Operator name:

Enter: int_E in the text field.

In the Settings Integration window > Source Selection,

Click > Geometric entity level:,

Select: Boundary from the pop-up menu.

Using the Clipboard,

Enter: 1 and 4 in the text field.

See Figure 8.11.

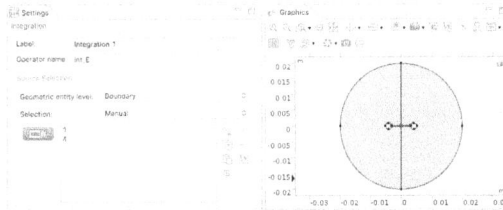

FIGURE 8.11 Settings integration 1 window parameters.

Figure 8.11 shows the Settings Integration 1 Window Parameters.

Integration 2 (intop2)

Click > Component 1 (comp1) > Definitions.

Right-Click > Definitions > Nonlocal Couplings,

Select: Integration.

In the Settings Integration window > Operator name:

Enter: int_H in the text field.

In the Settings Integration window > Source Selection,

Click > Geometric entity level:,

Select: Boundary from the pop-up menu.

Using the Clipboard,

Enter: 2, 3, 11, 12 in the text field.

NOTE *Because this model employs the PMC (Perfect Magnetic Conductor) boundary condition, there is no tangential H field on the outermost boundary (no leakage field). Thus, boundaries 11 and 12 could nominally be ignored.*

See Figure 8.12.

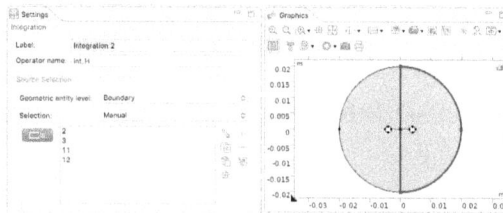

FIGURE 8.12 Settings integration 2 window parameters.

Figure 8.12 shows the Settings Integration 2 Window Parameters.

STUDY 1

Step 1: Mode Analysis

In Model Builder, Click > Study 1 twistie (as needed).

Click > Step 1: Mode Analysis.

In the Settings Mode Analysis window,

In Settings Mode Analysis > Study Settings,

Select: (by clicking) desired number of modes check box,

Enter: 1 in the associated text field.

See Figure 8.13.

FIGURE 8.13 Settings mode analysis window parameters.

Figure 8.13 shows the Settings Mode Analysis Window Parameters.

DEFINITIONS

Variables 1

Click > Definitions,

Right-Click > Definitions > Variables

TABLE 8.2 Parameters.

Name	Expression	Unit	Description
V	int_E(-emw.Ex*t1x-emw.Ey*t1y)	V	Voltage
I	-int_H(emw.Hx*t1x+emw.Hy*t1y)	A	Current
Z_model	V/I	Ohm	Characteristic impedance

See Figure 8.14.

FIGURE 8.14 Settings variables window parameters.

Figure 8.14 shows the Settings Variables Window Parameters.

NOTE

In the above Table 8.2, t1x and t1y are the tangential vector components along the integration boundaries (1 designates the boundary dimension). The emw. prefix indicates the correct physics-interface designator for the electric and magnetic field vector components.

ELECTROMAGNETIC WAVES, FREQUENCY DOMAIN (EMW)

Next, set up the Physics. The default boundary condition is as a Perfect Electric Conductor. Then, override the outermost boundaries with that of a Perfect Magnetic Conductor (PMC) condition to approximate an essentially infinite modeling space.

Perfect Magnetic Conductor 1

In the Model Builder, under Component 1 (comp1),

Right-Click > Electromagnetic Waves, frequency Domain (emw)

Select: Perfect Magnetic Conductor from the pop-up menu.

Enter: 5, 6, 11, 12 using the Clipboard.

See Figure 8.15.

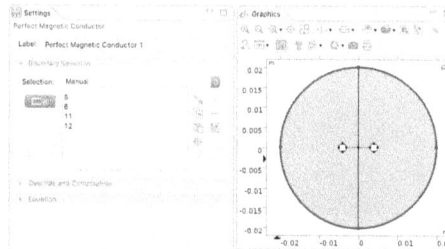

FIGURE 8.15 Settings PMC window parameters.

Figure 8.15 shows the Settings PMC Window Parameters.

MATERIALS

In Model Builder under Component 1 (comp1),

Click > Materials.

Right-Click > Materials,

Select: Add Material from Library.

In the Add Material window,

Click > Built-in twistie.

Select: Built-in > Air

See Figure 8.16.

FIGURE 8.16 Settings materials add air.

Figure 8.16 shows the Settings Materials Add Air.

Click > Add to Component.

Click > Add Material close button (hollow X).

MESH 1

NOTE *Next, set up the Mesh. The default mesh is suitable.*

In the Model Builder window, under Component 1 (comp1),

Click > Mesh 1.

In the Settings Mesh window, Verify: On the Physics-Controlled Mesh tab,

Contributor text box

Electromagnetic Waves, Frequency Domain (emw) check box > Selected (checked).

Click > Build All.

See Figure 8.17.

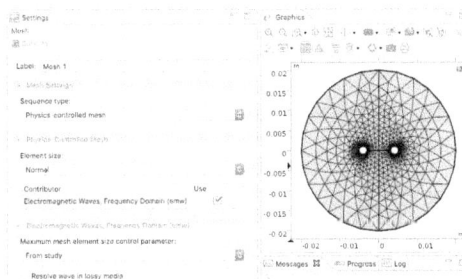

FIGURE 8.17 Settings mesh parameters.

Figure 8.17 shows the Settings Mesh Parameters.

NOTE *The Mesh contains 1326 domain elements and 108 boundary elements.*

STUDY 1

Click > Study 1.

Right-Click > Study 1,

Select: Compute from the pop-up menu.

See Figure 8.18.

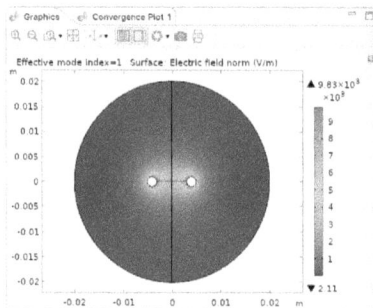

FIGURE 8.18 Default electric field plot.

Figure 8.18 shows the Default Electric Field Plot.

RESULTS

Electric Field (emw)

NOTE *The Default plot shows the distribution of the norm of the electric field. Next, an arrow plot will be added for the magnetic field.*

Arrow Surface 1

Under Results, in the Model builder,

Right-Click > Electric Field (emw),

Select: Arrow Surface from the pop-up menu.

In the Settings Arrow Surface > Arrow Positioning,

Enter: 30 in the X grid points > Points text field.

Enter: 30 in the Y grid points > Points text field.

See Figure 8.19.

FIGURE 8.19 Magnetic field plot grid parameters.

Figure 8.19 shows the Magnetic Field Plot Grid Parameters.

Right-Click > Electric Field (emw),

Select: Plot from the pop-up menu

In Results > Electric field (emw) > Arrow Surface,

Click > Arrow Surface.

In the Settings Arrow Surface window > Coloring and Style,

Click > Color

Select: Yellow from the pop-up menu.

Select the Scale factor Check Box,

Enter: 2e-4 in the associated text field.

See Figure 8.20.

FIGURE 8.20 Magnetic field arrow parameters.

Figure 8.20 shows the Magnetic Field Arrow Parameters.

NOTE *Next, an Arrow Plot is added for the Tangent Vector Field (TVF) along the boundaries to display the orientation of the TVF.*

Arrow Line 1

In Model Builder under Results, Click > Electric Field (emw).

Right-Click > Electric Field (emw),

Select: Arrow Line from the pop-up menu.

In the Settings Arrow Line window, on the Expression tab,

Click > Replace Expression,

Select: Component 1 > Geometry > tx, ty, Tangent.

In Model Builder under Results, Click > Electric Field (emw).

Right-Click > Electric Field (emw),

Select: Plot.

See Figure 8.21.

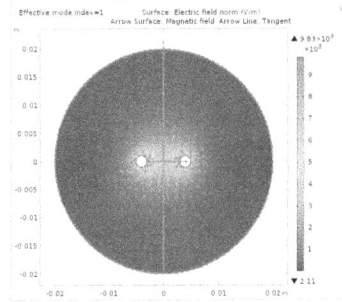

FIGURE 8.21 Magnetic field arrow line: tangent.

Figure 8.21 shows the Magnetic Field Arrow Line: Tangent.

In the Settings Arrow Line window, on the Arrow Positioning tab,

Enter: 100 in the Number of arrows: text field.

Right-Click > Electric Field (emw),

Select: > Plot.

See Figure 8.22.

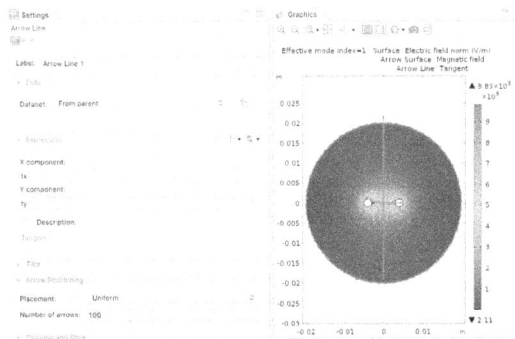

FIGURE 8.22 Magnetic field arrow line: number of arrows.

Figure 8.22 shows the Magnetic Field Arrow Line: Number of arrows.

NOTE

Next, as shown in Figure 8.1c, the line integral for voltage computes the potential difference $V_2 - V_1$. When computing the line integral for current, the right-hand (clockwise) rotational direction of the integration contour indicates a positive current in the negative Z direction (into the modeling plane). Thus, a minus sign is needed to reverse the current direction.

Derived Values

NOTE *Next, finish the Model by computing the characteristic impedance of the TWTL.*

Global Evaluation 1

In Model Builder,

Click > Results twistie (as needed).

Right-Click > Results > Derived Values,

Select: Global Evaluation.

TABLE 8.3 Characteristic Impedance Parameters.

Expression	Unit	Description
Z_model	[Omega]	Characteristic impedance

In the Settings Global Evaluation window,

Enter: The items found in Table 8.3 in the table on the Expressions tab.

Click > Evaluate in the Settings Global Evaluation toolbar.

See Figure 8.23.

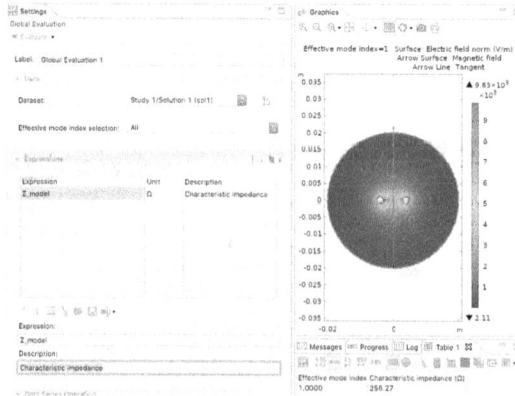

FIGURE 8.23 Characteristic impedance evaluation.

Figure 8.23 shows the Characteristic Impedance Evaluation.

NOTE *The characteristic impedance of the TWTL is evaluated to be approximately 256 Ohms. For an estimate that is more accurate, the size of the Air Domain can be expanded.*

2D Finding the Impedance of a Two (2) Wire, Parallel-Wire, Air-Dielectric, Transmission Line Model Summary and Conclusions

The Finding of the Impedance of a Two (2) Wire, Parallel-Wire, Air-Dielectric, Transmission Line Model is a powerful modeling tool that can be used to calculate the impedance of various two-wire transmission line configurations. With this model, the modeler can easily vary all of the geometric parameters and optimize the design before the first prototype transmission line is physically built. The technique has broad application support in industry.

Coaxial Cables

2D Finding the Impedance of a Concentric, Two (2) Wire, Transmission Line (Coaxial Cable)

In this section of Chapter 8, the Modeler will explore the characteristic impedance of a coaxial cable. Coaxial cables were first used in 1858 {8.12} in the first transatlantic cable. The theory explaining the behavior of coaxial cables was not available until twenty- two years later, when that theory was first developed by Oliver Heaviside {8.13} in 1880.

NOTE

In this complex mixed mode 2D model, the basics of a coaxial transmission line are presented. This 2D model (derived from COMSOL Model 12351) is somewhat more conceptually complex than the models that were previously presented in the earlier chapters of this text. The physics of a coaxial cable transmission line dictates that any applied signal must propagate in the TEM (Transverse Electromagnetic mode). That statement means that E_z and B_z both vanish uniquely (i.e. all the Electric (E) and Magnetic (B) fields lie in the X-Y Plane).

Building a Concentric, Two (2) Wire, Transmission Line (Coaxial Cable) Model

Figure 8.24 shows a cross-section of a Concentric, Two (2) Wire, Transmission Line (Coaxial Cable).

See Figure 8.24.

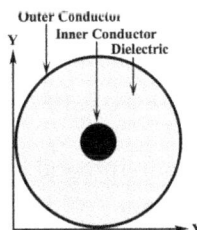

FIGURE 8.24 A concentric, two (2) wire, transmission line (Coaxial cable).

Figure 8.24 shows a Concentric, Two (2) Wire, Transmission Line (Coaxial Cable).

It is assumed herein, for the purposes of this model, that the dielectric surrounding the inner conductor is a lossless dielectric (no leakage current). It is also assumed that both the inner and outer conductors are perfect conductors.

Figure 8.25 shows how a Line-Integral {8.14} can be used to calculate the voltage between the inner and outer conductors.

See Figure 8.25.

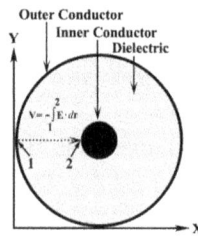

FIGURE 8.25 Voltage between the inner and outer conductors line integral.

Figure 8.25 shows Voltage Between the Inner and Outer Conductors Line Integral.

Figure 8.26 shows a Contour-Integral {8.15} being used to calculate the current flow through the coaxial cable.

See Figure 8.26.

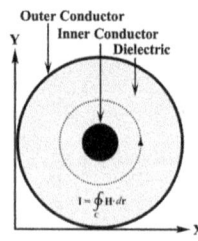

FIGURE 8.26 Contour-integral being used to calculate the current flow through the coaxial cable.

Figure 8.26 shows Contour-Integral being used to calculate the current flow through the coaxial cable.

Building the Concentric, Two (2) Wire, Transmission Line (Coaxial Cable) Model.

Startup 5.x, Click > Model Wizard button.

Select Space Dimension > 2D.

Click > Twistie for Radio Frequency.

Select: > Electromagnetic Waves, Frequency Domain (emw).

Click > Add.

Click > Study (Right Pointing Arrow button).

Select > Preset Studies for Selected Physics interfaces > Mode Analysis.

Click > Done.

Save > MM2E5X_2D_CCTL_RF_1.mph

See Figure 8.27.

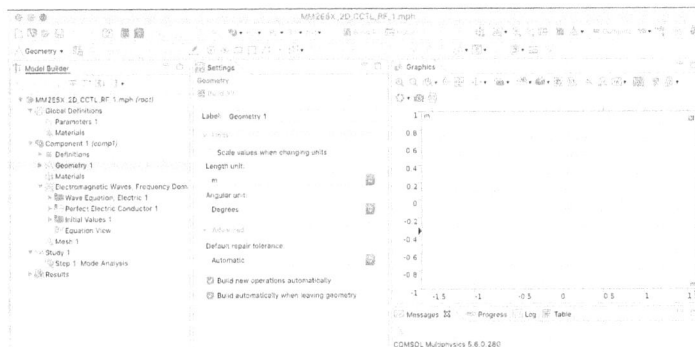

FIGURE 8.27 MM2E5X_2D_CCTL_RF_1 Desktop.

Figure 8.27 shows MM2E5X_2D_CCTL_RF_1 Desktop.

Global Definitions

Parameters 1

In the Model Builder window, under Global Definitions,

Click > Parameters 1.

In the Settings Parameters window,

Enter the Parameters contained in Table 8.4.

NOTE *Numerical parameter values are followed by [units] except when the numerical value is dimensionless. Then the numerical parameter is followed by [1].*

TABLE 8.4 Parameters.

Name	Expression	Value	Description
r_i	0.5[mm]	5e-4 m	Coax inner radius
r_o	3.43[mm]	0.00343 m	Coax outer radius
eps_r	2.4[1]	2.4	Relative dielectric constant
Z0_analytic	(Z0_const/(2*pi*sqrt(eps_r)))* log(r_o/r_i)	74.531 Ohms	Characteristic impedance, analytic

NOTE *Herein Z0_const is a predefined COMSOL constant for the characteristic impedance of vacuum (empty space), Z_0=sqrt(12.566e-7[Henries/m]/8.854e-12 [Farads/m]) {8.16}. Z_0 is approximately 377 Ohms.*

See Figure 8.28.

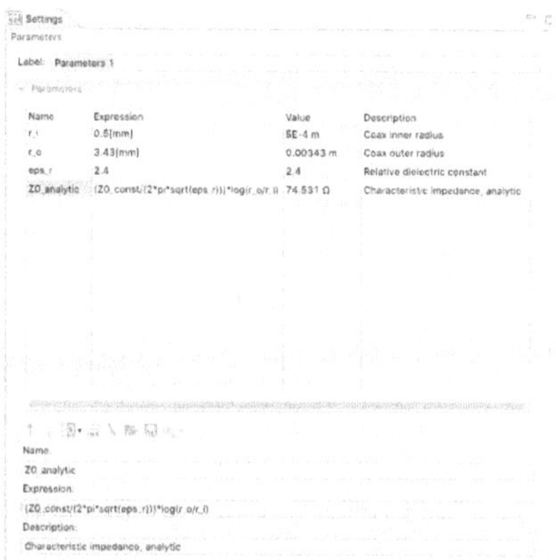

FIGURE 8.28 Settings parameters 1 entry table.

Figure 8.28 shows Settings Parameters 1 Entry Table.

Geometry 1

NOTE *A sequence of geometrical figures will now be selected to create the geometric elements of this 2D Model.*

Circle 1 (c1)

Right-Click > Geometry 1,

Select: Circle from the pop-up menu.

In the Settings Circle window,

Object Type tab,

Click > Type list.

Select: Curve from the pop-up menu.

In the Size and Shape tab,

Enter: r_o in the Radius text field.

Click > Layers tab twistie (as needed),

Enter: r_o-r_i in the Layer 1 Thickness column.

Click > Build Selected in the Settings Circle window tool bar.

See Figure 8.29.

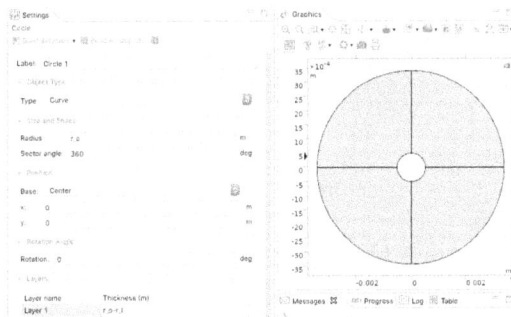

FIGURE 8.29 Circle 1 parameters.

Figure 8.29 shows Circle 1 Parameters.

NOTE *One of the advantages of using Layers to build this model is that Layers automatically generates a radial line, that can be used for the line integral of the electric field.*

MATERIALS

NOTE *Here, the Modeler defines a suitable dielectric material for the coaxial cable.*

In Model Builder,

under Component 1 (comp1),

Click > Materials.

Right-Click > Materials,

Select: Blank Material.

In the Settings Material window,

Enter: Insulator in the Label text field.

Enter: eps_r, 1, 0 in the Value column of the Settings Material > Material Contents tab > Property parameters table.

See Figure 8.30.

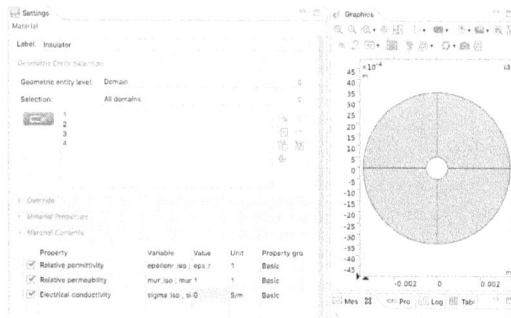

FIGURE 8.30 Dielectric material parameters.

Figure 8.30 shows Dielectric Material Parameters.

DEFINITIONS

NOTE

At this point, the Modeler needs to add a variable for the characteristic imped-ance that is computed by calculating the voltage between the inner and outer conductors divided by the current that is flowing through the cable. For this purpose, the Modeler needs to define two integration coupling operators.

Integration 1 (intop1)

In the Model Builder, under Component 1 (comp1),

Click > Definitions.

Right-Click > Definitions,

Select: Nonlocal Couplings > Integration from the pop-up menu.

In the Settings Integration window,

Enter: int_rad in the Operator name text field

In the Source Selection tab,

Click > Geometric entity level: list,

Select: Boundary from the pop-up list.

Using the Clipboard,

Enter: 1 in the text field.

See Figure 8.31.

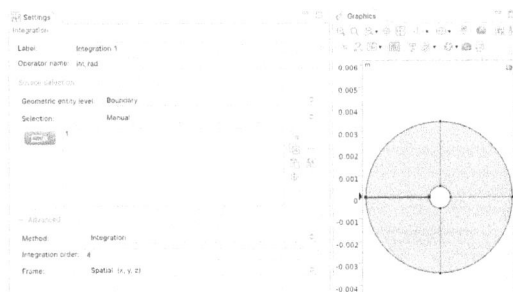

FIGURE 8.31 Integration 1 parameters.

Figure 8.31 shows Integration 1 Parameters.

Integration 2 (intop2)

In the Model Builder, under Component 1 (comp1),

Click > Definitions.

Right-Click > Definitions,

Select: Nonlocal Couplings > Integration from the pop-up menu.

In the Settings Integration window,

Enter: int_circ in the Operator name text field

In the Source Selection tab,

Click > Geometric entity level: list,

Select: Boundary from the pop-up list.

Using the Clipboard,

Enter: 5, 6, 9, 12 in the text field.

See Figure 8.32.

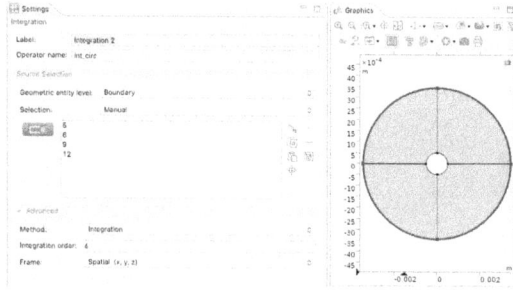

FIGURE 8.32 Integration 2 parameters.

Figure 8.32 shows Integration 2 Parameters.

Next, the Modeler needs to define the variable for the characteristic imped-ance that is computed using the Model.

Variables 1

In the Model Builder, under Component 1 (comp1),

Click > Definitions.

Right-Click > Definitions,

Select: Variables from the pop-up menu.

Enter: the variables shown in Table 8.5, in the Settings Variables > Variables 1 text field

TABLE 8.5 Variables 1.

Name	Expression	Unit	Description
V	int_rad(-emw.Ex*tlx-emw.Ey*tly)	V	Voltage
I	-int_circ(emw.Hx*tlx+emw.Hy*tly)	A	Current
Z0_model	V/I	Ohms	Characteristic impedance

See Figure 8.33.

FIGURE 8.33 Variables 1 parameters.

Figure 8.33 shows Variables 1 Parameters.

NOTE *The parameters t1x and t1y are the tangential vector components along the integration boundaries (where the number (e.g 1) refers to the particular boundary dimension).The signs in the variable parametric definitions are chosen such that the $V = V_1 - V_0$ and so that a positive current value indicates that current is flowing in the positive Z direction. The emw. prefix indicates the correct physics-interface region (scope) for the electric and magnetic field vector components.*

ELECTROMAGNETIC WAVES, FREQUENCY DOMAIN (EMW)

NOTE *Next, the Modeler needs to keep the default physics settings, which include the perfect electric conductor conditions for the outer boundaries.*

MESH 1

In the Model Builder, under Component 1 (comp1),

Click > Mesh 1.

In the Settings Mesh window, on the Physics-Controlled Mesh tab,

Verify that,

Electromagnetic Waves, Frequency Domain (emw) is in the Contributor text field,

And that the Use check box is checked.

See Figure 8.34.

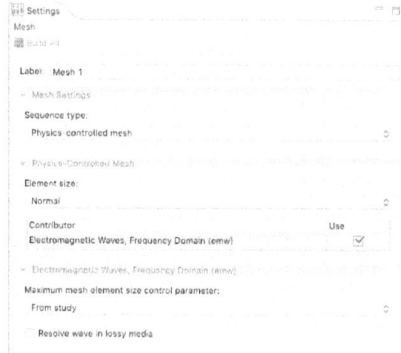

FIGURE 8.34 Mesh 1 parameters.

Figure 8.34 shows Mesh 1 Parameters.

Click > Build All.

See Figure 8.35.

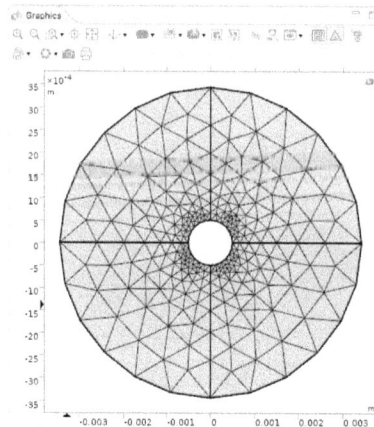

FIGURE 8.35 Mesh.

Figure 8.35 shows Mesh.

NOTE   *The built Mesh comprises 502 domain elements and 76 boundary elements.*

STUDY 1

Step 1: Mode Analysis

In the Model Builder, under Component 1 (comp1),

Click > Study 1 twistie (as needed).

Click > Step 1: Mode Analysis.

In Settings Mode Analysis, Study Settings tab,

Select: Desired number of modes Check box

Enter: 1 in the associated text field.

Select: Search for modes around Check box

Enter: sqrt(eps_r) in the associated text field.

See Figure 8.36.

FIGURE 8.36 Mode analysis parameters.

Figure 8.36 shows Mode Analysis Parameters.

Right-Click > Study 1,

Select: Compute from the pop-up menu.

See Figure 8.37.

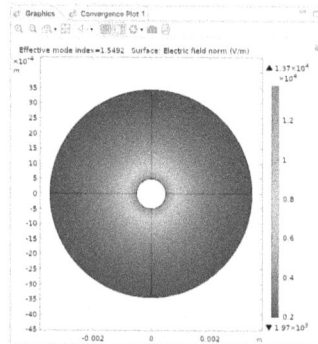

FIGURE 8.37 Default electric field plot.

Figure 8.37 shows Default Electric Field Plot.

RESULTS

Electric Field (emw)

NOTE *The Default Electric Field Plot shows the distribution of the norm of the electric field. Next, the Modeler will add an Arrow plot of the Magnetic Field.*

Arrow Surface 1

In the Model Builder, under Results,

Click > Electric Field (emw).

Right-Click > Electric Field (emw),

Select: Arrow Surface from the pop-up menu.

In the Settings Arrow Surface window, on the Arrow Positioning tab,

Enter: 21 in the X grid points > Points text field.

Enter: 21 in the Y grid points > Points text field.

In the Settings Arrow Surface window, on the Coloring and Style tab,

Select: Scale factor check box,

Enter: 7e-6 in the associated text field.

Click > Color list,

Select: Black from the pop-up list.

In the Model Builder, under Results,

Click > Electric Field (emw).

Click > Plot in the Electric Field (emw) toolbar.

See Figure 8.38.

FIGURE 8.38 Arrow surface parameters.

Figure 8.38 shows Arrow Surface Parameters.

See Figure 8.39.

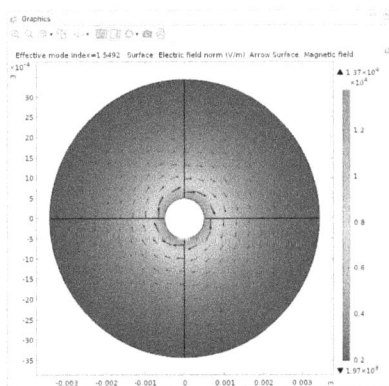

FIGURE 8.39 Electric and magnetic field plots.

Figure 8.39 shows Electric and Magnetic Field Plots.

NOTE *Next, the Modeler will add the integration contour orientations, by plotting the tangent vector (t1) along the boundaries.*

Arrow Line 1

In the Model Builder, under Results,

Click > Electric Field (emw).

Right-Click > Electric Field (emw),

Select: Arrow Line from the pop-up menu.

In the Settings Arrow Line window, in the Expression tab,

Click > Replace Expression (green triangle over red triangle),

Select: Component 1 (comp1) >Geometry > tx, ty – Tangent from the pop-up lists.

In the Settings Arrow Line window, in the Arrow Positioning tab,

Enter: 50 in the Number of arrows text field.

See Figure 8.40.

FIGURE 8.40 Arrow line parameters.

Figure 8.40 shows Arrow Line Parameters.

In the Model Builder, under Results,

Click > Electric Field (emw).

Click > Plot in the Electric Field (emw) toolbar.

See Figure 8.41.

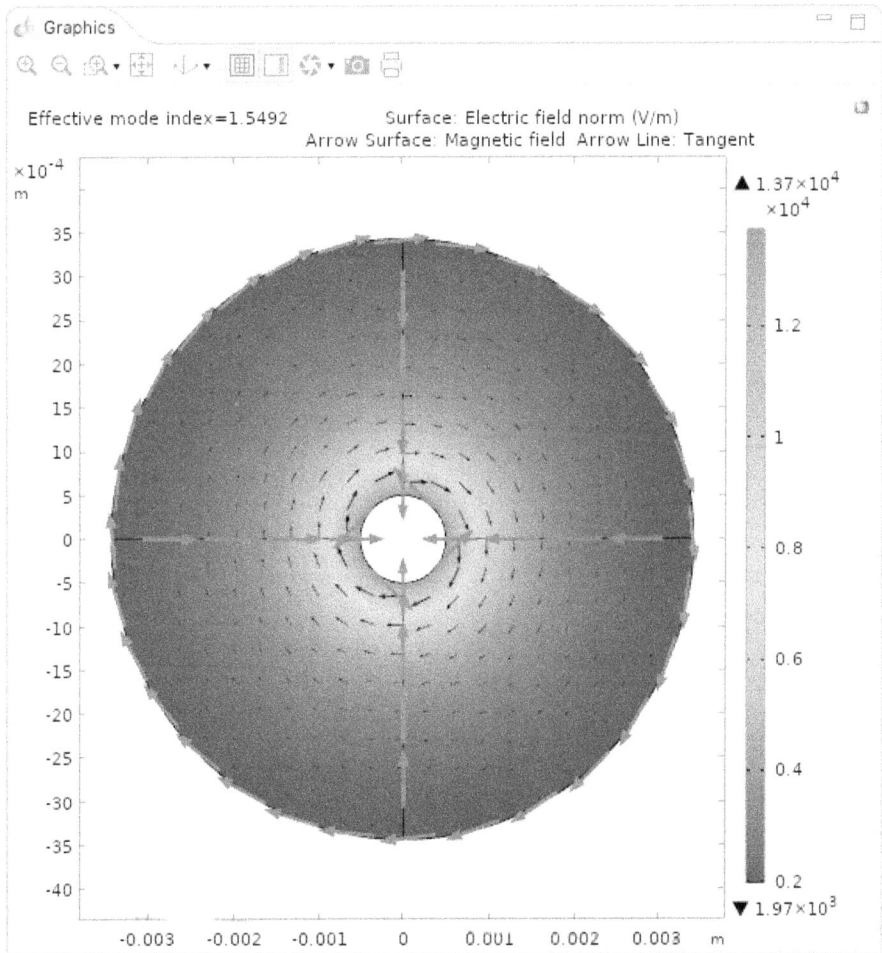

FIGURE 8.41 Arrow line plot.

Figure 8.41 shows Arrow Line Plot.

Right-Click > Arrow Line 1

Select: Disable from the pop-up menu, to retrieve the result plot.

In Model Builder,

Click > Results twistie (as needed).

Right-Click > Results > Derived Values,

Select: Global Evaluation.

In the Settings Global Evaluation window,

Click > Replace Expression,

Select: Component 1 > Definitions > Variables > Z0_model-Characteristic impedance – Ohms.

Enter: Z0_analytic found in Table 8.4 in the table on the Expressions tab.

Click > Evaluate in the Settings Global Evaluation toolbar.

See Figure 8.42.

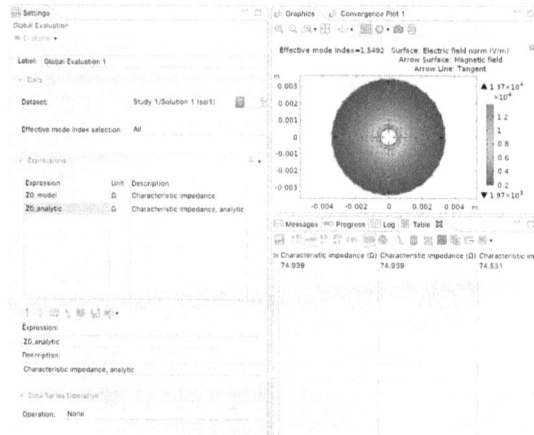

FIGURE 8.42 Characteristic impedance evaluation.

Figure 8.42 shows the Characteristic Impedance Evaluation.

NOTE   *The characteristic impedance of the CCTL is evaluated to be approximately 74.939 Ohms. The Z0_analytic is approximately 74.53 Ohms. The net difference is 0.5%.*

2D Finding the Impedance of a Concentric, Two (2) Wire

Transmission Line (Coaxial Cable) Model Summary and Conclusions

Finding the Impedance of a Concentric, Two (2) Wire, Transmission Line (Coaxial Cable) Model is a powerful modeling tool that can be used to calculate the impedance of various Coaxial Cable transmission line geometries. With this model, the modeler can easily vary all of the geometric parameters and optimize the design before the first prototype transmission line is physically built. The technique has broad application support in industry.

2D Axisymmetric Transient Modeling of a Coaxial Cable

In this section of Chapter 8, the Modeler will explore the transient response of a coaxial cable, using a 2D Axisymmetric Model. As mentioned earlier, coaxial cables were first used in 1858 {8.12} in the first transatlantic cable. The theory explaining the behavior of coaxial cables was not available until twenty-two years later when that theory was first developed by Oliver Heaviside {8.13} in 1880.

NOTE

In this complex mixed-mode 2D Axisymmetric model, the transient response of an air core, coaxial transmission line, under three different load conditions, is built. This 2D Axisymmetric model (derived from COMSOL Model 12349) is somewhat more conceptually complex than the models that were previously presented in the earlier chapters of this text. The physics of a coaxial cable transmission line dictates that any applied signal must propagate in the TEM (Transverse Electromagnetic mode). That statement means that E_z and B_z both vanish uniquely (i.e., all the Electric (E) and Magnetic (B) fields lie in the X-Y Plane).

Schematic of a 2D Axisymmetric, Concentric, Two (2) Wire, Transmission Line (Coaxial Cable) Model

Figure 8.43 shows a schematic of a Concentric, Two (2) Wire, Transmission Line (Coaxial Cable) segment, with a time variable voltage source and variable test load.

See Figure 8.43.

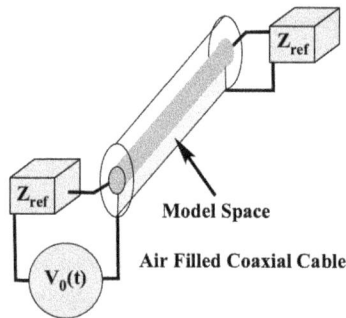

FIGURE 8.43 A concentric, two (2) wire, transmission line (Coaxial cable), segment, with a time variable voltage source and variable test load.

It is assumed herein, for the purposes of this model, that the dielectric surrounding the inner conductor (Air) is a lossless dielectric (no leakage current). It is also assumed that both the inner and outer conductors are perfect conductors.

Three different termination (load) conditions will be applied to the driven Coaxial Cable segment. They are:

1. Perfect Electric Conductor (PEC) - Simulation of a Short Circuit (no resistance).

2. Perfect Magnetic Conductor (PMC) - Simulation of an Open Circuit (no conduction).

3. Lumped Port – Simulation of a Matched Load.

Building the 2D Axisymmetric, Concentric, Two (2) Wire, Transmission Line (Coaxial Cable) Model.

Startup 5.x, Click > Model Wizard button.

Click > 2D Axisymmetric.

In the Select Physics window,

Click > Radio Frequency twistie,

Click > Electromagnetic Waves, Transient (temw).

Click > Add.

See Figure 8.44.

FIGURE 8.44 Select physics window.

Figure 8.44 shows the Select Physics Window.

Click > Study

In the Select Study window,

Click > General Studies > Time Dependent.

See Figure 8.45.

FIGURE 8.45 Select study window.

Figure 8.45 shows the Select Study Window.

Click > Done.

Save > MM2E5X_2DAxi_CCTR_RF_1.mph

See Figure 8.46.

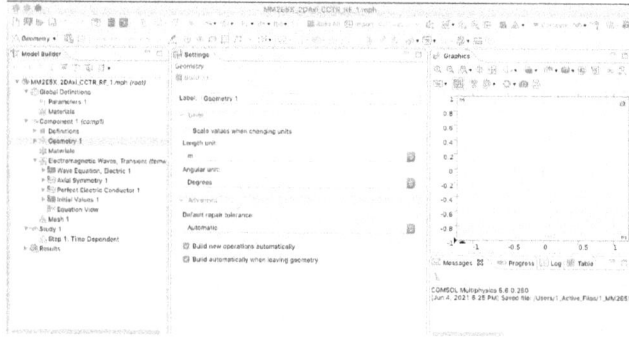

FIGURE 8.46 MM2E5X_2DAxi_CCTR_RF_1 Desktop.

Figure 8.46 shows MM2E5X_2DAxi_CCTR_RF_1 Desktop.

Global Definitions

Parameters 1

In the Model Builder window, under Global Definitions,

Click > Parameters 1.

In the Settings Parameters window, in the Parameters text table,

Enter the Parameter values found in Table 8.6.

TABLE 8.6 Parameters.

Name	Expression	Value	Description
Ir_coax	1[mm]	0.001 m	Coax inner radius
Or_coax	2[mm]	0.002 m	Coax outer radius
L_coax	40[mm]	0.04 m	Length of coax
P_freq	20[GHz]	2e10 Hz	Pulse frequency
W_long	c_const/P_freq	0.01499 m	Wavelength, free space
W_period	1/P_freq	5e-11 s	Period
H_max	min(W_long/8, (Or_coax-Ir_coax)/2)	5e-4 m	Maximum element size

See Figure 8.47.

FIGURE 8.47 Parameters 1, settings parameter table.

Figure 8.47 shows Parameters 1, Settings Parameter Table.

Save: Parameter 1 Table as MM2E5X_2DAxi_CCTR_RF_Param_1.txt

NOTE *Next, the driver excitation is defined in terms of a Gaussian pulse {8.17} and a Sine function {8.18}.*

Define a Gaussian pulse.

Gaussian Pulse 1 (gp1)

In the Model Builder,

Right-Click > Global Definitions,

Select: Functions > Gaussian Pulse from the pop-up menu.

In the Settings Gaussian Pulse window,

Enter: gauss_pulse in the Function name: text field.

In the Settings Gaussian Pulse window, on the Parameters tab,

Enter: 2*W_period in the Location text field.

Enter: W_period/2 in the Standard deviation text field.

See Figure 8.48.

FIGURE 8.48 Settings gaussian pulse parameters.

Figure 8.48 shows Settings Gaussian Pulse Parameters.

NOTE *Next, the Gaussian Pulse will be used to create an analytic function.*

Analytic 1 (an1)

In the Model Builder,

Right-Click > Global Definitions,

Select: Functions > Analytic from the pop-up menu.

In the Settings Analytic window,

Enter: V0 in the Function name: text field.

In the Settings Analytic window, Definition tab,

Enter: gauss_pulse(t)*sin(2*pi*P_freq*t) in the Expression text field.

Enter: t in the Arguments text field.

In the Settings Analytic window, Units tab,

Enter: s in the Arguments text field.

Enter: V in the Function text field.

In the Settings Analytic window, Plot Parameters tab,

Enter: t, 0[ns], 0.2[ns].

See Figure 8.49.

FIGURE 8.49 Settings analytic parameters.

Figure 8.49 shows Settings Analytic Parameters.

In the Settings Analytic toolbar,

Click > Plot

See Figure 8.50.

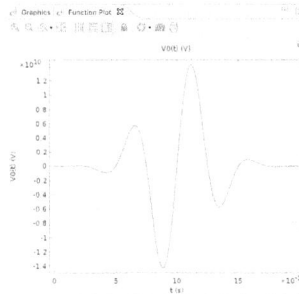

FIGURE 8.50 Pulse profile.

Figure 8.50 shows Pulse Profile.

NOTE *Next, the Modeler will create the Modeling Space.*

GEOMETRY 1

Rectangle 1 (r1)

Click > Geometry 1,

Right-Click > Geometry 1,

Select: Rectangle

In the Settings Rectangle window, Size and Shape tab,

Enter: Or_coax – Ir_coax in the Width text field.

Enter: L_coax in the Height text field.

Enter: Ir_coax in the r: text field on the Position tab.

Click > Build All Objects in the Settings Rectangle toolbar.

See Figure 8.51.

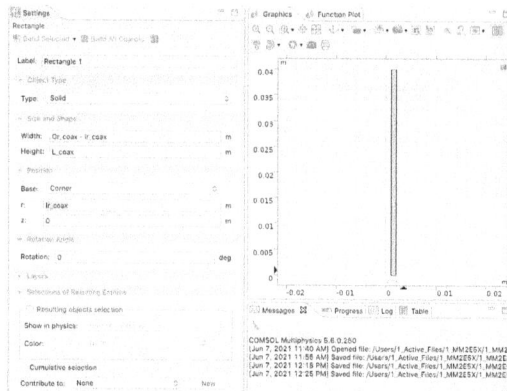

FIGURE 8.51 Settings rectangle parameters and rectangle 1 graphic.

Figure 8.51 shows Settings Rectangle Parameters and Rectangle 1 Graphic.

DEFINITIONS

NOTE

Next, the Modeler will create a point probe to allow plotting the electric field component Er, while the model is being solved. The Probe will also be used to obtain plotted data reflecting coax signal status under different cable termination conditions.

Domain Point Probe 1

In the Model Builder, under Component 1 (comp1),

Click > Definitions.

Right-Click > Definitions,

Select: Probes > Domain Point Probe.

In the Settings Domain Point Probe window,

Enter: Ir_coax in the Point Selection tab > Coordinates: > r: text field.

Select: the Snap to closest boundary check box (by Clicking).

See Figure 8.52.

FIGURE 8.52 Settings domain point probe parameters.

Figure 8.52 shows Settings Domain Point Probe Parameters.

Point Probe Expression 1 (ppb1)

In the Model Builder, under Component 1 (comp1),

Click > Domain Point Probe twistie.

Click > Point Probe Expression (ppb1).

In the Settings Point Probe Expression window,

Click > Replace Expression in the upper right corner of the Expression tab,

Select: Component 1 (comp1) > Electromagnetic Waves, Transient > Electric >

Electric field-V/m > temw.Er-Electric field, r component from the pop-up menus.

See Figure 8.53.

FIGURE 8.53 Settings domain point probe parameters.

Figure 8.53 shows Settings Domain Point Probe Parameters.

ELECTROMAGNETIC WAVES, TRANSIENT (TEMW)

NOTE *Next, the Modeler will set up the necessary physics conditions. The Modeler begins by defining the Lumped port input condition.*

Lumped Port 1

In the Model Builder, under Component 1 (comp1),

Click > Electromagnetic Waves, Transient (temw).

Right-Click > Electromagnetic Waves, Transient (temw),

Select: Lumped Port from the pop-up menu.

Enter: 2 in the Boundary Selection text field using the clip-board.

In the Settings Lumped Port window,

Enter: V0(t) in the V_0 text field, on the Lumped Port Properties tab.

Enter: (Z0_const*pi/2)*log(Or_coax/Ir_coax) in the Z_{ref} text field.

See Figure 8.54.

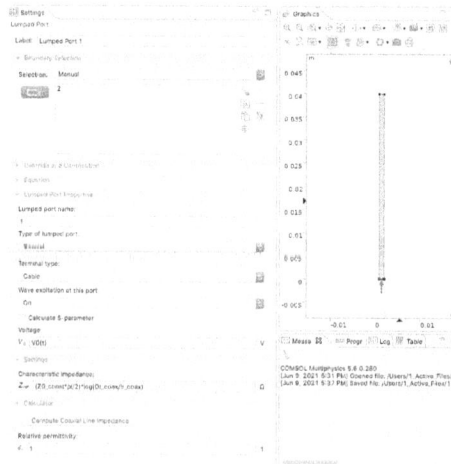

FIGURE 8.54 Settings lumped port parameters and graphic.

Figure 8.54 shows Settings Lumped Port Parameters and Graphic.

NOTE *Next, the Modeler will set up the necessary additional Lumped Port conditions. The Modeler begins by defining the PMC termination condition.*

Perfect Magnetic Conductor 1

Click > Electromagnetic Waves, Transient (temw).

Right-Click > Electromagnetic Waves, Transient (temw),

Select: Perfect Magnetic Conductor from the pop-up menu.

Enter: 3 in the Boundary Selection text field using the clip-board.

NOTE *Next, the Modeler will set up the necessary additional Lumped Port 2 conditions for a matched load termination.*

Lumped Port 2

Click > Electromagnetic Waves, Transient (temw).

Right-Click > Electromagnetic Waves, Transient (temw),

Select: Lumped Port from the pop-up menu.

Enter: 3 in the Boundary Selection text field using the clip-board.

Enter: 42[ohm] in the Z_{ref} text field.

See Figure 8.55.

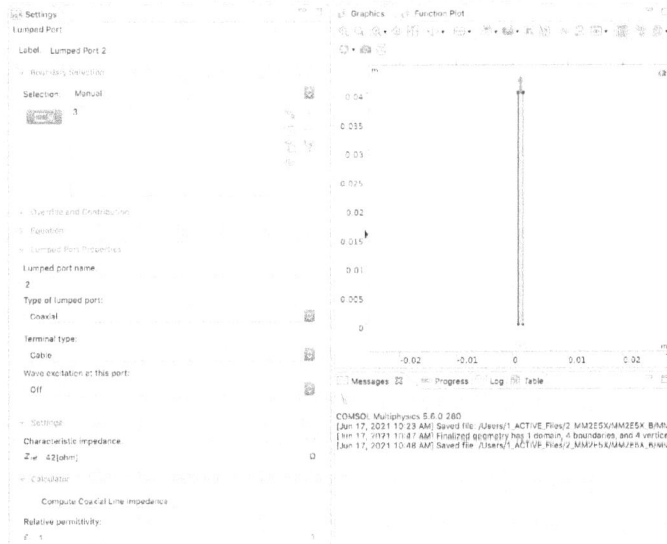

FIGURE 8.55 Settings lumped port 2 parameters and graphic.

Figure 8.55 shows Settings Lumped Port 2 Parameters and Graphic.

ADD MATERIAL

In the Model Builder, under Component 1 (comp1),

Right-Click > Materials,

Select: Add Material from Material Library on the pop-up menu.

Select: Built-in > Air from the Add Material window list.

Click > Add to Component.

Close the Add Material window.

MESH 1

Free Triangular 1

Right-Click > Mesh 1,

Select: Free Triangular.

Size

In the Model Builder window,

Click > Size.

In the Settings Size window,

Select: The Custom button (by clicking).

In the Settings Size window, in the Element Size Parameters tab,

Enter: H_max in the Maximum element size text field

See Figure 8.56.

FIGURE 8.56 Settings size mesh parameters.

Figure 8.56 shows Settings Size Mesh Parameters.

Click > Build All.

See Figure 8.57.

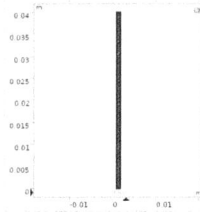

FIGURE 8.57 Mesh.

Figure 8.57 shows Mesh, which contains 482 domain elements and 164 boundary elements.

STUDY 1

Step 1: Time Dependent

In the Model Builder window,

Click > the Study 1 twistie (as needed).

Click > Step 1: Time Dependent.

In the Settings Time Dependent window,

On the Study Settings tab,

Click > Range button,

Enter: Start =0, Step=W_period/24, Stop=10*W_period

See Figure 8.58.

FIGURE 8.58 Range parameters.

Figure 8.58 shows Range Parameters.

Click > Replace button.

Click > Tolerance:

Select: User controlled.

Enter: 0.0001 in the Relative tolerance text field.

> **NOTE**
>
> *In order to study the "short" termination case, first the Modeler will need to disable the PMC and Lumped Port conditions. Once that is done, the default condition becomes PEC ("short") and is activated on the termination boundary.*

In the Settings Time Dependent window, on the Physics and Variables Selection tab,

Click > Modify model configuration for study step check box.

Right-Click > Perfect Magnetic Conductor,

Select: Disable from the pop-up menu.

Right-Click > Lumped Port 2,

Select: Disable from the pop-up menu.

See Figure 8.59.

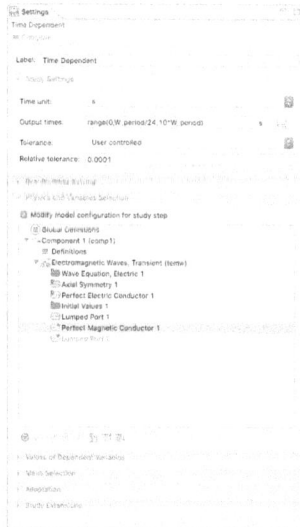

FIGURE 8.59 Settings time dependent parameters.

Figure 8.59 shows Settings Time Dependent Parameters.

Click > Compute in Settings Time Dependent window toolbar.

RESULTS

2D Plot Group 1

Click > Probe tab in the Desktop Message Window Area (as needed).

See Figure 8.60.

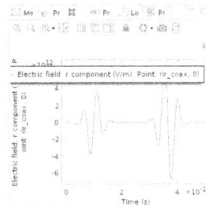

FIGURE 8.60 Probe plot.

Figure 8.60 shows Probe Plot.

In the Model Builder,

Probe Plot Group 2

Click > Probe Plot Group 2

In the Settings 1D Plot Group,

Click > Legend > Position

Select: Lower Right from the pop-up menu.

See Figure 8.61.

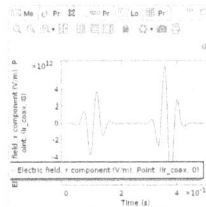

FIGURE 8.61 Settings 1D plot group probe plot group 2 and pulse profile.

Figure 8.61 shows Settings 1D Plot Group Probe Plot Group 2 and Pulse Profile.

2D Plot Group 1

NOTE *In order to see a more interesting Surface: Electric field norm (V/m) plot. The Modeler will now adjust the plot time shown.*

In the Model Builder window,

Click > 2D Plot Group 1.

In the Settings 2D Plot Group window,

Click > Time (s) pull-down list,

Select: 1.5E-10.

See Figure 8.62.

FIGURE 8.62 Settings 2D plot group 1 parameters.

Figure 8.62 shows Settings 2D Plot Group 1 Parameters.

Click > Plot in the Settings 2D Plot Group toolbar.

See Figure 8.63.

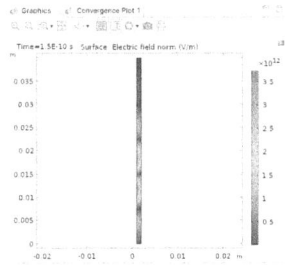

FIGURE 8.63 Electric field norm at 1.5E-10 seconds for coaxial cable "Short" termination.

Figure 8.63 shows Electric Field Norm at 1.5E-10 Seconds for Coaxial Cable "Short" Termination.

NOTE *The Modeler will now adjust the model to view the case of the "Open" (No Termination) termination.*

DEFINITIONS

Point Probe Expression 1 (ppb1)

In the Model Builder, under Component 1 (comp1),

Click > the Definitions twistie.

Click > the Domain Point Probe 1 twistie (as needed),

Click > Point Probe Expression 1 (ppb1).

See Figure 8.64.

FIGURE 8.64 Point probe expression 1 selection.

Figure 8.64 shows Point Probe Expression 1 Selection.

In the Settings Point Probe Expression window,

Click > the Table and Window Settings twistie.

Click > Output table: pull-down list,

Select: New table.

See Figure 8.65.

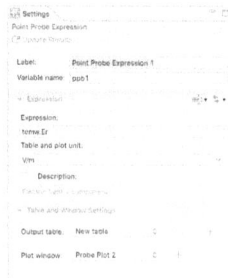

FIGURE 8.65 Settings point probe expression window.

Figure 8.65 shows Settings Point Probe Expression Window.

NOTE *The Modeler will now get a plot that contains information for both the "Short" and "Open" terminations..*

ELECTROMAGNETIC WAVES, TRANSIENT (TEMW)

Perfect Magnetic Conductor 1

In the Model Builder,

Click > the Study 1 twistie,

Click > Step 1: Time Dependent.

In the Settings Time Dependent window, on the Physics and Variables selection tab,

In the Modify model configuration for study step window,

Right-Click > Perfect Magnetic Conductor 1,

Select: Enable.

NOTE *The Perfect Magnetic Conductor becomes enabled when the asterisk is removed from the symbol.*

See Figure 8.66.

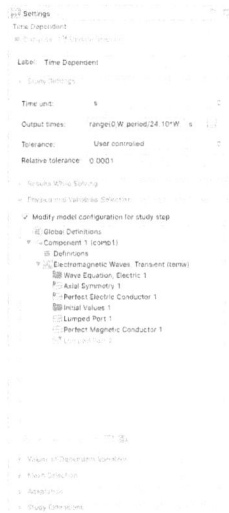

FIGURE 8.66 Settings time dependent window.

Figure 8.66 shows Settings Time Dependent Window.

STUDY 1

In the Model Builder window,

Click > Study 1.

In the Settings Study window,

Clear the check box for Generate Default Plots.

See Figure 8.67.

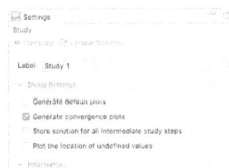

FIGURE 8.67 Settings study window parameters.

Figure 8.67 shows Settings Study Window Parameters.

In the Home toolbar,

Click > Compute.

See Figure 8.68.

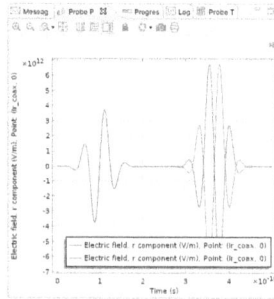

FIGURE 8.68 Reflected waves for "Short" and "Open" coax terminations.

Figure 8.68 shows Reflected Waves for "Short" and "Open" Coax Terminations.

NOTE *Next, The Modeler will plot the results for all three loads.*

DEFINITIONS

Point Probe Expression 1 (ppb1)

In the Model Builder, under Component 1 (comp1),

Click > the Definitions twistie.

Click > the Domain Point Probe 1 twistie,

Click > Point Probe Expression 1 (ppb1).

See Figure 8.69.

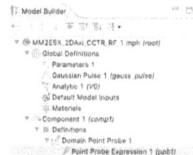

FIGURE 8.69 Point probe expression 1 selection.

Figure 8.69 shows Point Probe Expression 1 Selection.

In the Settings Point Probe Expression window,

Click > the Table and Window Settings twistie (as needed).

Click > Output table: pull-down list,

Select: New table.

See Figure 8.70.

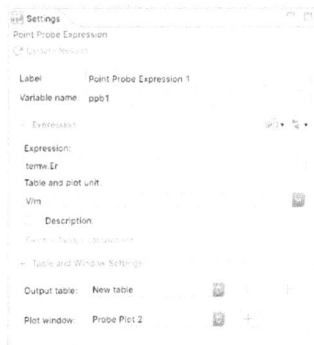

FIGURE 8.70 Settings point probe expression window.

Figure 8.70 shows Settings Point Probe Expression Window.

Lumped Port 2

In the Model Builder,

Click > the Study 1 twistie,

Click > Step 1: Time Dependent.

In the Settings Time Dependent window, on the Physics and Variables selection tab,

In the Modify model configuration for study step window,

Right-Click > Lumped Port 2,

Select: Enable.

NOTE *The Lumped Port 2 becomes enabled when the asterisk is removed from the symbol.*

See Figure 8.71.

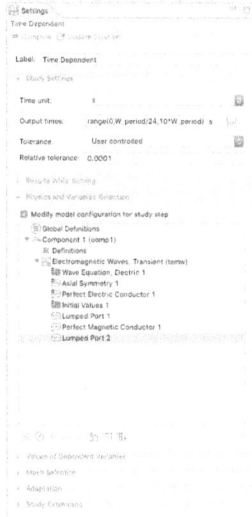

FIGURE 8.71 Settings time dependent window.

Figure 8.71 shows Settings Time Dependent Window.

STUDY 1

In the Home toolbar,

Click > Compute.

RESULTS

Probe Plot Group 2

In the Model Builder window, under Results,

Click > Probe Plot Group 2.

In the Settings 1D Plot Group window,

Click > the x-axis label: check box on the Plot Settings tab.

Enter: t (s) in the associated text field.

Click > the y-axis label: check box on the Plot Settings tab.

Enter: Electric field, r component (V/m) in the associated text field.

Click > Plot in the Settings 1D Plot Group window toolbar.

See Figure 8.72.

FIGURE 8.72 Settings 1D plot group window.

Figure 8.72 shows Settings 1D Plot Group Window.

Probe Table Graph 1

In the Model Builder window,

Click > the Probe Plot Group 2 twistie (as needed),

Click > Probe Table Graph 1

In the Settings Table Graph window,

Click > the Legends tab twistie,

Click > Legends pull-down list,

Select: Manual.

Enter: PEC (short) in the Legends text field.

See Figure 8.73.

FIGURE 8.73 Probe table graph 1 parameters.

Figure 8.73 shows Probe Table Graph 1 Parameters.

Probe Table Graph 2

In the Model Builder window,

Click > the Probe Plot Group 2 twistie (as needed),

Click > Probe Table Graph 2

In the Settings Table Graph window,

Click > the Legends tab twistie (as needed),

Click > Legends pull-down list,

Select: Manual.

Enter: PMC (open) in the Legends text field.

See Figure 8.74.

FIGURE 8.74 Probe table graph 2 parameters.

Figure 8.74 shows Probe Table Graph 2 Parameters.

Probe Table Graph 3

In the Model Builder window,

Click > the Probe Plot Group 2 twistie (as needed),

Click > Probe Table Graph 3

In the Settings Table Graph window,

Click > the Legends tab twistie (as needed),

Click > Legends pull-down list,

Select: Manual.

Enter: Matched Load in the Legends text field.

See Figure 8.75.

FIGURE 8.75 Probe table graph 3 parameters.

Figure 8.75 shows Probe Table Graph 3 Parameters.
See Figure 8.76.

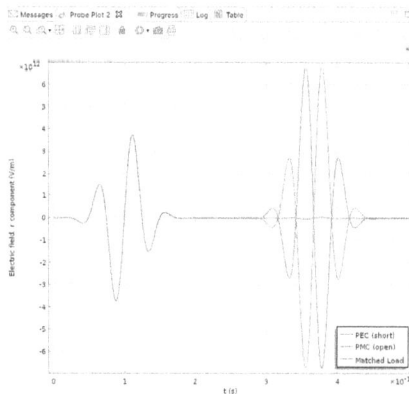

FIGURE 8.76 Transient graph all conditions.

Figure 8.76 shows the Transient Graph All Conditions.

FIRST PRINCIPLES AS APPLIED TO 2D COMPLEX MIXED MODE MODEL DEFINITION

First Principles Analysis derives from the fundamental laws of nature. In the case of models using this Classical Physics Analysis approach, the laws of conservation in physics require that what goes in (as mass, energy, charge, etc.) must come out (as mass, energy, charge, etc.) or must accumulate within the boundaries of the model.

The careful modeler must be knowledgeable of the implicit assumptions and default specifications that are normally incorporated into the COMSOL Multiphysics software model when a model is built using the default settings.

Consider, for example, the two 2D models and one 2D Axisymmetric model developed in this chapter. In these models, it is implicitly assumed there are no thermally related changes (mechanical, electrical, etc.). It is also assumed the materials are homogeneous and isotropic, except as specifically indicated and there are no thin insulating contact barriers at the thermal junctions. None of these assumptions are typically true in the general case. However, by making such assumptions, it is possible to easily build a 2D Complex Mixed Mode First Approximation Model.

NOTE

A First Approximation Model is one that captures all the essential features of the problem that needs to be solved, without dwelling excessively on all of the small details. A good First Approximation Model will yield an answer that enables the modeler to determine if he needs to invest the time and the resources required to build a more highly detailed model. Also, the modeler needs to remember to name model parameters carefully as pointed out in Chapter 1.

REFERENCES

1. RFModuleUsersGuide, p. 158

2. *https://en.wikipedia.org/wiki/Twin-lead*

3. *https://en.wikipedia.org/wiki/Characteristic_impedance*

4. *https://en.wikipedia.org/wiki/Frequency_response*

5. *https://en.wikipedia.org/wiki/Radio_frequency*

6. *https://en.wikipedia.org/wiki/Radio_wave*

7. *https://en.wikipedia.org/wiki/James_Clerk_Maxwell*

8. *https://en.wikipedia.org/wiki/Guglielmo_Marconi*

9. *https://en.wikipedia.org/wiki/Radio-frequency_engineering*

10. *https://en.wikipedia.org/wiki/Radio_Frequency_Systems*

11. *https://en.wikipedia.org/wiki/Impedance_of_free_space*

12. *https://en.wikipedia.org/wiki/Coaxial_cable*

13. *https://en.wikipedia.org/wiki/Oliver_Heaviside*

14. *https://en.wikipedia.org/wiki/Line_integral*

15. *https://en.wikipedia.org/wiki/Contour_integration*

16. *https://en.wikipedia.org/wiki/Impedance_of_free_space*

17. *https://en.wikipedia.org/wiki/Pulse_(signal_processing)*

18. *https://en.wikipedia.org/wiki/Sine*

SUGGESTED MODELING EXERCISES

1. Build, mesh, and solve the 2D Finding the Impedance of a Two (2) Wire, Parallel-Wire, Transmission Line as presented earlier in this chapter.

2. Build, mesh, and solve the 2D Finding the Impedance of a Concentric, Two (2) Wire, Transmission Line (Coaxial Cable), as presented earlier in this chapter.

3. Build, mesh, and solve the 2D Axisymmetric Transient Modeling of a Coaxial Cable, as presented earlier in this chapter.

4. Change the values of the materials parameters and then build, mesh, and solve the 2D Finding the Impedance of a Two (2) Wire, Parallel-Wire, Transmission Line as presented earlier in this chapter, as an example problem.

5. Change the value of the materials parameters and then build, mesh, and solve the 2D Finding the Impedance of a Concentric, Two (2) Wire, Transmission Line (Coaxial Cable), as an example problem.

6. Change the value of the Model geometries and then build, mesh, and solve the 2D Axisymmetric Transient Modeling of a Coaxial Cable, as presented earlier in this chapter, as an example problem.

CHAPTER 9

3D MODELING USING COMSOL MULTIPHYSICS 5.x

In This Chapter

- Guidelines for 3D Modeling in 5.x
  - 3D Modeling Considerations
- 3D Models
  - 3D Spiral Coil Microinductor Model
  - 3D Linear Microresistor Beam Model
- First Principles as Applied to 3D Model Definition
- References
- Suggested Modeling Exercises

GUIDELINES FOR 3D MODELING IN 5.x

NOTE

In this chapter, 3D models will be presented. Such 3D models are typically more conceptually and physically complex than the models that were presented in earlier chapters of this text. 3D models have proven to be very valuable to the science and engineering communities as first-cut evaluations of potential systemic physical behavior under the influence of mixed external stimuli. 3D model responses and other such ancillary information can be gathered and screened early in a project for a first-cut evaluation of the physical behavior of a planned prototype. The calculated model (simulation) information can be used in the prototype fabrication stage as guidance in the selection of prototype geometry and materials.

Since the models in this and subsequent chapters are more complex and more difficult to solve than the models presented thus far, it is important that the modeler have available the tools necessary to most easily utilize the powerful capabilities of the 5.x software. In order to do that, if you have not done this previously, the modeler should go to the main 5.x toolbar, Click > Options – Preferences – Show More Options. When the Preferences – Show More Options edit window is shown, Select > Equation view checkbox and Equation Section checkbox. Click > OK. {9.1}.

3D Modeling Considerations

3D models are typically difficult, simply based on their mathematical size, if nothing else. The modeler needs to ensure that all the appropriate assumptions have been made and that the correct materials properties have been incorporated in all dimensions. 3D models also allow the use of vector and tensor materials properties {9.2, 9.3}. Thus, depending upon the nature of the materials properties, the behavior of the material in the model may not be linear or homogeneous or isotropic or well behaved, in any sense of the word. However, if the materials are properly characterized and the modeler is sufficiently patient, a first approximation model can be built and converged to yield a first approximation answer (solution) to a difficult, non-analytic problem.

NOTE   *An Analytic Function has a Closed-form Solution which can be written as a bounded number of well-known functions {9.4}. A non-analytic function cannot be so written.*

In compliance with the laws of physics, a 3D model initially implicitly assumes that energy flow, materials properties, environment, and all other conditions and variables that are of interest are homogeneous, isotropic, and/or constant, unless otherwise specified, throughout the entire domain of interest both within the model and through the boundary conditions and in the environs of the model.

It is the responsibility of the modeler to ensure that any implicit assumptions that need to be adjusted are. Especially in 3D models, the modeler needs to bear the above-stated conditions in mind and carefully ensure that all of the modeling conditions and associated parameters (default settings) in each model created are properly considered, defined, verified, and/or set to the appropriate values.

It is always mandatory that the modeler be able to accurately anticipate the expected results of the model and accurately specify the manner in which those results will be presented. Never assume that any of the default values that are present when the model is created necessarily satisfy the needs or conditions of a particular model.

NOTE   *Always verify that any parameters employed in the model are of the correct value needed for that model. Calculated solutions that significantly deviate from the anticipated solution or from a comparison of values to those measured in an experimentally derived realistic model are probably indicative of one or more modeling errors either in the original model design, in the earlier model analysis, in the understanding of the underlying physics, or are simply due to human error.*

3D Coordinate System

In a 3D model, if the parameters can only vary as a function of the position, the (x), (y), and (z) coordinates, then such a 3D model represents the parametric condition of the model in a time-independent mode (stationary). In a time-dependent study or frequency domain study model, parameters can vary both with position in (x), (y), and/or (z) and with time (t).

See Figure 9.1.

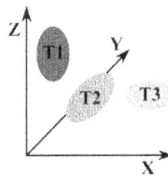

FIGURE 9.1 Parametric variations in a 3D coordinate system, plus time.

Figure 9.1 shows parametric variations in a 3D coordinate system, plus time.

In this chapter, the modeler is introduced to three new modeling concepts: the Terminal boundary condition, lumped parameters {9.5}, and coupled thermal, electrical, and structural multiphysics analysis. The Terminal boundary condition and the lumped parameter concepts are employed in the solution of the 3D Spiral Coil Microinductor Model (Derived from COMSOL Model 129). The fully coupled multiphysics solution is employed in the 3D Linear Microresistor Beam Model.

The lumped parameter (lumped element) modeling approach approximates a spatially distributed collection of diverse physical elements by a collection of topologically (series and/or parallel) connected discrete elements. This technique is commonly employed for first approximation models in electrical, electronic, mechanical, heat transfer, acoustic, and other physical systems.

Inductance Theory

Michael Faraday {9.6} discovered electromagnetic induction {9.7} in 1831. Joseph Henry {9.8} independently discovered electromagnetic induction during approximately the same era; however, he chose not to publish his discovery until later. Oliver Heaviside coined the name inductance in 1886 {9.9}.

Inductance (a lumped parameter) is the physical property associated with electrical components configured in specific circuit configurations. There are basically two commonly employed configurations for inductance: self-inductance and mutual inductance. Self-inductance (L) is defined as the inductance of an individual coil in a circuit and is mathematically defined as follows:

$$v = L \frac{di}{dt} \tag{9.1}$$

Where v = induced voltage in Volts [V].

L = inductance in Henries [H].

i = current flowing in the circuit in Amperes [A].

The configuration for a self-inductive coil is as shown in Figure 9.2.

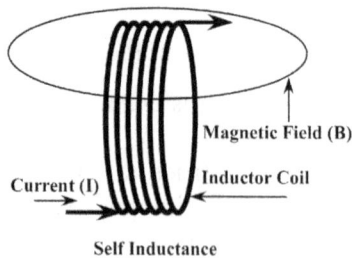

Self Inductance

FIGURE 9.2 Self inductance circuit.

Figure 9.2 shows a Self Inductance circuit.

Mutual inductance is somewhat more complex. Mutual inductance involves the coupling between at least two discrete coils through an intermediate medium. The mutual inductance for two coils is typically defined as follows:

$$M_{\alpha\beta} = k_{\alpha\beta}(L_{\alpha} \cdot L_{\beta})^{\frac{1}{2}} \qquad (9.2)$$

Where $M_{\alpha\beta}$ = the mutual inductance in Henries [H].

$k_{\alpha\beta}$ = coupling coefficient through the intermediate media $0 \leq k_{\alpha\beta} \leq 1$ and is unitless [1].

L_{α} = inductance of the first coil in Henries [H].

L_{β} = inductance of the second coil in Henries [H].

The configuration for two mutual inductive coils is as shown in Figure 9.3.

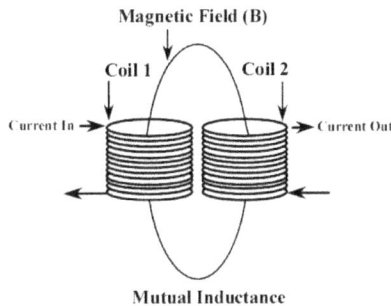

FIGURE 9.3 Mutual inductance circuit.

Figure 9.3 shows a Mutual Inductance circuit.

Inductance can be calculated in the general case by using Maxwell's equations {9.10}. Expressions (formulas) for the inductance of most common inductor configurations (geometries) have been calculated and are available in the form of tables and handbooks {9.11}.

NOTE

The easiest modeling method to employ in the solution of a lumped parameter is the use of an impedance matrix {9.12, 9.13}, which is merely an expansion of Ohm's Law {9.14}.

When current flows in an electrical circuit, energy is stored in the magnetic field {9.15}. The energy stored in the magnetic field of an inductor {9.16, 9.17} is proportional to the inductance and to the square of the current as follows:

$$E_m = \frac{1}{2} L I^2 \tag{9.3}$$

Where E_m = stored magnetic energy in joules [J].

L = inductance in Henries [H].

I = current in amperes [A].

NOTE — *In a circuit that contains an inductive component L, the diagonal term L_{11} of the impedance matrix is the magnitude of the self-inductance L.*

3D MODELS

3D Spiral Coil Microinductor Model

Building the 3D Spiral Coil Microinductor Model

Startup 5.x, Click > Model Wizard button.

Select Space Dimension > 3D.

Click > Twistie for AC/DC Interface.

Click > Electromagnetic Fields twistie.

Click > Vector Formulations twistie.

Click > Magnetic and Electric Fields (*mef*).

Click > Add button.

Click > Study (Right Pointing Arrow button).

Select > General Studies > Stationary.

Click > Done (Checked Box button).

Click > Save As.

Enter: MM2E5X_3D_SCM_1.mph.

Click > Save.

See Figure 9.4.

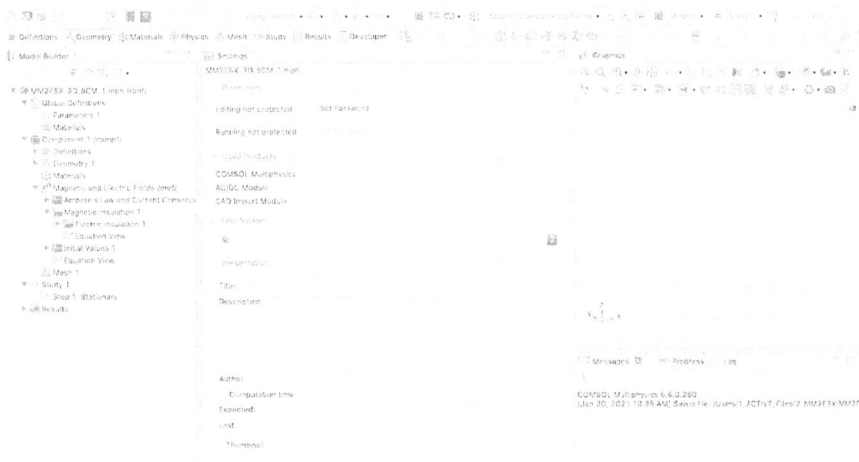

FIGURE 9.4 Desktop display for the MM2E5X_3D_SCM_1.mph model.

Figure 9.4 shows the Desktop Display for the MM2E5X_3D_SCM_1.mph model.

Geometry

Spiral Coil

Block 1

In Model Builder,

Right-Click > Model Builder > Component 1 (*comp1*) > Geometry 1.

Select > Block from the Pop-up menu.

Enter > 60E-6[m] in the Settings – Block – Size and Shape – Width entry window.

Enter > 20E-6[m] in the Settings – Block – Size and Shape – Depth entry window.

Enter > 20E-6[m] in the Settings – Block – Size and Shape – Height entry window.

Click: Base pull-down menu on Settings Block Position tab (as needed),

Select: Corner from the pull-down menu.

Enter > 0E-6[m] in the Settings Block Position tab x text field.

Enter > 0E-6[m] in the Settings Block Position tab y text field.

Enter > 0E-6[m] in the Settings Block Position tab z text field.

Click > Build All Objects in the Settings Block toolbar.

See Figure 9.5.

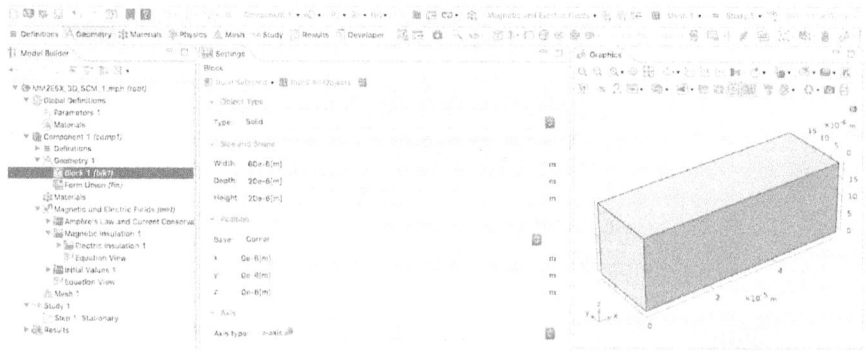

FIGURE 9.5 Settings block window > Block 1 parameters and the built block 1.

Figure 9.5 shows the Settings Block Window > Block 1 Parameters and the Built Block 1.

NOTE *The remaining portions of the geometry will be built without explicitly showing the results incrementally. The total result will be shown, once completed.*

Build the remainder of the Spiral Coil Blocks 2 – 15, by following the procedure employed to build Block 1.

Click > Zoom Extents after each block is built.

TABLE 9.1 Spiral Coil Elements.

	Block Dimensions			Corner Position		
Block	Width	Depth	Height	x	y	z
2	20E-6	80E-6	20E-6	40E-6	20E-6	0E-6
3	200E-6	20E-6	20E-6	40E-6	100E-6	0E-6
4	20E-6	140E-6	20E-6	220E-6	−40E-6	0E-6
5	160E-6	20E-6	20E-6	80E-6	−60E-6	0E-6
6	20E-6	100E-6	20E-6	80E-6	−40E-6	0E-6

7	120E-6	20E-6	20E-6	80E-6	60E-6	0E-6
8	20E-6	60E-6	20E-6	180E-6	0E-6	0E-6
9	80E-6	20E-6	20E-6	120E-6	−20E-6	0E-6
10	20E-6	20E-6	20E-6	120E-6	0E-6	0E-6
11	40E-6	20E-6	20E-6	120E-6	20E-6	0E-6
12	20E-6	20E-6	20E-6	140E-6	20E-6	20E-6
13	140E-6	20E-6	20E-6	140E-6	20E-6	40E-6
14	20E-6	20E-6	20E-6	260E-6	20E-6	20E-6
15	40E-6	20E-6	20E-6	260E-6	20E-6	0E-6

Right-Click > Model Builder > Component 1 (*comp1*) > Geometry 1.

Select: Booleans and Partitions > Union from the pop-up menu.

Uncheck: Keep interior boundaries Checkbox.

Click > Select All Button in the Graphics Toolbar.

Click > Build Selected.

See Figure 9.6.

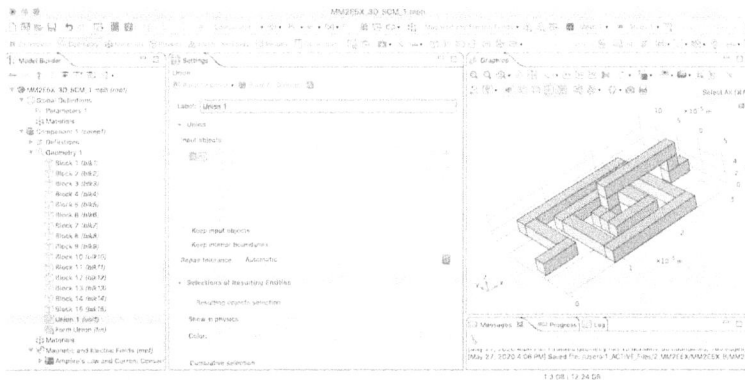

FIGURE 9.6 Graphics window – Built spiral coil.

Figure 9.6 shows the Graphics Window – Built Spiral Coil.

NOTE *The Spiral Coil needs to have an environment as part of the model. The next step adds a Block to define the Spiral Coil environment.*

Surrounding Environment

Right-Click > Model Builder > Component 1 (*comp1*) > Geometry 1.

Select: Block from the Pop-up menu.

Enter > 300E-6[m] in the Settings Block > Size and Shape tab >Width text field.

Enter > 300E-6[m] in the Settings Block > Size and Shape – Depth text field.

Enter > 200E-6[m] in the Settings – Block > Size and Shape – Height text field.

Verify > Corner from the Settings > Block > Position Pull-down menu.

Enter > 0E-6[m] in the Settings Block > Position tab > x text field.

Enter > –120E-6[m] in the Settings Block > Position tab > y text field.

Enter > –100E-6[m] in the Settings Block >Position tab > z text field.

Click > Build Selected.

Click > Zoom Extents.

Click > Wireframe Rendering in the Graphics Window Toolbar.

See Figure 9.7.

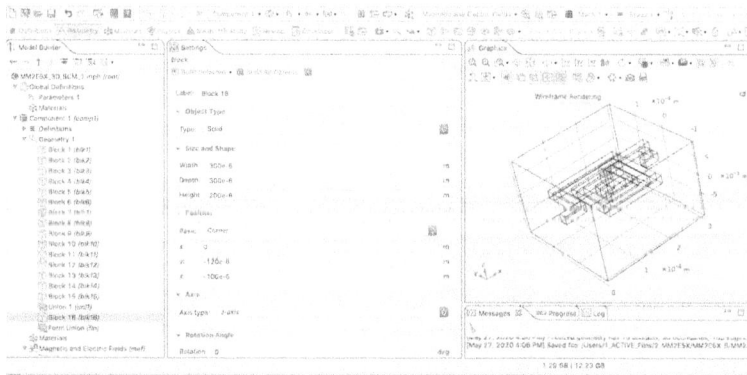

FIGURE 9.7 Graphics window – Built spiral coil plus environmental box.

Figure 9.7 shows the Graphics Window – Built Spiral Coil plus Environmental Box.

The Spiral Coil now has the necessary environment to complete the Geometry of the model. The next step defines the materials of the Spiral Coil and the Environmental Box.

Materials

Material 1

In this model, the materials properties will be entered by the modeler.

Right-Click > Model Builder > Component 1 (*comp1*) > Materials.

Select > Blank Material from the Pop-up menu.

Right-Click > Material 1.

Select > Rename from the Pop-up menu.

Enter > Conductor in the New Name edit window.

Click > OK.

Click > 1 in Settings Material window > Geometric Entity Selection edit window.

Click > Remove from Selection (Minus Sign) in Settings – Material – Geometric Scope.

Enter: 1e0 in Settings Material window,

Material Contents tab,

mur Value text field.

Enter: 1e6 in Settings Material window,

Material Contents tab,

sigma Value text field.

Enter: 1e0 in Settings Material window,

Material Contents tab,

epsilonr Value text field.

See Figure 9.8.

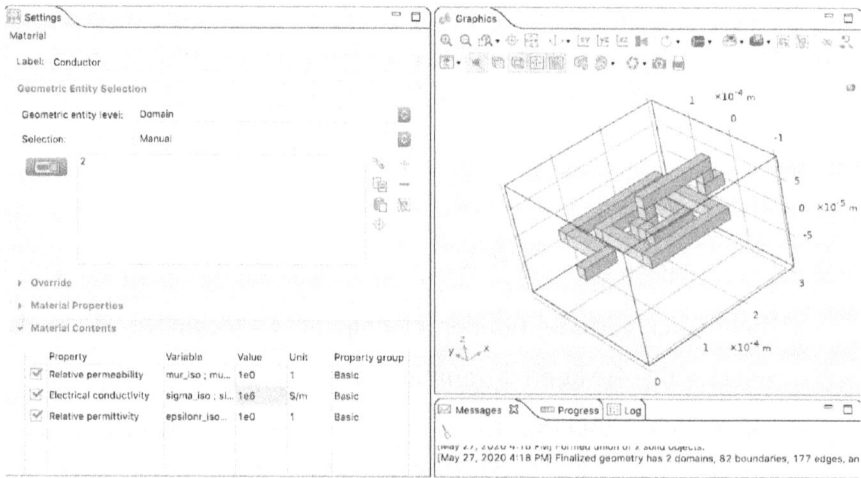

FIGURE 9.8 Settings – Material contents (Properties) conductor windows.

Figure 9.8 shows the Settings – Material Contents (Properties) Conductor Windows.

Material 2

Right-Click > Model Builder > Component 1 (*comp1*) > Materials.

Select > Blank Material from the Pop-up menu.

Right-Click > Material 2.

Select > Rename from the Pop-up menu.

Enter > Air in the New Name edit window.

Click > OK.

Click > Domain 1 (Environmental Box) in the Graphics window.

Enter: 1e0 in Settings Material window,

Material Contents tab,

mur Value text field.

Enter: 1e-6 in Settings Material window,

Material Contents tab,

sigma Value text field.

Enter: 1e0 in Settings Material window,

Material Contents tab,

epsilonr Value text field.

NOTE *The value chosen for sigma Air (1E-6) needs to be sufficiently small, but not zero (0), in order for the model to converge to solution without incurring a divide by zero problem.*

See Figure 9.9.

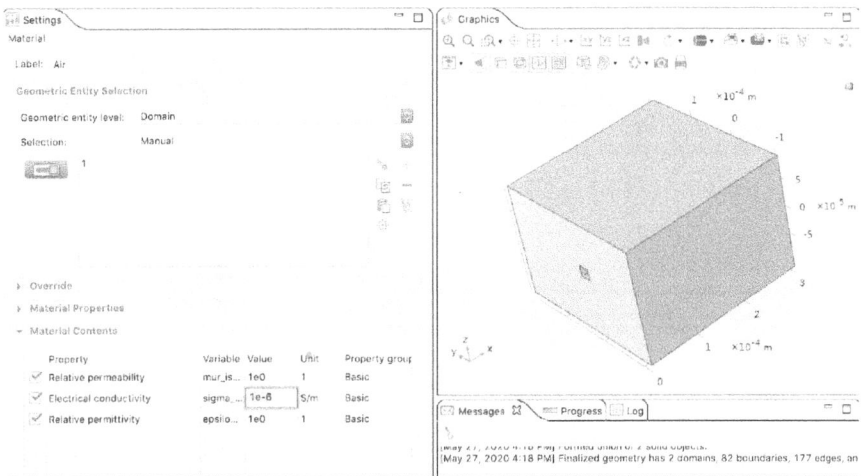

FIGURE 9.9 Settings – Material contents (Properties) air windows.

Figure 9.9 shows the Settings – Material Contents (Properties) Air Windows.

Magnetic and Electric Fields (mef)

Terminal 1

Click > Model Builder > Component 1 (*comp1*) >Twistie for the Magnetic and Electric Fields (mef).

Right-Click > Model Builder > Component 1 (*comp1*) > Magnetic and Electric Fields (mef) > Magnetic Insulation 1.

Select > Terminal from the Pop-up menu.

Click > Terminal 1 (as needed).

Enter: 5 in the Clipboard,

Click > OK (this selects the left end of the coil).

Enter > 1e0 in Settings Terminal window > Terminal tab > I_0 text field.

See Figure 9.10.

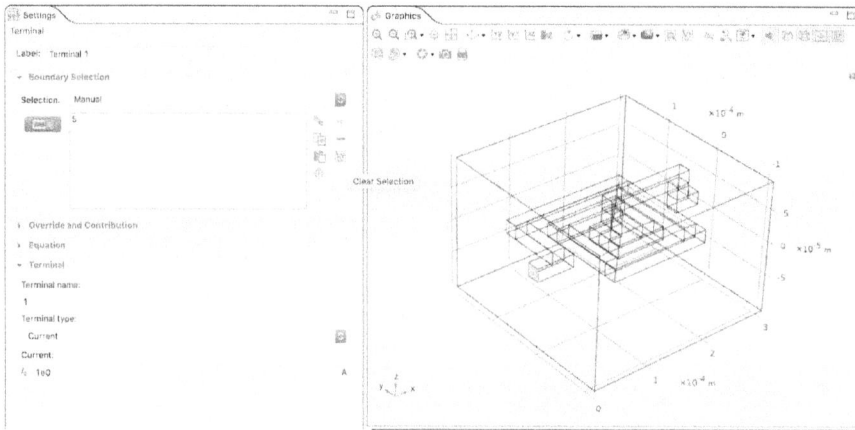

FIGURE 9.10 Settings – Terminal 1 edit windows.

Figure 9.10 shows the Settings – Terminal 1 edit windows.

Ground 1

Right-Click > Model Builder > Component 1 (*comp1*) > Magnetic and Electric Fields (mef) > Magnetic Insulation 1.

Select > Ground from the upper section of the Pop-up menu.

Click > Ground 1.

Select > All Boundaries from the Boundary Selection Pull-down menu.

Select > 1-80 Settings Ground window,

Boundaries text field.

Click > Remove from Selection (minus sign).

(Boundaries 81 and 82 remain).

See Figure 9.11.

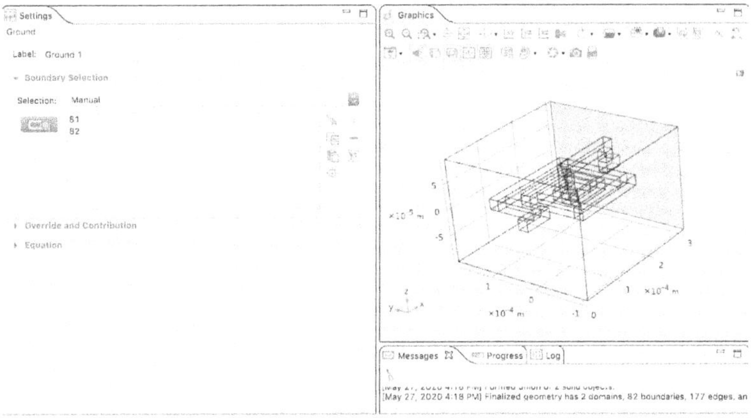

FIGURE 9.11 Settings – Ground 1 edit window.

Figure 9.11 shows the Settings – Ground 1 edit window.

Mesh 1

Right-Click > Model Builder > Component 1 (*comp1*) > Mesh 1.

Select > Free Tetrahedral from the Pop-up menu.

Click > Model Builder > Component 1 (*comp1*) > Mesh 1 > Size.

Click > Predefined Pull-down list in Settings Size window > Element Size tab,

Select: Coarse.

See Figure 9.12.

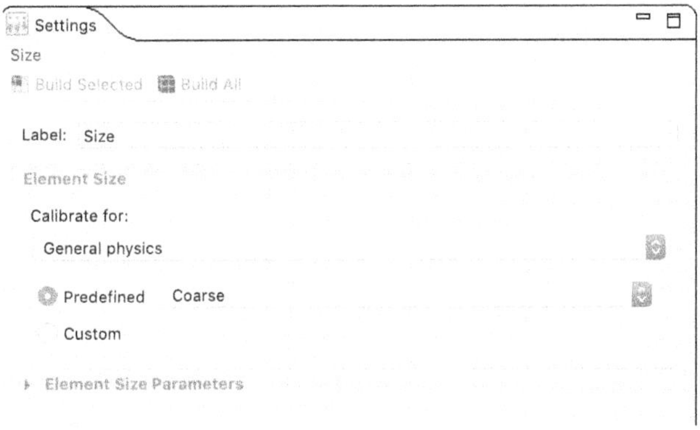

FIGURE 9.12 Settings size window pull-sown list.

Figure 9.12 shows the Settings Size Window Pull-Down List.

Click > Build All in the Settings Size toolbar.

See Figure 9.13.

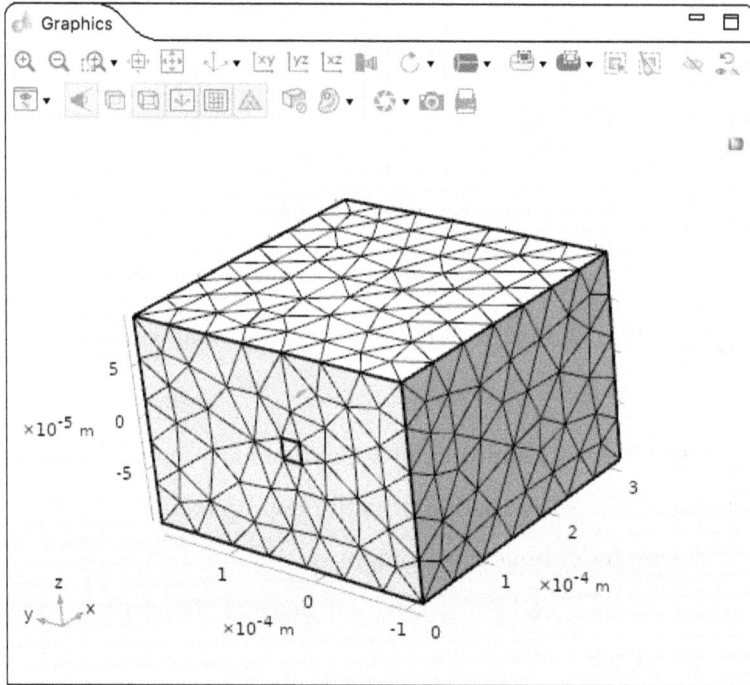

FIGURE 9.13 Graphics window, mesh.

Figure 9.13 shows the Graphics Window, Mesh.

NOTE *After the mesh is built, the model should have 3621 domain elements, 914 boundary elements and 307 edge elements.*

Study 1

In Model Builder, Right-Click Study 1 > Select > Compute.

Computed results, using the default display settings, are shown in Figure 9.14.

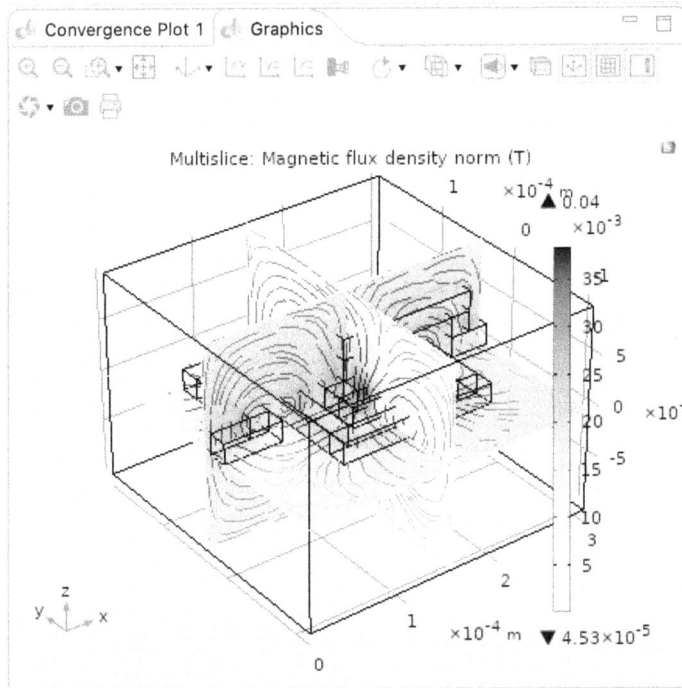

FIGURE 9.14 Graphics window, default display solution.

Figure 9.14 shows the Graphics Window, Default Display Solution.

> *The modeler should note that the computed results, using the default display settings, initially display the model solution. However, with some additional display parameter adjustments in the way that the data are presented, the model solution presentation will be significantly enhanced.*

NOTE

Results

Color and Style

Right-Click > Model Builder > Results

Select: 3D Plot Group.

Right-Click > Model Builder > Results > 3D Plot Group 2

Select: Surface.

Click > Surface 1.

Select: Thermal from the Settings Surface window > Coloring and Style tab > Color table Pull-down menu.

Click > Plot in the Settings Surface toolbar.

See Figure 9.15.

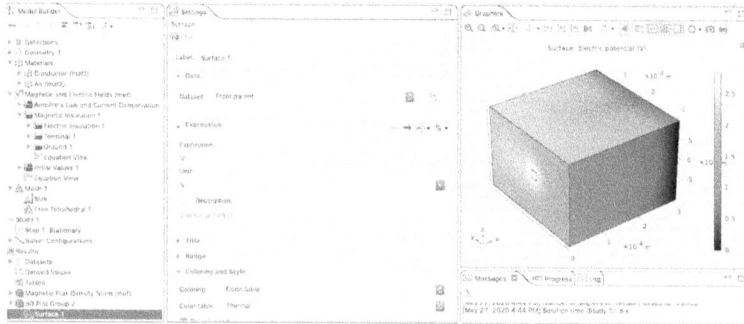

FIGURE 9.15 Color and style plot settings.

Figure 9.15 shows the Color and Style Plot Settings.

Datasets

Click > Model Builder > Results > Datasets twistie.

Right-Click > Model Builder > Results > Datasets > Study 1 / Solution 1.

Select: Selection from the Pop-up menu.

NOTE

To select all the boundaries of the Spiral Coil, first, select all of the boundaries of the model and then Remove from Selection all of the boundaries of the Air Box.

Click > Settings Selection window > Geometric Entity Selection > Geometric entity level.

Select: Boundary from the Pop-up menu.

Click > Settings – Selection > Geometric Entity Selection > Selection.

Select: All boundaries from the Pop-up menu.

Remove Boundaries: 1, 2, 3, 4, 10, 81 (first by Selection of each and then Clicking the minus (−) sign).

NOTE *When selected boundaries are removed from the selection window, the process becomes manual.*

See Figure 9.16.

FIGURE 9.16 Selected spiral coil boundaries.

Figure 9.16 shows the Selected Spiral Coil Boundaries.

3D Plot Group 2

Right-Click > Model Builder > Results > 3D Plot Group 2.

Select: Streamline from the Pop-up menu.

Click > Replace Expression in Settings Streamline window > Expression tab,

Select > Component 1 > Magnetic and Electric Fields > Magnetic > Magnetic flux density (mef.Bx, …, mef.Bz) from the Pop-up menu.

Click > Positioning Pull-down menu in the Settings Streamline window > Streamline Positioning.

Select: Starting - point controlled.

Click > Entry method pull-down menu in the Settings Streamline window > Streamline Positioning tab.

Select: Number of points.

Enter: 3 in the Settings Streamline window > Streamline Positioning > Points: text field.

Click > In the Settings Streamline window > Coloring and Style tab > Line style > Type:,

Select: Tube from the Pull-down menu.

Enter: 1E-6 in the Settings Streamline window > Coloring and Style > Tube radius expression text field.

Click > Radius scale factor check box.

Enter > 1 in the Radius scale factor text field (as needed).

Click > Settings Streamline window > Quality tab twistie.

Click > Settings Streamline window > Quality tab > Resolution pop-up menu.

Select: Finer from the Pop-up menu.

See Figure 9.17.

FIGURE 9.17 Streamline settings.

Figure 9.17 shows the Streamline Settings.

Click > Plot in the Settings Streamline toolbar.

See Figure 9.18.

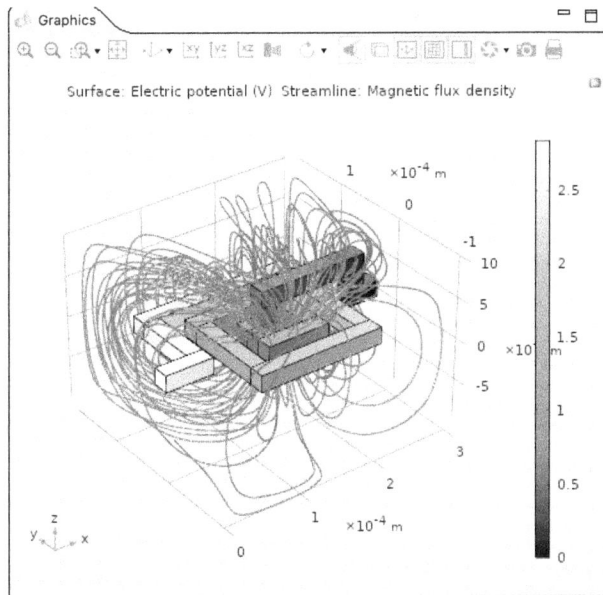

FIGURE 9.18 Spiral coil magnetic field.

Figure 9.18 shows the Spiral Coil Magnetic Field.

NOTE *Because the Spiral Coil is contained within the Air Box and the Air Box is magnetically insulated, the field lines follow the contour of the box. The limited size of the box introduces a small error in the calculated value of the inductance. The error can be roughly estimated from the relative magnitude of the magnetic field at the boundary of the box.*

Right-Click > Model Builder > Results > 3D Plot Group 2 – Streamline 1.

Select: Color Expression from the Pop-up menu.

Click > Replace Expression in Settings Color Expression window > Expression tab,

Select: Component 1 > Magnetic and Electric Fields > Magnetic > Magnetic flux density norm (mef.normB).

Click > Plot in the Settings Color Expression window toolbar.

See Figure 9.19.

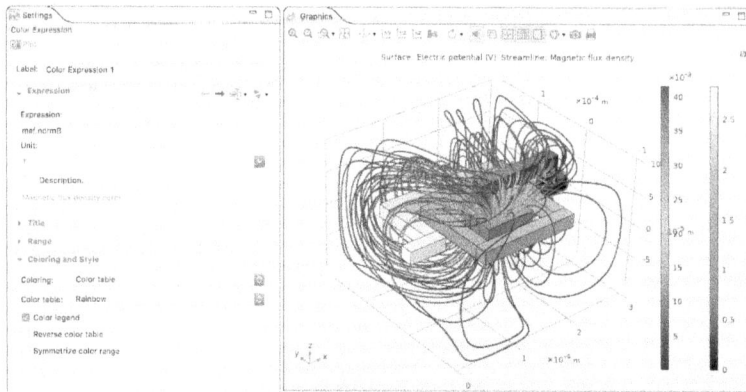

FIGURE 9.19 Spiral coil magnetic field intensity.

Figure 9.19 shows the Spiral Coil Magnetic Field Intensity.

NOTE

The modeler should note that the magnetic field intensity in the center of the Spiral Coil is approximately 40 times stronger than the magnetic field intensity at the boundary of the Air Box (that is the ratio of the strongest magnetic field in the box to that of the weakest magnetic field in the box).

Calculating Inductance

NOTE

The Plot command was just executed to ensure that the software was in the correct place to execute the next step.

Right-Click > Model Builder > Results > Derived Values.

Select: Global Evaluation from the pop-up menu.

Click > Settings Global Evaluation > Expression tab > Replace Expression.

Select: Component 1 > Magnetic and Electric Fields > Terminals > (mef.L11) Inductance.

Click > Evaluate (Equals Sign (=)) in the Settings Toolbar.

NOTE

The calculated value for the Spiral Coil Inductance is shown in the Results window below the Graphics window. The value of L for this model is approximately 5.2528 E-10H (nanoHenry [nH]).

See Figure 9.20.

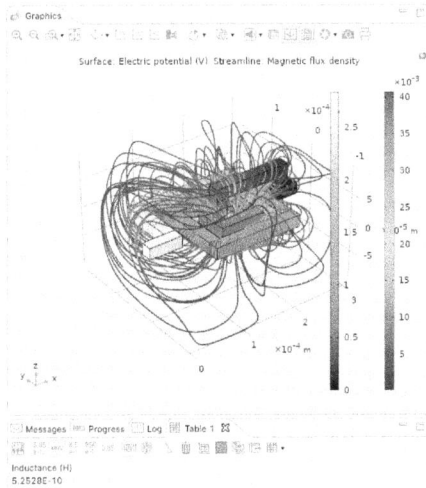

FIGURE 9.20 Spiral coil inductance.

Figure 9.20 shows the Spiral Coil Inductance.

3D Spiral Coil Microinductor Model Summary and Conclusions

The 3D Spiral Coil Microinductor Model is a powerful modeling tool that can be used to calculate the inductance of spiral coils of various different sizes. With this model, the modeler can easily vary all of the deposition parameters and optimize the design before the first prototype is physically built. These types of spiral coils are widely used in industry.

3D Linear Microresistor Beam Model

In this model (derived from COMSOL Model 366), three different areas of physical analysis are combined and solved simultaneously. Those three areas of physics are electrical, thermal, and structural analysis. The analysis of this problem begins explicitly with the use of Ohm's Law {9.18}.

Which is:

$$V = I * R \qquad (9.4)$$

Where V = Applied voltage [V].

I = Current in amperes [A].

R = Resistance in ohms [ohms].

When current flows through a resistance, power is dissipated and then Joule's First Law {9.19} applies. James Prescott Joule {9.20} discovered his power (first) law in the 1840s.

That Law is:

$$Q = I^2 * R * t \qquad (9.5)$$

Where Q = Heat Generated in Joules [J].

I = Current in Amperes [A] (assumed to be constant).

R = Resistance in Ohms [ohms] (assumed to be constant).

t = Time in Seconds [s].

Using Ohm's Law and substituting for the resistance R, yields:

$$Q = I^2 * \frac{V}{I} * t = I * V * t \qquad (9.6)$$

Where Q = Heat Generated in joules [J].

I = Current in amperes [A].

V = Voltage in volts [V].

t = Time in seconds [s].

Dividing the heat generated by the time yields:

$$\frac{Q}{t} = I * V = P \qquad (9.7)$$

Where Q = Heat Generated in joules [J].

I = Current in amperes [A].

V = Voltage in volts [V].

t = Time in seconds [s].

P = Power in watts [J/s].

Where P is the instantaneous rate of power dissipation.

It has long been observed that most materials change volume in either a positive (expand) or a negative (shrink) fashion when heated or cooled.

Quantification of this anecdotal evidence began in approximately 1650 {9.21} when Otto von Guericke {9.22} built the first vacuum pump {9.23}. Subsequent work by Nicolas Leonard Sadi Carnot {9.24}, William Thomson (Lord Kelvin) {9.25}, and many others have brought thermodynamics to a very high level of development. Of interest in this particular model is the change in the resistance and shape of the copper beam, due to thermal expansion, from the resistive heating caused by current flow through the beam.

NOTE

As current flows through the beam, heat is deposited. As heat is deposited, the copper expands. As the copper expands, the beam bends upward. The upward bending is caused by the beam becoming longer with the beam ends fixed in place.

Building the 3D Linear Microresistor Beam Model

Startup 5.x, Click > Model Wizard button.

Select Space Dimension > 3D.

Click > Twistie for Structural Mechanics Interface,

Click > Twistie for Thermal Structure Interaction,

Click > Joule Heating and Thermal Expansion.

Click > Add button.

Click > Study (Right Pointing Arrow button).

Select: General Studies > Stationary.

Click > Done (Checked Box button).

Click > Save As.

Enter MM2E5X_3D_MRTE_1.mph.

Click > Save.

See Figure 9.21.

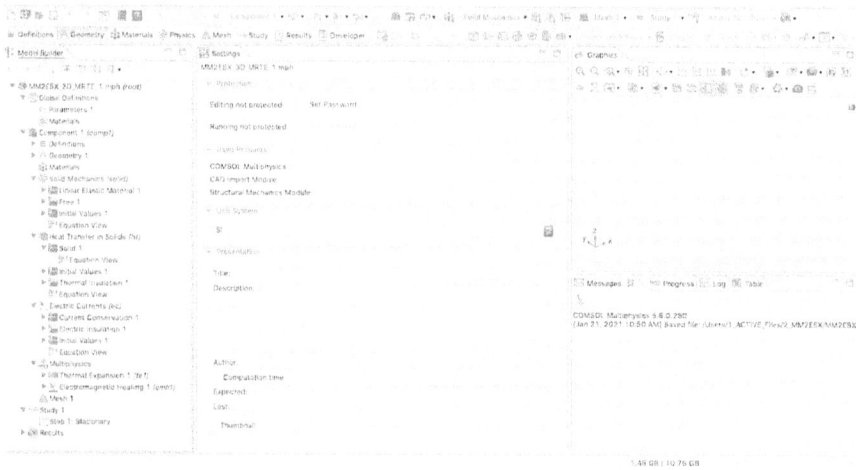

FIGURE 9.21 Desktop display for the MM2E5X_3D_MRTE_1.mph model.

Figure 9.21 shows the Desktop Display for the MM2E5X_3D_MRTE_1. mph model.

Global Definitions – Parameters

Click > Model Builder > Global Definitions > Parameters 1.

In the Settings Parameters, Parameters Table,

Enter: The Parameters from Table 9.2.

TABLE 9.2 Microresistor Beam Parameters.

Name	Expression	Description
V0	0.2[V]	Applied voltage
T0	323[K]	Base temperature
Text	298[K]	Room temperature
ht	5[W/(m^2*K)]	Heat transfer coefficient

NOTE *The Parameter "Room temperature" applies to the temperature of the environment (air, gas, liquid, etc.) into which the heat is being transferred.*

See Figure 9.22.

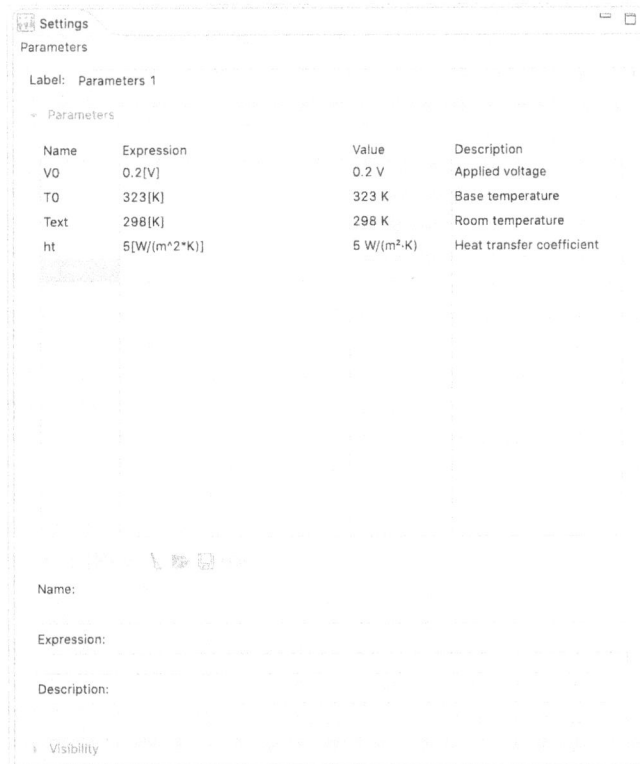

FIGURE 9.22 Filled settings – Parameters > Parameters edit window.

Figure 9.22 shows the Filled Settings – Parameters > Parameters edit window.

Geometry

Geometry 1

Click > Model Builder > Component 1 (*comp1*) > Geometry 1.

Click > Length unit: selection bar in Settings Geometry window > Units tab.

Select: μm from the list on the Pull-down menu.

See Figure 9.23.

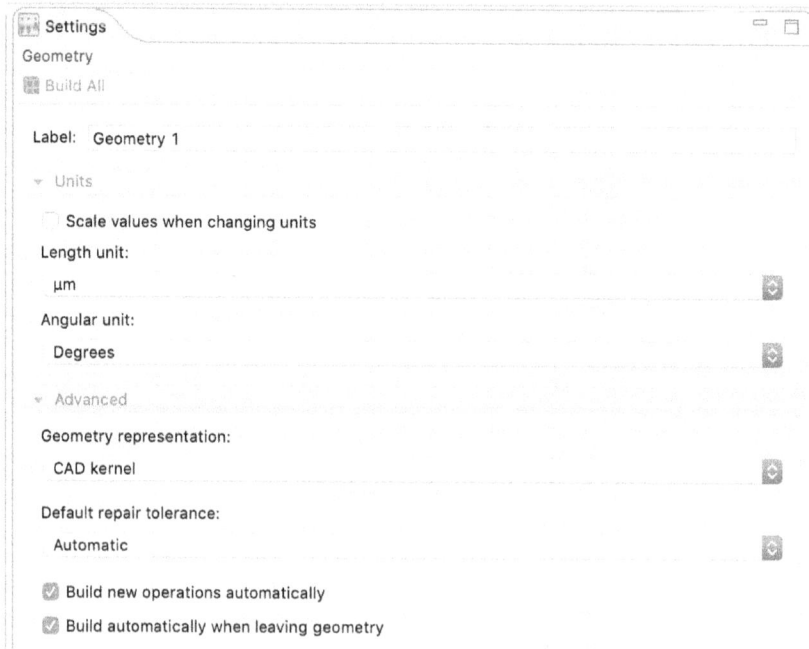

FIGURE 9.23 Settings – Geometry > Units > Length unit setting.

Figure 9.23 shows the Settings – Geometry > Units > Length unit setting.

NOTE *In the following portion of the development of this model, a powerful tool is introduced. That tool is the Work Plane. The Work Plane allows the modeler to create the outline of a 3D object in 2D and then to add the third dimension by extrusion {9.26}.*

Work Plane 1

Right-Click > Model Builder > Component 1 (*comp1*) > Geometry 1.

Select > Work Plane from the Pop-up menu.

Polygon 1

Right-Click > Model Builder > Component 1 (*comp1*) > Geometry 1 > Work Plane 1 > Plane Geometry.

Select: Polygon 1 (*pol1*) from the Pop-up menu.

Click > In Settings Polygon window, in the Coordinates tab > Data source:,

Select: Vectors from the pull-down menu.

In the xw text field,

Enter: 0 5 5 18 18 23 23 23 23 18 18 5 5 0 0 0.

In the yw text field,

Enter: 0 1.5 1.5 1.5 1.5 0 0 4 4 2.5 2.5 2.5 2.5 4 4 0.

See Figure 9.24.

FIGURE 9.24 Settings polygon coordinates entry window.

Figure 9.24 shows the Settings Polygon Coordinates Entry Window.

Click > Build Selected.

Click > Zoom Extents.

See Figure 9.25.

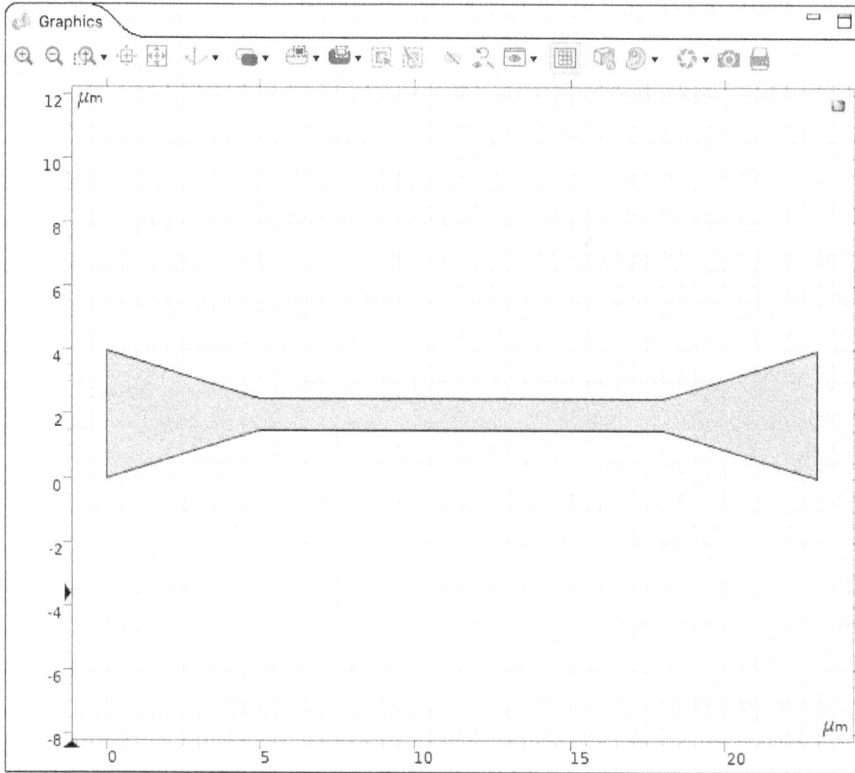

FIGURE 9.25 Polygon settings and built curve.

Figure 9.25 shows the Polygon Settings and Built Curve.

Extrude 1

Right-Click > Model Builder > Component 1 (*comp1*) > Geometry 1 > Work Plane 1.

Select > Extrude from the Pop-up menu.

Enter > 3 in Settings Extrude – Distances tab > Distances text field.

Click > Build All Objects.

See Figure 9.26.

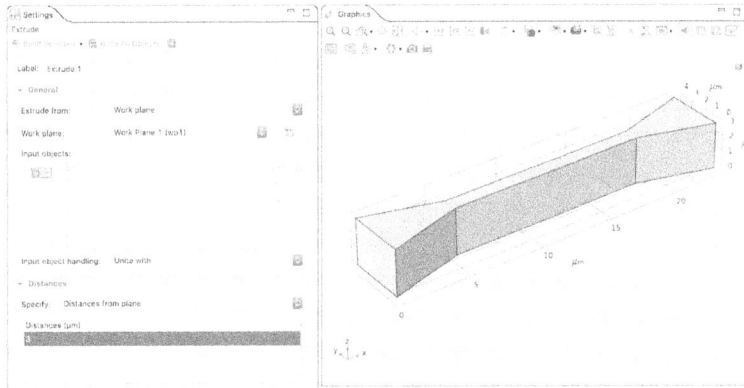

FIGURE 9.26 Polygon settings and built curve extruded.

Figure 9.26 shows the Polygon Settings and Built Curve Extruded.

Work Plane 2

Right-Click > Model Builder > Component 1 (*comp1*) > Geometry 1.

Select > Work Plane from the Pop-up menu.

In Settings Work Plane,

Click > Plane type in Settings,

Select: Face parallel from the Pop-up menu.

Click > Boundary 6 on ext 1 in the Graphics window (see Figure 9.27).

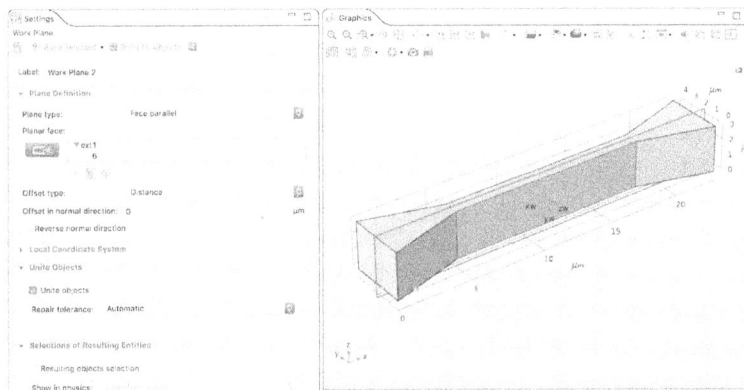

FIGURE 9.27 Boundary 6 of ext 1 in graphics window.

Figure 9.27 shows the Boundary 6 of ext 1 in Graphics window.

See Figure 9.28.

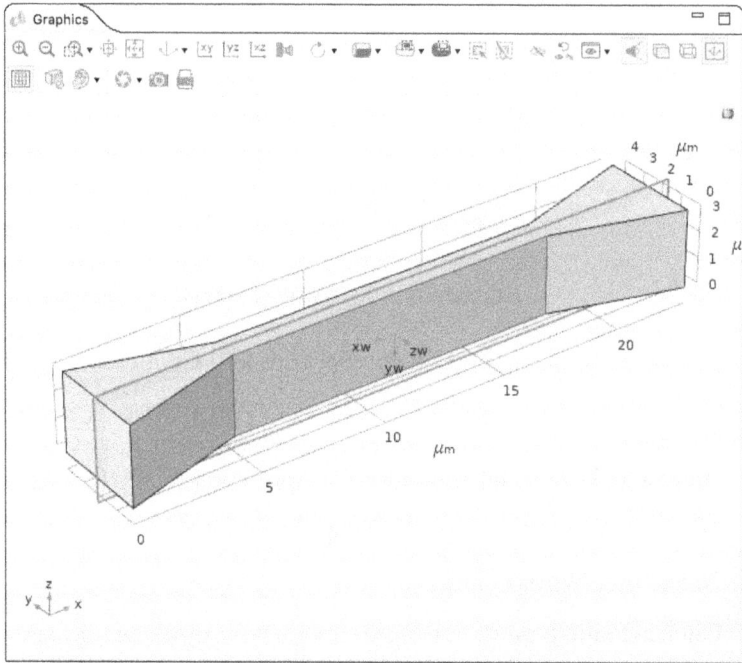

FIGURE 9.28 Work plane 2 before offset graphic window only.

Figure 9.28 shows Work Plane 2 before offset Graphic Window Only..

NOTE *In the following model building steps, the location of the work plane and the local axis will be changed. These changes are made so that when the final solid figure is extruded, it will be created in the correct location.*

Next,

Enter: –1.5 in Settings Work Plane window > Plane Definition tab > Offset in normal direction text field.

Click > Reverse normal direction check box in Settings Work Plane window > Plane Definition tab.

See Figure 9.29.

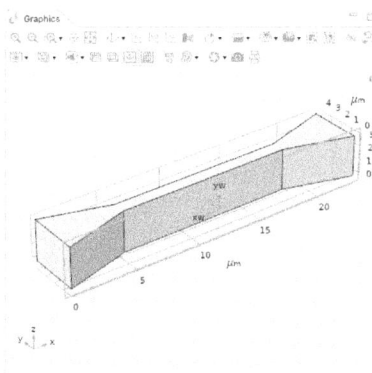

FIGURE 9.29 Work plane 2 after offset.

Figure 9.29 shows Work Plane 2 after offset.

Click > Work Plane 2 (wp2) twistie,

Click > Plane Geometry.

In the Settings Plane Geometry window,

Un-Check > Intersection (green) check box on the Visualization tab.

Un-Check > Coincident entities (blue) check box on the Visualization tab.

Polygon 1

Right-Click > Model Builder > Component 1 (*comp1*) > Geometry 1 > Work Plane 2 > Plane Geometry.

Select: Polygon 1 (*pol1*) from the Pop-up menu.

Click > In Settings Polygon window, in the Coordinates tab > Data source:,

Select: Vectors from the pull-down menu.

In the xw text field,

Enter: −11.5 −6.3 −6.3 −6.3 −6.3 6.3 6.3 6.3 6.3 11.5 11.5 6.5 6.5 −6.5 −6.5 −11.5.

In the yw text field,

Enter: −1.5 −1.5 −1.5 0.5 0.5 0.5 0.5 −1.5 −1.5 −1.5 −1.5 1.5 1.5 1.5 1.5 −1.5.

Click > Zoom Extents in the Graphics toolbar.

See Figure 9.30.

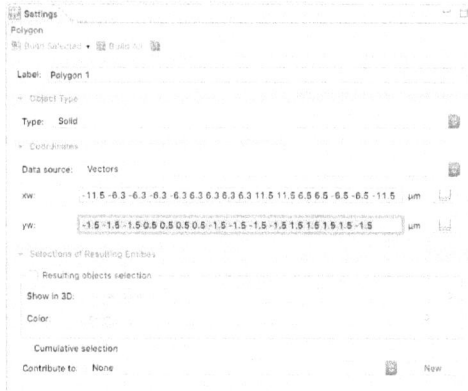

FIGURE 9.30 Settings polygon edit window.

Figure 9.30 shows the Settings Polygon edit window.

Click > Build Selected.

Click > Zoom Extents.

See Figure 9.31.

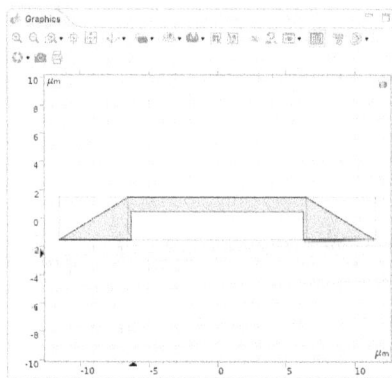

FIGURE 9.31 Settings polygon edit window and closed polygon.

Figure 9.31 shows the Settings Polygon edit window and Closed Polygon.

Fillet 1

Right-Click > Model Builder > Component 1 (*comp1*) > Geometry 1 > Work Plane 2 > Plane Geometry.

Select: Fillet from the Pop-up menu.

Click > Points 4 and 6 in the Graphics window (left inner corner (4), right inner corner (6)).

Enter: 0.3 in Settings Fillet window > Radius tab > Radius: text field.

Click > Build All in the Settings Fillet toolbar.

See Figure 9.32.

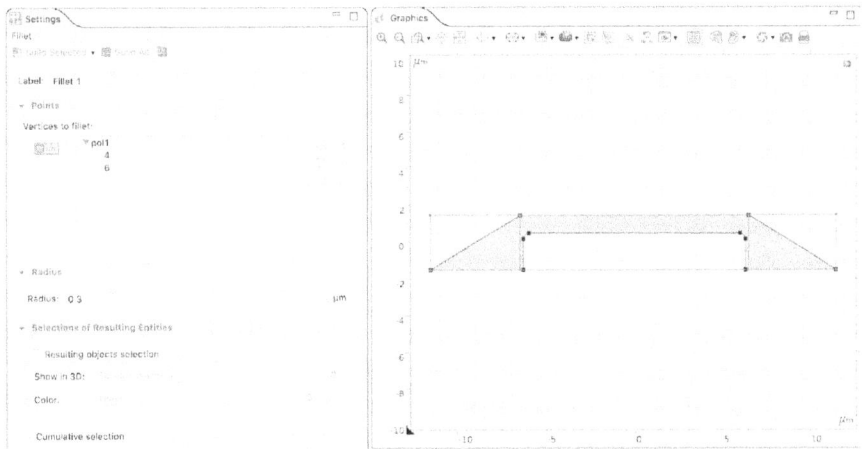

FIGURE 9.32 Settings polygon with fillets.

Figure 9.32 shows the Settings Polygon with Fillets.

Extrude 2

Right-Click > Model Builder > Component 1 (*comp1*) > Geometry 1 > Work Plane 2.

Select: Extrude from the Pop-up menu.

Enter: 4 in Settings Extrude window > Distances tab > Specify: Distances from plane (as needed) > Distances text field.

Click > Build All Objects.

See Figure 9.33.

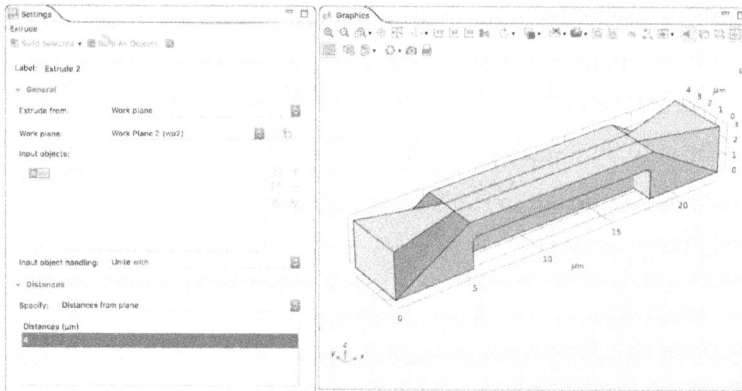

FIGURE 9.33 Work plane 2 > Settings extrude > Settings and graphical results.

Figure 9.33 shows the Work Plane 2 > Settings Extrude > Settings and Graphical Results.

Intersection 1

Right-Click > Model Builder > Component 1 (*comp1*) > Geometry 1.

Select: Boolean and Partitions > Intersection from the Pop-up menu.

Select: Objects ext1 and ext2 in the Graphics window.

See Figure 9.34.

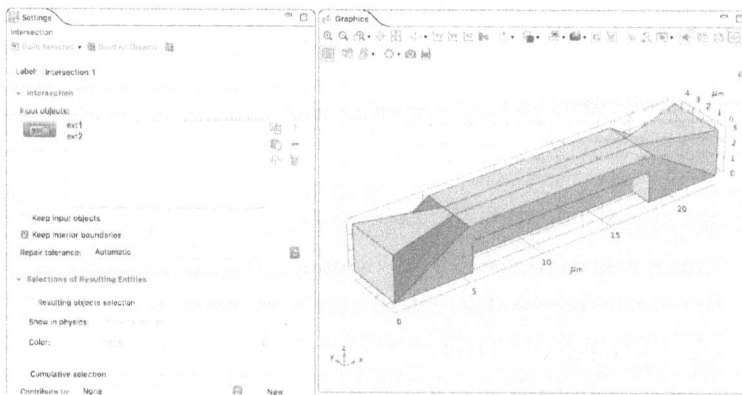

FIGURE 9.34 Work plane 2 settings and extrusion.

Figure 9.34 shows the Work Plane 2 Settings and Extrusion.

Click > Build Selected.

Form Union

Click > Model Builder > Component 1 (*comp1*) > Geometry 1 > Form Union (*fin*).

Click > Build Selected.

See Figure 9.35.

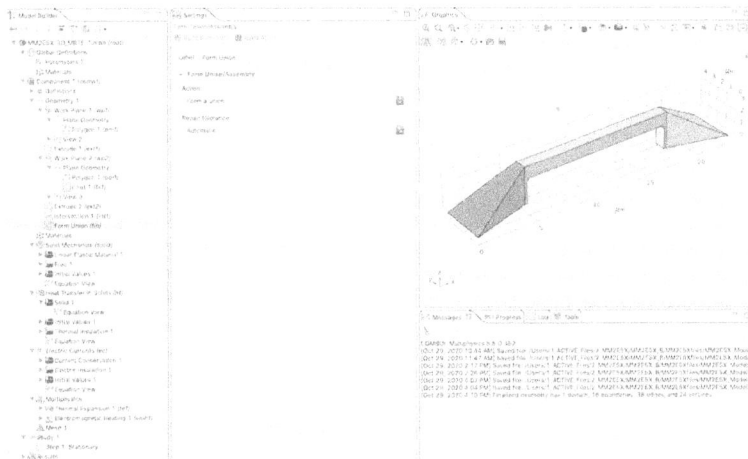

FIGURE 9.35 Desktop display of the completed microresistor beam.

Figure 9.35 shows the Desktop Display of the Completed Microresistor Beam.

Definitions

NOTE *Now that the Microresistor Beam has been built, some definitions of collections of boundaries are going to be added to make the setting of boundary conditions easier later in the model.*

Selection 1

Right-Click > Model Builder > Component 1 (*comp1*) > Definitions.

Select: Selections >Explicit.

Right-Click > Explicit 1.

Select > Rename.

Enter > connector 1 in the Rename Selection – New label text field.

Click > OK.

Click > Geometric entity level in Settings Explicit > Input Entities tab > Geometric entity level,

Select: Boundary from the Pull-down menu.

The easiest method of picking a single boundary or a small set of boundaries is by Clicking the All boundaries check box first and then Clicking the All boundaries check box a second time. All the modeler then has to do is remove (Remove from Selection (Minus sign)) the boundaries that are not needed

NOTE *from the edit window.*

If the modeler wants to view the selected boundary in the Graphics window to verify his choice and the selected boundary does not appear to be visible in that view, rotate the graphic in the Graphics window.

Select: Boundary 1 only.

Selection 2

Right-Click > Model Builder > Component 1 (*comp1*) > Definitions.

Select > Selections > Explicit.

Right-Click > Explicit 2.

Select > Rename.

Enter > connector 2 in the Rename Selection – New label text field.

Click > OK.

Click > Geometric entity level in Settings Explicit window > Input Entities > Geometric entity level,

Select: Boundary from the Pull-down menu.

Select > Boundary 13 only.

Selection 3

> Right-Click > Model Builder > Component 1 (*comp1*) > Definitions.
>
> Select > Selections > Explicit.
>
> Right-Click > Explicit 3.
>
> Select > Rename.
>
> Enter > connectors in the Rename Selection – New label text field.
>
> Click > OK.
>
> Click > Geometric entity level in Settings Explicit window > Input Entities > Geometric entity level,
>
> Select: Boundary from the Pull-down menu.
>
> Select: Boundaries 1 and 13 only.

Materials

Material 1

> Right-Click > Model Builder > Component 1 (*comp1*) > Materials.
>
> Select: Add Material from Library in the Pop-up menu.
>
> Click > Add Material > MEMS > Metals twistie.
>
> Click > Add Material > MEMS > Metals > Cu.
>
> Right-Click > Add Material > MEMS > Metals > Cu.
>
> Select: Add Material to Component 1.
>
> Click > Close on the Add Material window (hollow X on Add Materials tab).

Material - Cu

> Click > Model Builder > Component 1 (*comp1*) > Materials > Cu.
>
> Enter > 1 in the Settings Material window > Material Contents tab > Relative Permittivity (epsilonr) Value text field.
>
> See Figure 9.36.

FIGURE 9.36 Cu settings material.

Figure 9.36 shows the Cu Settings Material.

Click > Settings Material window > Material Properties tab twistie > Electromagnetic Models twistie.

Click > Settings Material window > Material Properties tab > Electromagnetic Models > Linearized Resistivity twistie.

Right-Click > Settings Material window > Material Properties tab > Electromagnetic Models > Linearized Resistivity > Reference Resistivity (rho0).

Select: Add to Material (plus sign) (The added materials will be found by scrolling to the Material Contents tab.).

Enter: Values as indicated in Table 9.3.

TABLE 9.3 Cu Reference Resistivity Parameters.

Property	Name	Value
Reference Resistivity	rho0	1.72E-8[ohm*m]
Resistivity temperature coefficient	alpha	3.9E-3[1/K]
Reference temperature	Tref	293[K]

See Figure 9.37.

FIGURE 9.37 Linearized cu settings material.

Figure 9.37 shows the Linearized Cu Settings Material.

Electric Currents (ec)

Click > Model Builder > Component 1 (*comp1*) > Electric Currents (*ec*) twistie.

Click > Model Builder > Component 1 (*comp1*) > Electric Currents (*ec*) > Current Conservation 1.

Click > Conduction model: Pull-down menu in Settings Current Conservation > Constitutive Relation Jc-E tab.

Select: Linearized resistivity from the Pull-down menu.

NOTE *The model is initially solved for a temperature independent resistivity to allow comparison of the temperature dependent and temperature independent resistive behavior.*

Click > Resistivity temperature coefficient in Settings Current Conservation,

Select > User defined from the Pull-down menu.

Enter > 0 in Resistivity temperature coefficient edit window in Settings Conduction Current (as needed, default value).

See Figure 9.38.

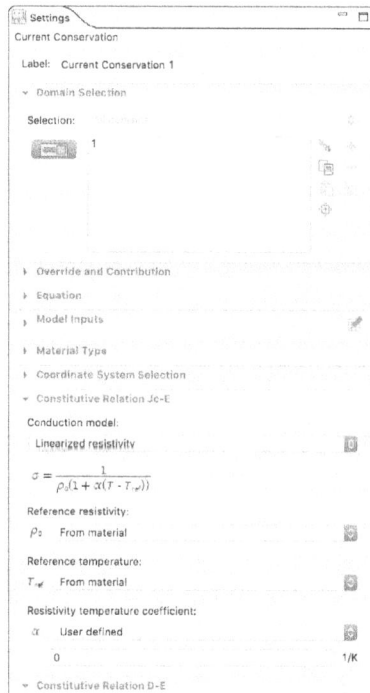

FIGURE 9.38 Settings current conservation.

Figure 9.38 shows the Settings Current Conservation.

Ground 1

NOTE *The Pop-up menu for the Physics Interfaces is divided into hierarchical sections {9.27}. First are Domains. Second are Boundaries, etc.*

Right-Click > Model Builder > Component 1 (*comp1*) > Electric Currents (*ec*).

Select > Ground from the boundary condition section of the Pop-up menu.

Click > Ground 1.

Click > Selection: pull-down menu bar in Settings Ground > Boundary Selection tab.

Select: connector 2 from the Pop-up menu.

See Figure 9.39.

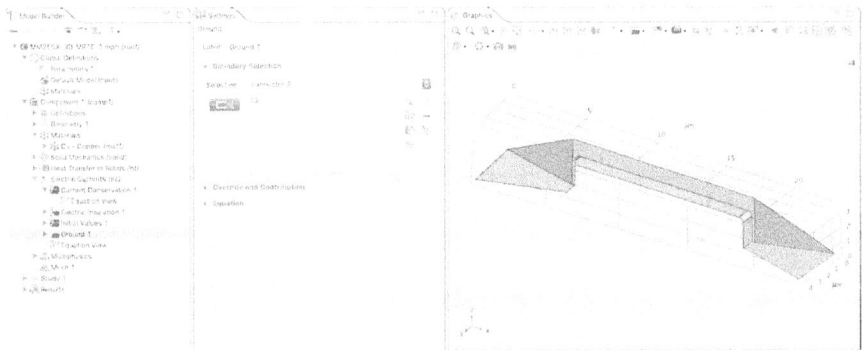

FIGURE 9.39 Settings ground window > Ground boundary selection.

Figure 9.39 shows the Settings Ground window > Ground Boundary Selection.

Electric Potential 1

Right-Click > Model Builder > Component 1 (*comp1*) > Electric Currents (*ec*).

Select: Electric Potential from the boundary condition section of the Pop-up menu.

Click > Electric Potential 1.

Click > Selection in Settings Electric Potential > Boundary Selection tab,

Select: connector 1 from the Pop-up menu.

Enter > V0 in the Settings Electric Potential > Electric Potential tab > Voltage V_0 text field.

See Figure 9.40.

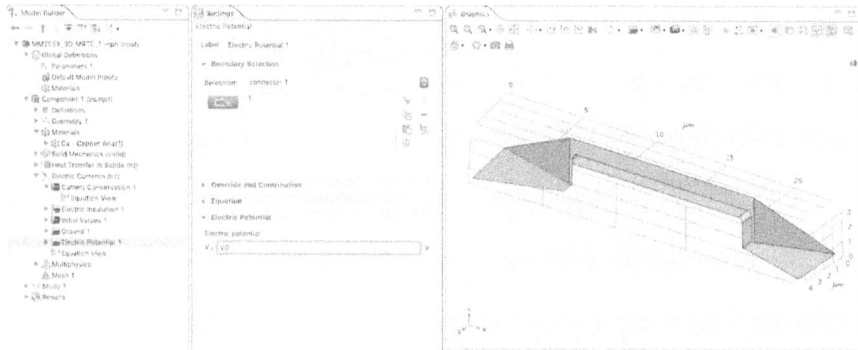

FIGURE 9.40 Electric potential boundary specification.

Figure 9.40 shows the Electric Potential Boundary Specification.

Multiphysics Thermal Linear Elastic 1 (*te*1)

Click > Model Builder > Component 1 (*comp1*) > Multiphysics twistie (as needed) > Thermal Expansion 1 (*te1*).

Click > In Settings Thermal Expansion > Model Input tab > Volume reference temperature: pull-down menu,

Select: User defined,

Enter: Text (a Global Parameter) in the associated Tref text field.

See Figure 9.41.

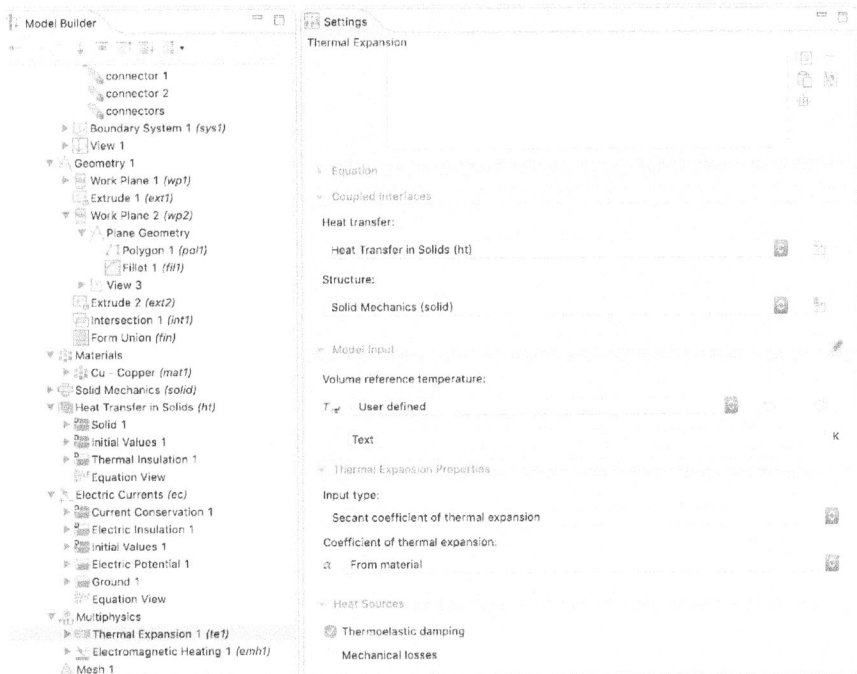

FIGURE 9.41 Settings thermal expansion window > Parameters.

Figure 9.41 shows the Settings Thermal Expansion Window > Parameters.

Heat Transfer in Solids (*ht*)

Initial Values 1

Click > Model Builder > Component 1 (*comp1*) > Heat Transfer in Solids (*ht*) twistie (as needed) > Initial Values 1.

Enter: T0 in Settings Initial Values window > Initial Values tab, Temperature T associated text field.

See Figure 9.42.

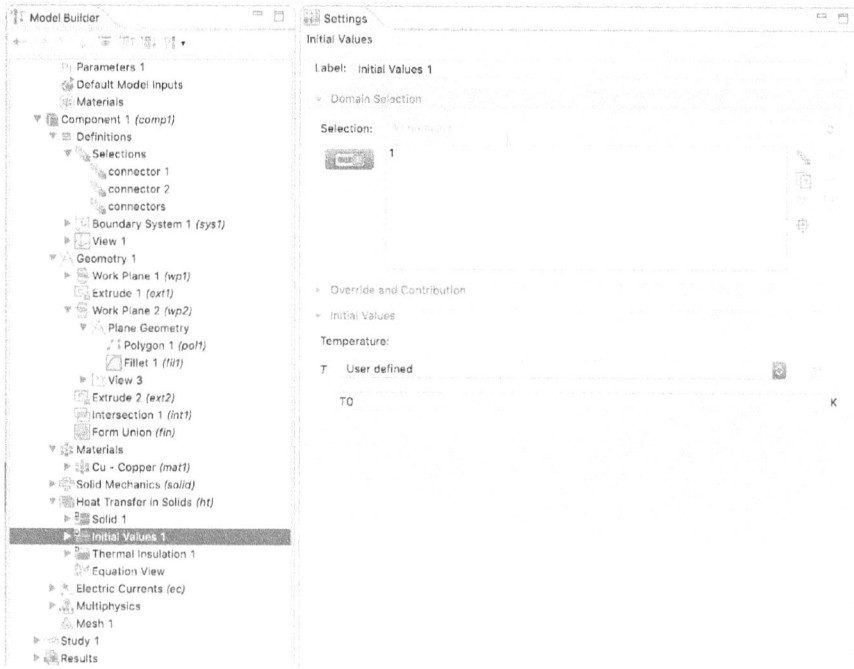

FIGURE 9.42 Initial values 1 settings.

Figure 9.42 shows the Initial Values 1 Settings.

Heat Flux 1

Right-Click > Model Builder > Component 1 (*comp1*) > Heat Transfer in Solids (*ht*).

Select: Heat Flux from boundary condition section of the Pop up menu.

Click > Heat Flux 1.

Click > Selection in Settings Heat Flux window > Boundary Selection tab,

Select: All boundaries from the Pop-up menu.

Click > Settings Heat Flux window > Heat Flux tab,

Select: Convective heat flux by clicking the radio button associated.

Enter: ht in the Settings Heat Flux window > Heat Flux tab > Heat transfer coefficient h associated text field.

Enter: Text in the Settings Heat Flux window > Heat Flux tab > External temperature Text associated text field.

See Figure 9.43.

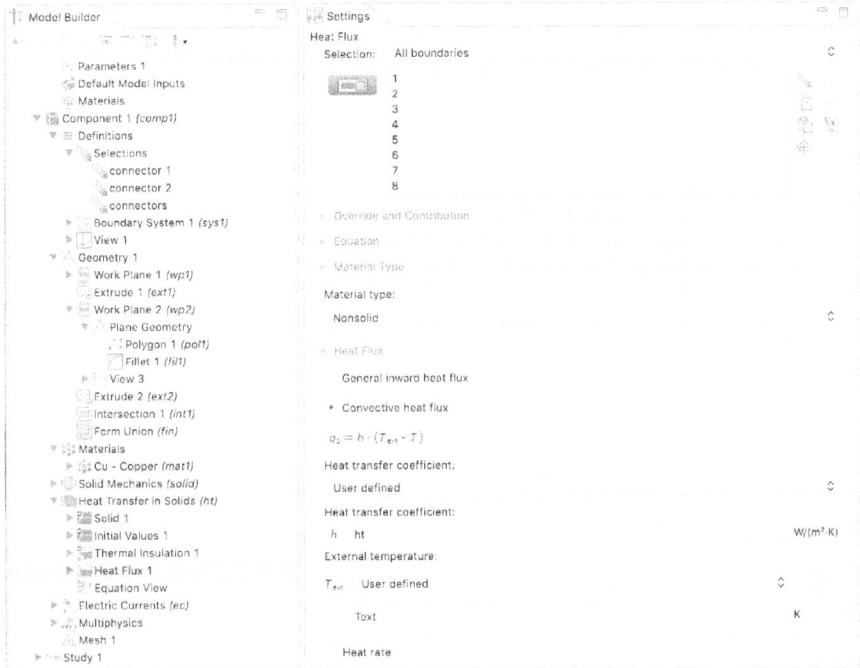

FIGURE 9.43 Heat flux boundary settings.

Figure 9.43 shows the Heat Flux Boundary Settings.

Temperature 1

Right-Click > Model Builder > Component 1 (*comp1*) > Heat Transfer in Solids (*ht*).

Select: Temperature from boundary condition section of the Pop-up menu.

Click > Temperature 1.

Click > Selection pop-up menu bar in Settings Temperature > Boundary Selection tab,

Select: connectors from the Pop-up menu.

Enter: T0 in the Settings Temperature window > Temperature tab > Temperature T_0 associated text field.

See Figure 9.44.

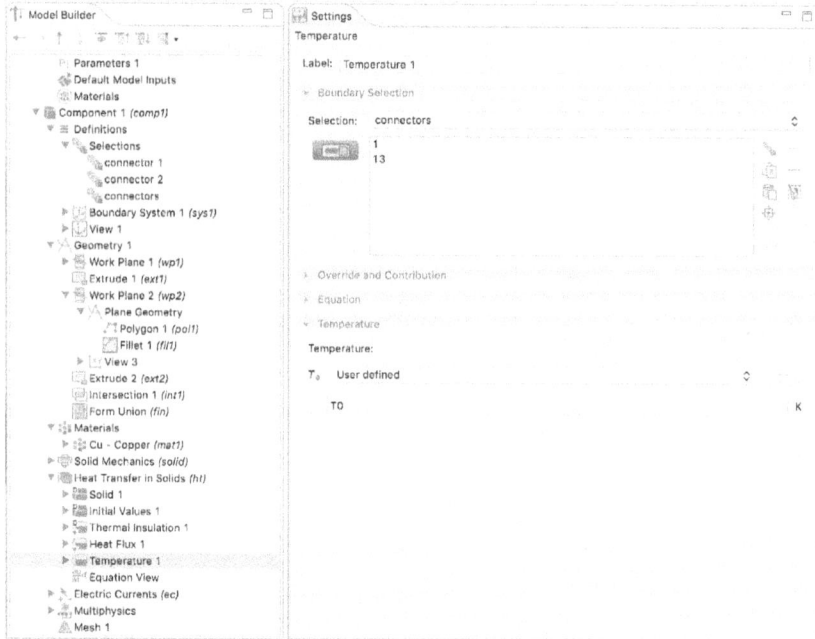

FIGURE 9.44 Temperature 1 boundary settings.

Figure 9.44 shows the Temperature 1 Boundary Settings.

Solid Mechanics (solid), Fixed Constraint 1

Right-Click > Model Builder > Component 1 (*comp1*) > Solid Mechanics (*solid*).

Select: Fixed Constraint from boundary condition section of the Pop-up menu.

Click > Fixed Constraint 1.

Click > Selection pop-up menu bar in Settings Fixed Constraint window > Boundary Selection tab,

Select: connectors from the Pop-up menu.

See Figure 9.45.

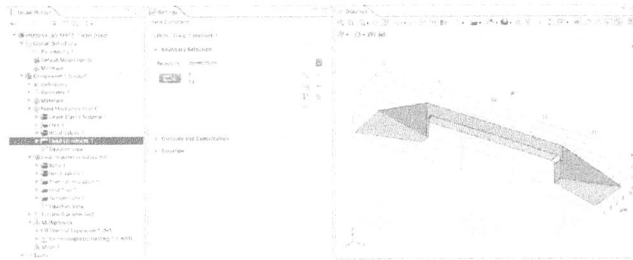

FIGURE 9.45 Fixed constraint 1 boundary settings.

Figure 9.45 shows the Fixed Constraint 1 Boundary Settings.

Mesh 1

Right-Click > Model Builder > Component 1 (*comp1*) > Mesh 1.

Select: Free Tetrahedral from the Pop-up menu.

Click > Model Builder > Component 1 (*comp1*) > Mesh 1 > Size.

Click > Predefined pull-down list in Settings Size window > Element Size tab,

Select > Finer.

See Figure 9.46.

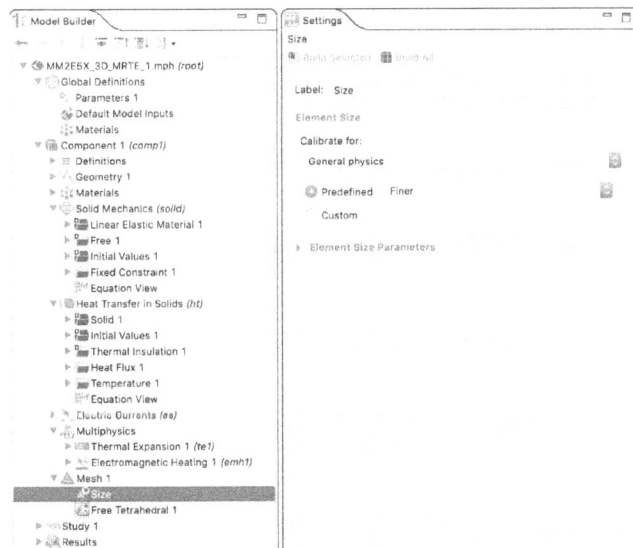

FIGURE 9.46 Mesh size settings.

Figure 9.46 shows the Mesh Size Settings.

Click > Build All in the Settings Size toolbar.

See Figure 9.47.

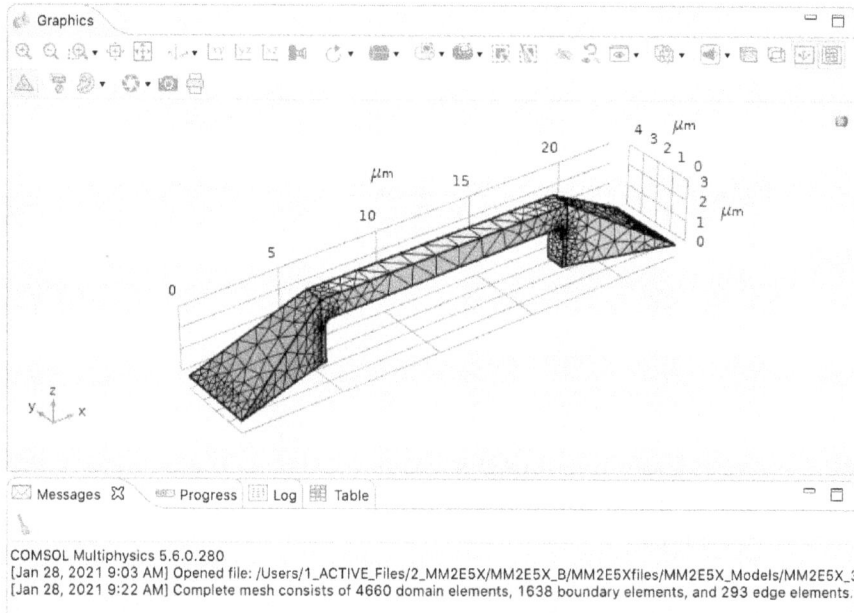

FIGURE 9.47 Graphics window, mesh.

Figure 9.47 shows the Graphics Window, Mesh.

NOTE
After the mesh is built, the model should have 4660 domain elements, 1638 boundary elements and 293 edge elements.

Study 1

In Model Builder,

Right-Click > Study 1.

Select > Compute from the Pop-up menu.

Computed results, using the default display settings, are shown in Figure 9.48.

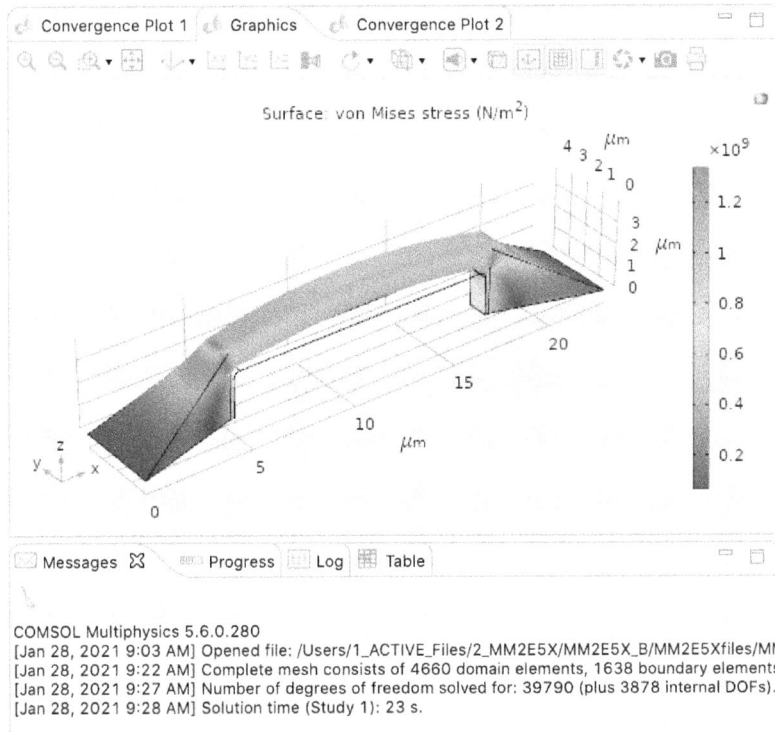

FIGURE 9.48 Converged microresistor beam model using the default von Mises stress plot.

Figure 9.48 shows the Converged Microresistor Beam Model using the Default von Mises Stress Plot.

NOTE *The modeler should note that the computed results, using the default display settings display four (4) different plots from the model solution. The first is the von Mises stress plot, shown in Figure 9.48.*

Results

Additional Results Plots

Click > Results > Temperature (ht).

See Figure 9.49.

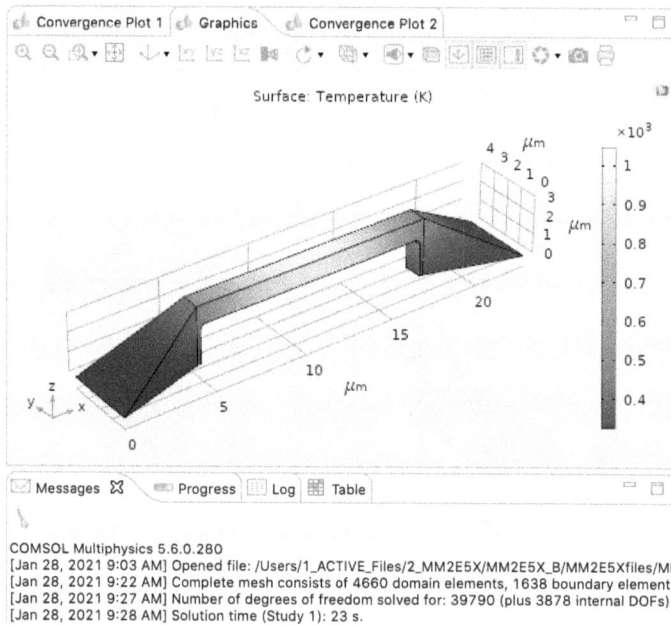

FIGURE 9.49 Linearized resistivity temperature profile.

Figure 9.49 shows the linearized resistivity temperature profile.

NOTE *The modeler should note that the computed results for the maximum temperature, using the copper linearized resistivity, is approximately 1048 K.*

NOTE *The modeler executes the following steps to ensure that the model is properly initialized, before proceeding to make modifications in the postprocessing display parameters.*

Displacement

Right-Click > Model Builder > Results > Stress (solid).

Select: Duplicate.

Click > Model Builder > Results > Stress (solid) 1.

Right-Click > Model Builder > Results > Stress (solid) 1.

Select: Rename.

Enter: Displacement, Study 1 in the Rename edit window.

Click > Displacement, Study 1 twistie.

Click > Surface 1.

Click > Settings Surface > Expression tab > Replace Expression,

Select > Component 1 > Solid Mechanics > Displacement > solid.disp – Displacement magnitude - m.

Click > Settings Surface > Expression tab > Unit:,

Select: nm from the Unit: pull-down menu.

Click > Plot in the Settings Surface window toolbar.

See Figure 9.50.

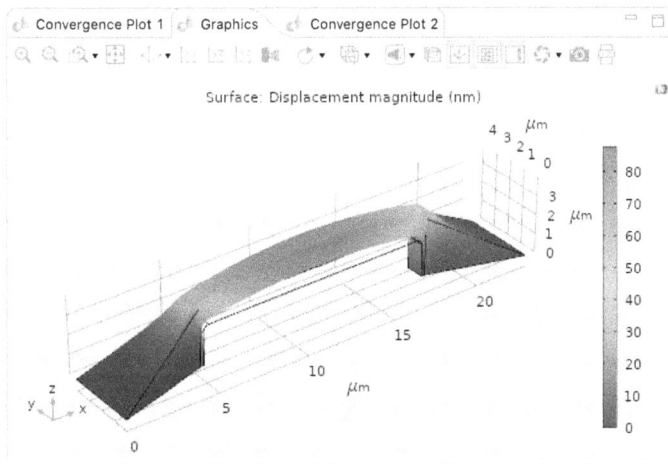

FIGURE 9.50 Thermal expansion displacement.

Figure 9.50 shows the Thermal Expansion Displacement.

NOTE *The modeler should note that the maximum thermal displacement under the linearized resistivity condition is approximately 88 nm.*

NOTE *Now that the modeler has computed the model with the linearized resistivity, it will be interesting to do the same model with a thermally dependent resistivity.*

Add a Study

Click > Model Builder > Component 1 (*comp1*) > Electric Currents (ec) twistie.

Click > Current Conservation 1.

Click > Settings Current Conservation > Constitutive Relation Jc-E tab > Resistivity temperature coefficient.

Select > From material for the α Pop-up menu.

See Figure 9.51.

FIGURE 9.51 Resistivity temperature coefficient settings.

Figure 9.51 shows the Resistivity Temperature Coefficient Settings.

Study 2

In Model Builder,

Right-Click > MM2E5X_3D_MRTE_1.mph (*root*).

Select > Add Study.

In the Add Study window,

Click > General Studies > Stationary.

Click > Add Study (Plus sign).

Close > Add Study window (X).

Right-Click > Study 2,

Select > Compute.

Results

Study 2

Computed results, using the thermally dependent resistivity settings, are shown in Figure 9.52 (Click > Model Builder > Results > Temperature (ht)1).

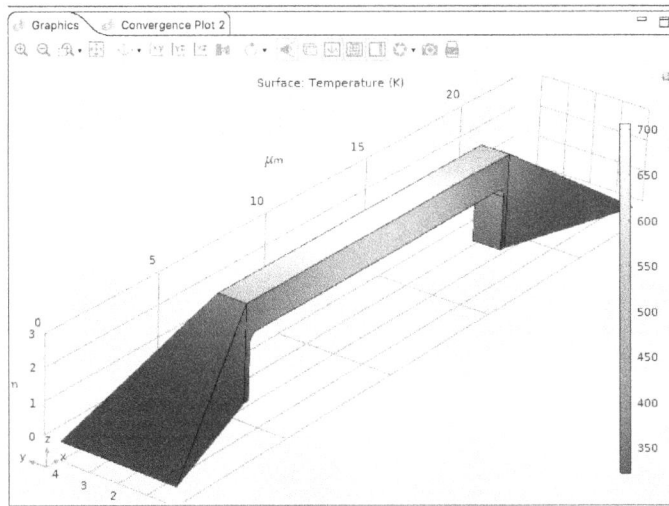

FIGURE 9.52 Converged thermally dependent microresistor beam model.

Figure 9.52 shows the Converged Thermally Dependent Microresistor Beam Model.

NOTE *The modeler should note that the maximum temperature has dropped from approximately 1048 K to approximately 710 K.*

Displacement 2

Right-Click > Model Builder > Results > Stress (solid) 1.

Select: Duplicate.

Click > Model Builder > Results > Stress (solid) 1.1.

Right-Click > Model Builder > Results > Stress (solid) 1.1.

Select: Rename.

Enter: Displacement, Study 2 in the Rename edit window.

Click > Displacement, Study 2 twistie.

Click > Surface 1.

Click > Settings Surface > Expression tab > Replace Expression,

Select > Component 1 > Solid Mechanics > Displacement > solid.disp – Displacement magnitude - m.

Click > Settings Surface > Expression tab > Unit:,

Select: nm from the Unit: pull-down menu.

Click > Plot in the Settings Surface window toolbar.

Computed deformation results, using the thermally dependent resistivity settings, are shown in Figure 9.53.

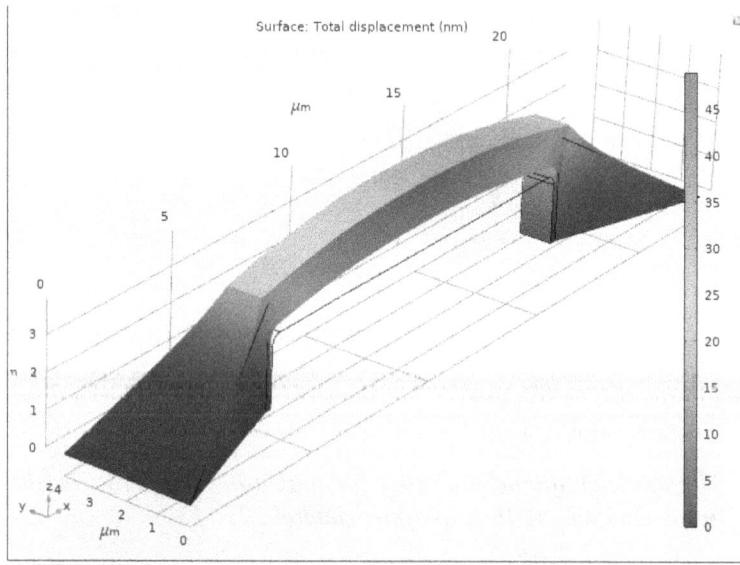

FIGURE 9.53 Converged thermally deformation in the microresistor beam model.

Figure 9.53 shows the Converged Thermally Deformation in the Microresistor Beam Model.

NOTE *The modeler should note that the maximum deformation has dropped from approximately 88 nm to approximately 48 nm.*

FIRST PRINCIPLES AS APPLIED TO 3D MODEL DEFINITION

First Principles Analysis derives from the fundamental laws of nature. In the case of models using this Classical Physics Analysis approach, the laws of conservation in physics require that what goes in (as mass, energy, charge, etc.) must come out (as mass, energy, charge, etc.) or must accumulate within the boundaries of the model.

The careful modeler must be knowledgeable of the implicit assumptions and default specifications that are normally incorporated into the COMSOL Multiphysics software model when a model is built using the default settings.

Consider, for example, the two 3D models developed in this chapter. In these models, it is implicitly assumed that there are no thermally related changes (mechanical, electrical, etc.), except as specified. It is also assumed the materials are homogeneous and isotropic, except as specifically indicated and there are no thin insulating contact barriers at the thermal junctions. None of these assumptions are typically true in the general case. However, by making such assumptions, it is possible to easily build 3D First Approximation Models.

NOTE

A First Approximation Model is one that captures all the essential features of the problem that needs to be solved, without dwelling excessively on all of the small details. A good First Approximation Model will yield an answer that enables the modeler to determine if he needs to invest the time and the resources required to build a more highly detailed model.

Also, the modeler needs to remember to name model parameters carefully as pointed out in Chapter 1.

REFERENCES

1. COMSOL Reference Manual, p. 254

2. *https://en.wikipedia.org/wiki/Vector_(mathematics_and_physics)*

3. *https://en.wikipedia.org/wiki/Tensor*

4. *https://en.wikipedia.org/wiki/Closed-form_expression*

5. *https://en.wikipedia.org/wiki/Lumped_element_model*

6. *https://en.wikipedia.org/wiki/Michael_Faraday*

7. *https://en.wikipedia.org/wiki/Electromagnetic_induction*

8. *https://en.wikipedia.org/wiki/Joseph_Henry*

9. *https://en.wikipedia.org/wiki/Oliver_Heaviside*

10. *https://en.wikipedia.org/wiki/Maxwell%27s_equations*

11. *https://en.wikipedia.org/wiki/Inductance*

12. *https://en.wikipedia.org/wiki/Impedance_parameters*

13. MEMS Module Users Guide, p. 47

14. *https://en.wikipedia.org/wiki/Ohm%27s_law*

15. *https://en.wikipedia.org/wiki/Magnetic_field*

16. *https://en.wikipedia.org/wiki/Magnetic_energy*

17. *https://en.wikipedia.org/wiki/Inductor*

18. *https://en.wikipedia.org/wiki/Ohm%27s_law*

19. *https://en.wikipedia.org/wiki/Joule_effect*

20. *https://en.wikipedia.org/wiki/James_Prescott_Joule*

21. *https://en.wikipedia.org/wiki/Thermodynamics*

22. *https://en.wikipedia.org/wiki/Otto_von_Guericke*

23. *https://en.wikipedia.org/wiki/Vacuum_pump*

24. *https://en.wikipedia.org/wiki/Nicolas_Léonard_Sadi_Carnot*

25. *https://en.wikipedia.org/wiki/William_Thomson,_1st_Baron_Kelvin*

26. COMSOL Reference Manual, p. 351

27. COMSOL Reference Manual, p. 564

SUGGESTED MODELING EXERCISES

1. Build, mesh, and solve the 3D Spiral Coil Microinductor Model as presented earlier in this chapter.

2. Build, mesh, and solve the 3D Linear Microresistor Beam Model as presented earlier in this chapter.

3. Change the values of the materials parameters and then build, mesh, and solve the 3D Spiral Coil Microinductor Model as an example problem.

4. Change the values of the materials parameters and then build, mesh, and solve the 3D Linear Microresistor Beam Model as an example problem.

5. Change the value of the electrical parameters and then build, mesh, and solve the 3D Spiral Coil Microinductor Model as an example problem.

6. Change the value of the geometries of the microresistor and then solve the 3D Linear Microresistor Beam Model as an example problem.

CHAPTER 10

PERFECTLY MATCHED LAYER MODELS USING COMSOL MULTIPHYSICS 5.x

In This Chapter

- Guidelines for Perfectly Matched Layer (PML) Modeling in 5.x
  - PML Modeling Considerations
- Perfectly Matched Layer Models
  - 2D Concave Metallic Mirror PML Model
  - 2D Energy Concentrator PML Model
- First Principles as Applied to PML Model Definition
- References
- Suggested Modeling Exercises

GUIDELINES FOR PERFECTLY MATCHED LAYER (PML) MODELING IN 5.x

NOTE

In this chapter, two 2D PML models will be presented. PML models have proven to be very valuable in the study and application of wave propagation for the science and engineering communities, both in the past and currently. Such models serve as first-cut evaluations of potential systemic physical behavior under the influence of complex external stimuli. PML model responses and other such ancillary information can be gathered and screened early in a project for a first-cut evaluation of the physical behavior of a planned prototype. PML models are typically more conceptually and physically complex than the models that were presented in earlier chapters of this text. The calculated model (simulation) information can be used in the prototype fabrication stage as guidance in the selection of prototype geometry and materials.

Since the models in this and subsequent chapters are more conceptually complex and are potentially more difficult to solve than the models presented thus far, it is important that the modeler have available the tools necessary to most easily utilize the powerful capabilities of the 5.x software. In order to do that, if you have not done this previously, the modeler should go to the main 5.x toolbar Click > Options – Preferences – Show More Options. When the Preferences – Show More Options edit window is shown, Select > Equation view checkbox and Equation Section checkbox. Click > OK.{10.1}.

Perfectly Matched Layer (PML) Modeling Guidelines and Coordinate Considerations

PML Theory

One of the fundamental difficulties underlying electromagnetic wave equation calculations (Maxwell's Equations {10.2}) is dealing with a propagating wave after the wave interacts with a boundary (reflection). If the boundary of a model domain is terminated in the typical fashion {10.3}, unwanted reflections will typically be incorporated into the solution, potentially creating undesired and possibly erroneous model solution values. Fortunately, for the modeler of today, there is a methodology that works sufficiently well that it essentially eliminates reflection problems at the domain boundary. That methodology is the Perfectly Matched Layer.

The Perfectly Matched Layer (PML) {10.4} is an approximation methodology originally developed in 1994 by Jean-Pierre Berenger for use with FDTD {10.5} (Finite-Difference Time-Domain) electromagnetic modeling calculations. The PML technique has now been adapted and applied to other calculational techniques that have similar domain-mediated needs (e.g. FEM and others) {10.6}. The PML methodology can be applied to a large variety of diverse wave equation problems {10.7}. Herein, however, it is only applied to electromagnetic problems within the context of the COMSOL RF Module {10.8}.

NOTE   *For more detailed applications and a history of the PML methodology in other types of wave problems, the modeler is referred to the literature.*

The PML methodology functions by adding anisotropic attenuating domains (layers) outside the modeled domain. The anisotropic attenuating domains (PMLs) create for the modeled domain a set of essentially reflectionless boundaries.

Examples of modeling domains with PMLs can be found in the COMSOL Multiphysics literature {10.9}. The coordinate systems employed with the domain structures are those that are associated with their respective geometries.

NOTE

In order to achieve the desired behavior of the wave equation PDE, the entire model domain, including the Perfectly Matched Layers, is transformed to a complex coordinate system. For a Cartesian System (x, y, z), the transformation occurs as follows:

$$\frac{\partial}{\partial x} \rightarrow \frac{1}{1 + \dfrac{i\sigma(x)}{\omega}} \frac{\partial}{\partial x} : \quad \frac{\partial}{\partial y} \rightarrow \frac{1}{1 + \dfrac{i\sigma(y)}{\omega}} \frac{\partial}{\partial y} : \quad \frac{\partial}{\partial z} \rightarrow \frac{1}{1 + \dfrac{i\sigma(z)}{\omega}} \frac{\partial}{\partial z} \qquad (10.1)$$

Where: $\sigma(x, y, z) =$ is a step function that is zero inside the solution domain and a positive real number or an appropriate function of the designated coordinate variable (x, y, z), outside the solution domain and inside the PML.

The transformation of the PDE in the above fashion results in a solution with a multiplicative term that is, in general, as follows:

$$F(x,y,z) = f(x,y,z) * e^{-\frac{k\sigma(x,y,z)}{\omega}} \qquad (10.2)$$

Where $F(x,y,z) = f(x,y,z)*e^{-0}$ (The Solution inside the domain).

$F(x,y,z) = f(x,y,z)*e^{-\kappa\sigma(x,y,z)/\omega}$ (The Decaying Solution within the PML domain).

At the outer PML boundary, the preferred boundary condition is the Scattering boundary condition. However, if the attenuation of the propagating wave, at the outer boundary of the PML, is sufficient, then the particular boundary condition invoked is largely irrelevant. This pertains, since the amplitude of the reflected wave will be sufficiently small as not to contribute to the final solution.

Figures 10.1 and 10.2 show examples of wavefront behavior within a domain and within a PML. For an example of the wavefront inside the modeling domain see Figure 10.1. For an example of the wavefront inside the PML domain see Figure 10.2.

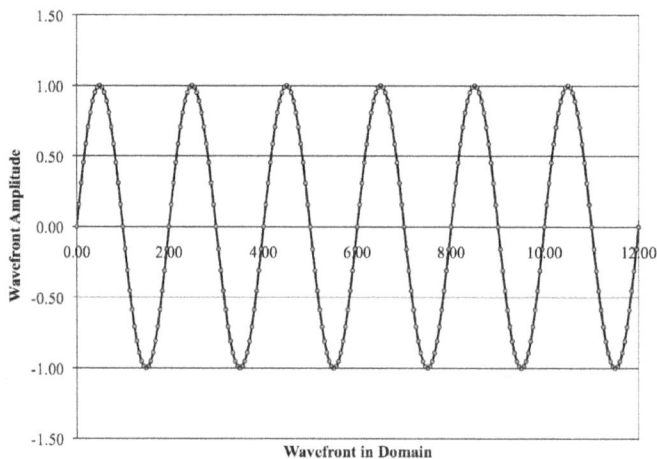

FIGURE 10.1 Wave equation solution example inside the modeling domain.

Figure 10.1 shows the wave equation solution example inside the modeling domain.

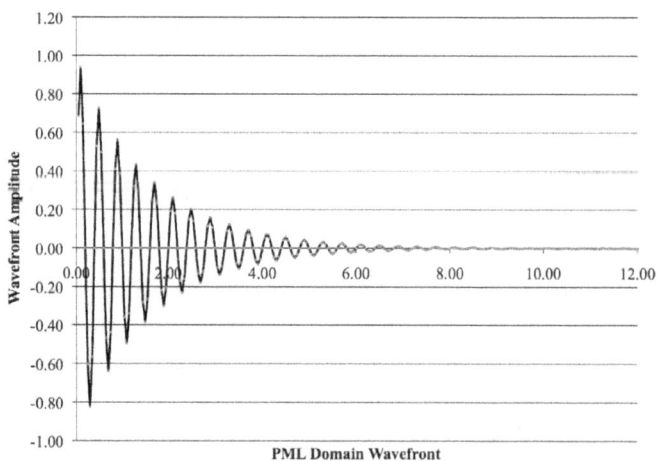

FIGURE 10.2 Wave equation solution example inside the PML domain.

Figure 10.2 shows the wave equation solution example inside the PML domain.

PERFECTLY MATCHED LAYER MODELS

NOTE *The two models presented in this chapter are derived from the COMSOL PML tutorial model "Radar Cross Section."*

Building the 2D Concave Metallic Mirror PML Model

The Concave Metallic Mirror is a concept widely utilized in Optical Physics. In this application, the principles of Optics are applied to lower frequency electromagnetic waves to focus an impinging wavefront into the region of a sensor. The act of focusing the wavefront effectively increases the magnitude of the impinging signal (i.e. adds Gain {10.10}) in the region of the sensor, thus making the focused signal more easily detectable.

NOTE

The focusing concept can also be utilized to concentrate large-area, diffuse energy from a renewable energy source (e.g. solar), into a smaller area, higher energy density source for more convenient application. An example of that concept will be demonstrated in the 2D Energy Concentrator PML Model.

Startup 5.x, Click > Model Wizard button.

Select Space Dimension > 2D.

Click > Twistie for the Radio Frequency interface.

Click > Electromagnetic Waves, Frequency Domain (emw).

Click > Add button.

Click > Study (Right Pointing Arrow button).

Select > General Studies > Frequency Domain.

Click > Done (Checked Box button).

Click > Save As.

Enter MM2E5X_2D_PML_CMM_1.mph.

Click > Save.

See Figure 10.3.

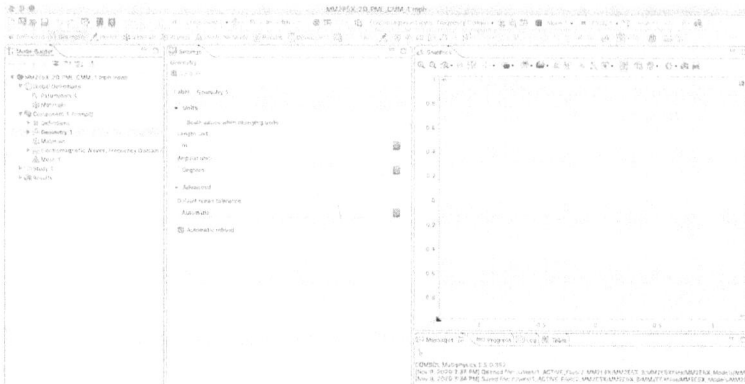

FIGURE 10.3 Desktop display for the MM2E5X_2D_PML_CMM_1.mph model.

Figure 10.3 shows the Display for the MM2E5X_2D_PML_CMM_1.mph model.

Global Definitions

Parameters

In Model Builder > under Global Definitions,

Click > Parameters 1.

In the Settings Parameters window > in the Parameters table, on the Parameters tab,

Enter: all the Parameters as shown, in Table 10.1.

NOTE

The speed of light in free space (vacuum) is actually 299,792,458 m/s (exactly) {10.11}. However, for use in this model, the approximate value of 3.0e8 m/s will be used, as a good first approximation.

The wave number is proportional to the reciprocal of wavelength {10.12}.

TABLE 10.1 CMM Parameters.

Name	Expression	Description
f_a	100[MHz]	Frequency
c_L	3.0e8[m/s]	Speed of light in free space
k_0	2*pi*f_a/c_L	Free space wave number
E_z	1[V/m]	Electric field amplitude
me_s	c_L/f_a/6	Maximum element size

See Figure 10.4.

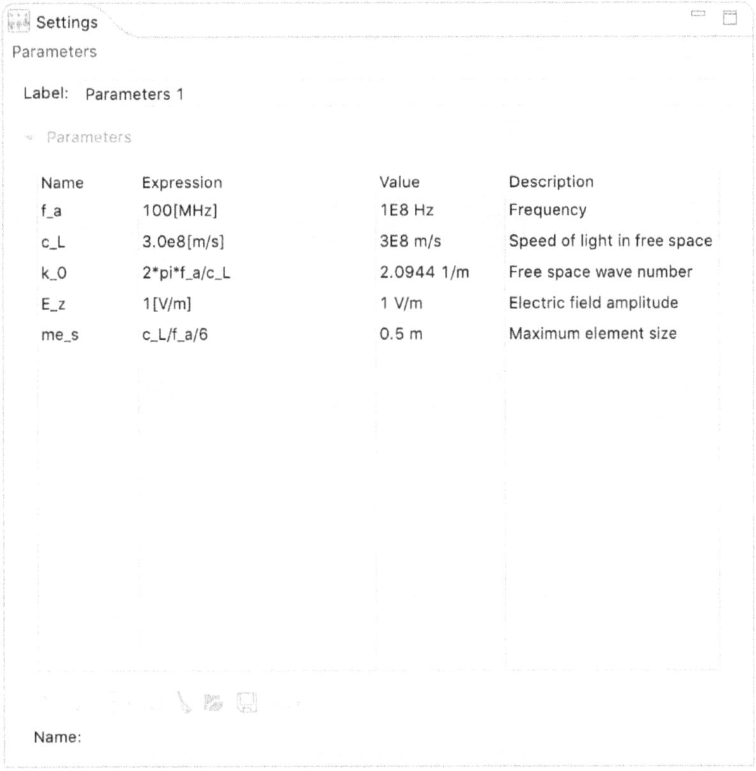

FIGURE 10.4 Settings parameters window – Parameters table.

Figure 10.4 shows the Settings Parameters Window – Parameters Table.

Variables

In Model Builder,

Right-Click > Model Builder > Global Definitions,

Select: Variables from the Pop-up menu.

In Settings Variable window, Variables tab,

Enter: all the CMM Variables into the Variables text field, as shown, in Table 10.2.

TABLE 10.2 CMM Variables.

Name	Expression	Description
phideg	0	Initial angle
phi	phideg*pi/180	Angle of incidence, radians
E_b	exp(j*k_0*(x*cos(phi)+y*sin(phi)))*E_z	Electric field
dm_x	x*cos(phi)	Destination map x expression
dm_y	y*sin(phi)	Destination map y expression
Efar	1	Reciprocal backscatter

See Figure 10.5.

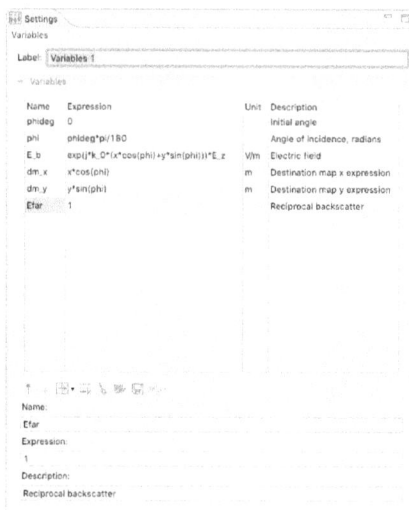

FIGURE 10.5 Settings variables > Variables edit window.

Figure 10.5 shows the Settings Variables > Variables edit window.

Geometry

PML Domain

In Model Builder

Right-Click > Model Builder > Component 1 (*comp1*) > Geometry 1.

Select: Circle from the Pop-up menu.

Enter the coordinates shown in Table 10.3.

Click as instructed.

Repeat the sequence until completed.

TABLE 10.3 PML Domain.

Circle	Radius	Click
1	15[m]	Build Selected
2	12[m]	Build All Objects

See Figure 10.6.

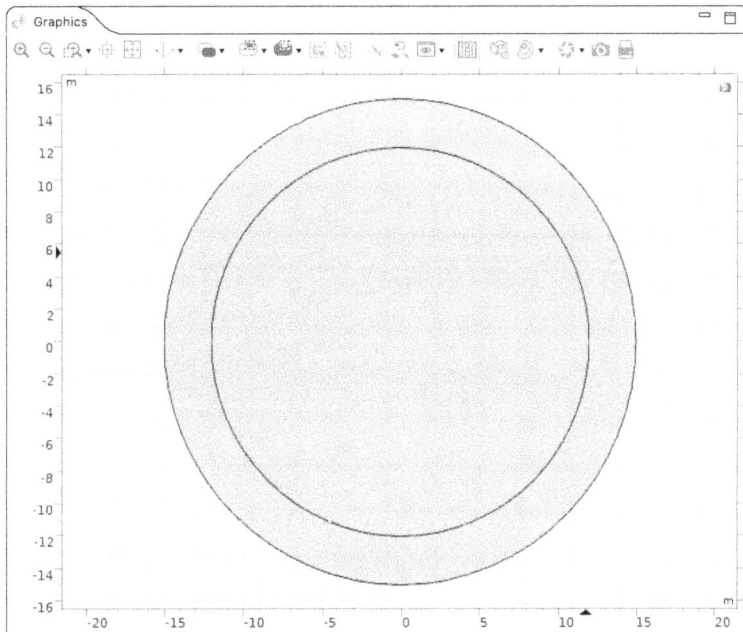

FIGURE 10.6 PML in graphics edit window.

Figure 10.6 shows the PML domain in the Graphics edit window.

CMM Domain

In Model Builder,

Right-Click > Model Builder > Component 1 (*comp1*) > Geometry 1.

Select: Circle from the Pop-up menu.

Enter the coordinates shown in Table 10.4.

Click as instructed.

Repeat the sequence until completed.

TABLE 10.4 CMM Domain.

Circle	Radius	Click
3	3.0[m]	Build Selected
4	2.9[m]	Build All Objects

See Figure 10.7.

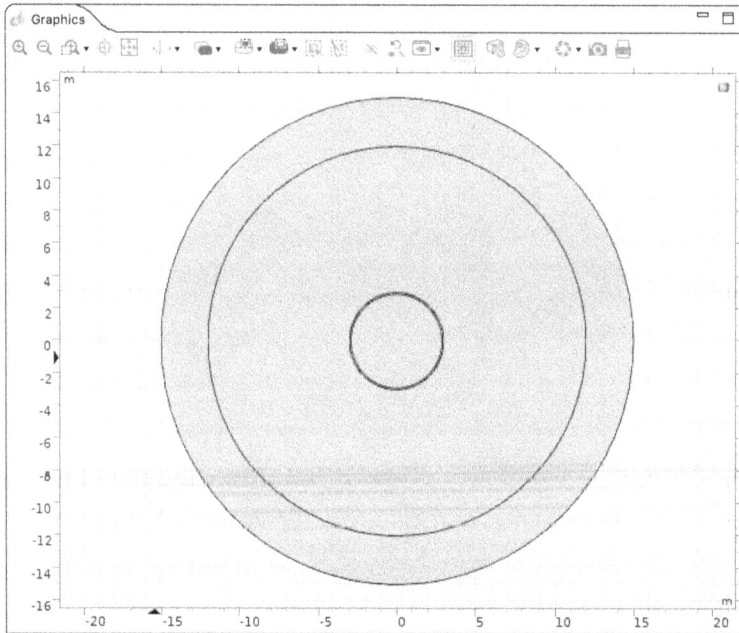

FIGURE 10.7 CMM circles in the graphics edit window.

Figure 10.7 shows the CMM Circles in the Graphics edit window.

NOTE *In order to build a metallic mirror using the CMM Circles, the modeler needs to perform two Boolean Difference operations.*

CMM Formation

Right-Click > Model Builder > Component 1 (*comp1*) > Geometry 1.

Select: Booleans and Partitions > Difference from the Pop-up menu.

Use the Clipboard to insert c3 in the Objects to add text field.

See Figure 10.8.

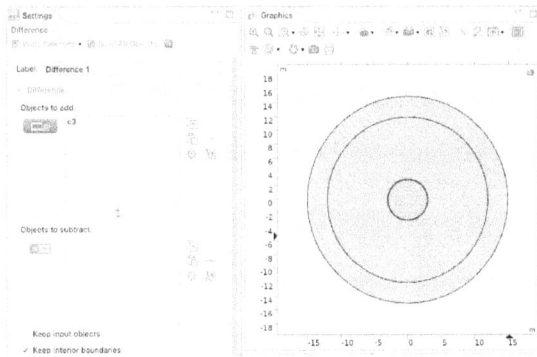

FIGURE 10.8 CMM circle 3 in graphics edit window.

Figure 10.8 shows the selected portion of the CMM Circle 3 in the Graphics edit window.

Click > Activate button on the Objects to subtract window.

Use the Clipboard to insert c4 in the Objects to subtract text field.

See Figure 10.9.

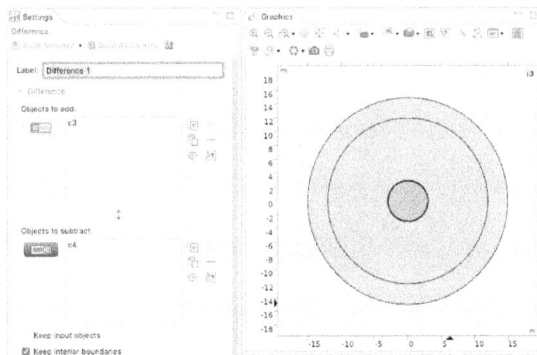

FIGURE 10.9 Settings difference > Difference prepared for the build selected operation.

Figure 10.9 shows Settings Difference >Difference Prepared for the Build Selected operation.

Click > Build Selected in the Settings Difference Toolbar.

Rectangle

NOTE *A Rectangle is needed to remove the Right Half of the difference object formed in the last step.*

Right-Click > Model Builder > Component 1 (*comp1*) > Geometry 1.

Select: Rectangle from the Pop-up menu.

Enter > Width = 3[m], Height = 6[m], Base = Corner, x = 0[m], and y = −3[m] in Settings Rectangle > Size and Shape tab and Position tab text field windows.

Click > Build Selected.

Click > Zoom In.

Click > Zoom in a second time.

See Figure 10.10.

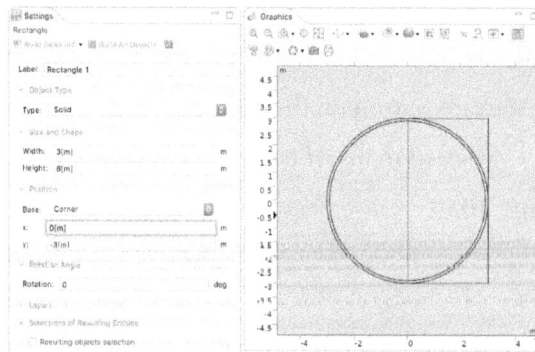

FIGURE 10.10 Graphics components prepared for build selected.

Figure 10.10 shows Graphics Components Prepared for Build Selected.

Right-Click > Model Builder > Component 1 (*comp1*) > Geometry 1.

Select > Booleans and Partitions > Difference from the Pop-up menu.

Use the Clipboard to insert dif1 in the Objects to add text field.

Click > Activate Selection in Settings Difference > Difference > Objects to subtract.

Use the Clipboard to insert r1 in the Objects to subtract text field.

See Figure 10.11.

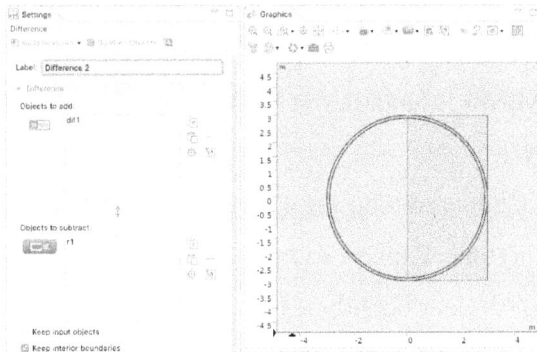

FIGURE 10.11 Settings – Difference – Difference prepared for the build selected operation.

Figure 10.11 shows Settings – Difference – Difference Prepared for the Build Selected operation.

Click > Build Selected in the Settings Toolbar.

Click > Zoom Extents.

See Figure 10.12.

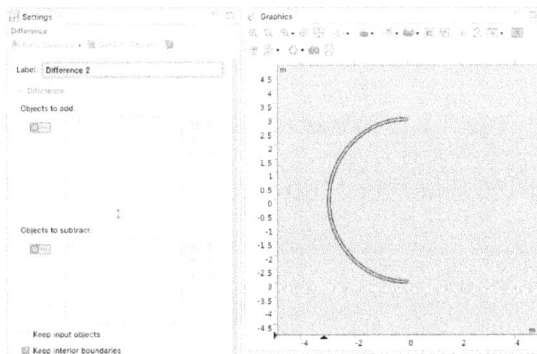

FIGURE 10.12 CMM in the graphics window.

Figure 10.12 shows the CMM in the Graphics window.

Selection 1

Right-Click > Model Builder > Component 1 (*comp1*) > Definitions.

Select: Selections > Explicit from the Pop-up menu.

Click > Model Builder > Component 1 (*comp1*) > Definitions > Explicit 1.

Right- Click > Model Builder > Component 1 (*comp1*) > Definitions > Explicit 1,

Select: Rename from the Pop-up menu.

Enter: Mirror Boundaries in the Rename Selection text field.

Click > OK.

Use the Clipboard to insert 2 in the Input Entities text field.

Click > Settings Explicit > Output Entities > Selection Popup-menu.

Select > Adjacent boundaries from the Pop-up menu.

See Figure 10.13.

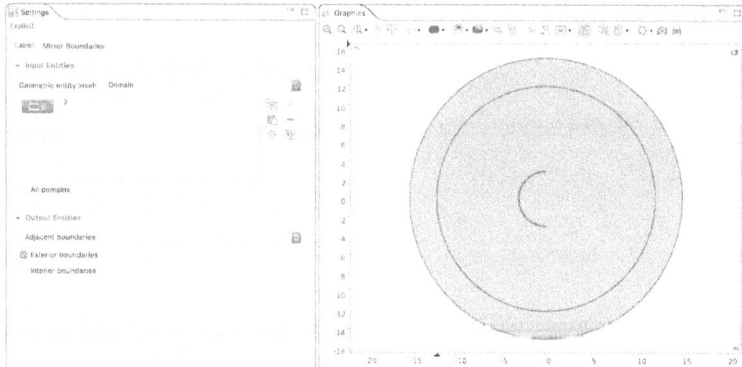

FIGURE 10.13 Mirror boundaries selection.

Figure 10.13 shows the Mirror Boundaries Selection.

Materials

Material 1

Right-Click > Model Builder > Component 1 (*comp1*) > Materials.

Select: Add Material from Library from the Pop-up menu.

Click > In the Add Material window > Built-in twistie,

Click > Air.

Right- Click > Air.

Select: Add to Component 1.

Click > Domain 2 in Settings Material > Geometric Entity Selection text field,

Click > Remove (minus sign).

Click > Close on the Add Material window (hollow X on tab).

Material 2

Right-Click > Model Builder > Component 1 (*comp1*) > Materials.

Select: Add Material from Library from the Pop-up menu.

Click > In the Add Material window > Built-in twistie.

Click > Aluminum.

Right-Click > Aluminum.

Select > Add to Component 1.

Click > Model Builder > Component 1 (*comp1*) > Materials > Aluminum (*mat2*).

Click > Geometric entity level in Settings Material > Geometric Entity Selection.

Select: Boundary from the Pop-up menu.

Selection

Click > Selection in Settings Material > Geometric Entity Selection.

Select: Mirror Boundaries from the Pop-up menu.

Close > Add Material window.

Electromagnetic Waves (emw)

Domains

Click > Model Builder > Component 1 (*comp1*) > Electromagnetic Waves, Frequency Domain (emw).

Click > 2 in Settings Electromagnetic Waves, Frequency Domain > Domain Selection > Selection edit window.

Click > Remove from Selection button (Minus sign) in Settings – Electromagnetic Waves – Domains.

This model uses the boundaries of the mirror to represent the mirror. Because that is the case, the field inside the mirror is identically zero and requires no solution inside. This approach saves both time and effort.

Settings Electromagnetic Waves, Frequency Domain

Click > Formulation pull-down menu in Settings Electromagnetic Waves, Frequency Domain > Formulation,

Select: Scattered field from the Pull-down menu.

Enter E_b in the z position of the Background electric field text field in Settings Electromagnetic Waves, Frequency Domain > Formulation tab.

See Figure 10.14.

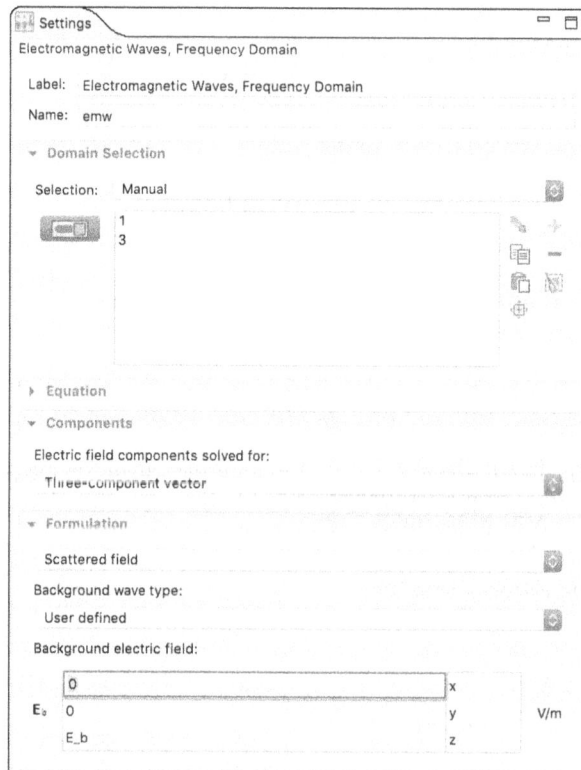

FIGURE 10.14 Electromagnetic waves settings selection.

Figure 10.14 shows the Electromagnetic Waves Settings Selection.

Perfectly Matched Layers 1

The Perfectly Matched Layer simulates an infinitely distant boundary layer. That eliminates possible interference of a back scattered wave with the waves in the region of interest at the center of the model.

Right-Click > Model Builder > Component 1 (*comp1*) > Definitions.

Select > Perfectly Matched Layer.

Click > Perfectly Matched Layer 1 (*pml1*).

Use the Clipboard to enter 3 in the Domain Selection text field.

Coordinates

The coordinates used in this 2D model are Cylindrical. They are set in that mode below.

Click > In Settings Perfectly Matched Layer > Geometry tab, > Type,

Select: Cylindrical from the Type Pull-down menu.

See Figure 10.15.

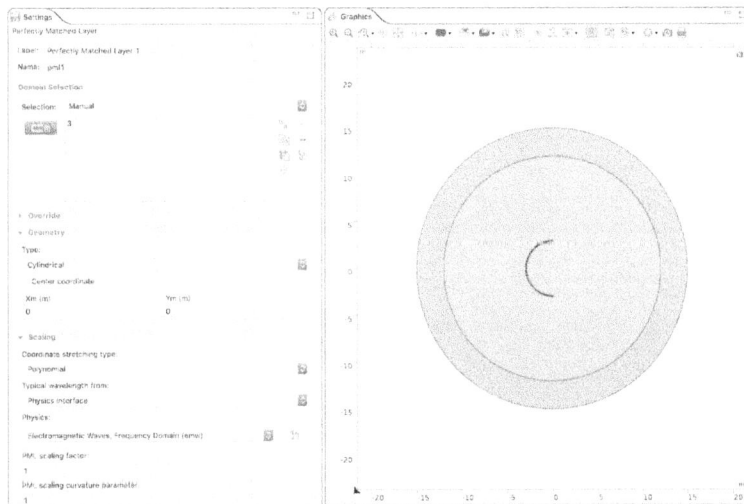

FIGURE 10.15 Perfectly matched layer settings selection.

Figure 10.15 shows the Perfectly Matched Layer Settings Selection.

Impedance Boundary Condition 1

The Impedance Boundary Condition assumes the skin depth in the material is significantly less than the material thickness. In this case it is on the order of microns.

Right-Click > Model Builder > Component 1 (*comp1*) > Electromagnetic Waves, Frequency Domain (emw).

Select: Impedance Boundary Condition.

Click > Impedance Boundary Condition 1.

Click > Selection in Settings Impedance Boundary Condition window > Boundary Selection tab,

Select: Mirror Boundaries from the Pull-down menu.

See Figure 10.16.

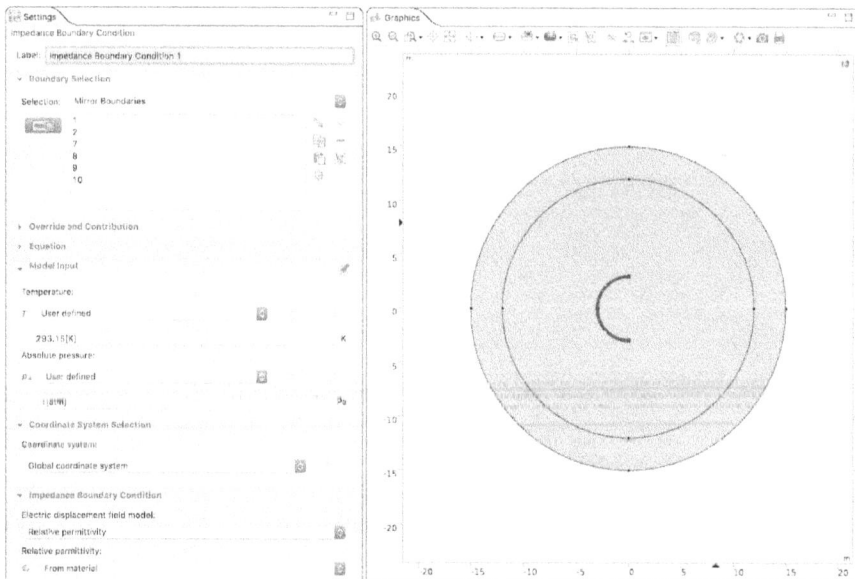

FIGURE 10.16 Impedance boundary condition settings selection.

Figure 10.16 shows the Impedance Boundary Condition Settings Selection.

Far-Field Calculation 1

The requirement for the Far-Field Calculation is that all reflecting surfaces are surrounded, which they are.

Right-Click > Model Builder > Component 1 (*comp1*) > Electromagnetic Waves, Frequency Domain (emw).

Select: Far-Field Domain.

Click > Far-Field Domain 1.

Verify that Domain 1 is Selected, if not use the Clipboard to insert 1 into the Domain Selection text field.

See Figure 10.17a.

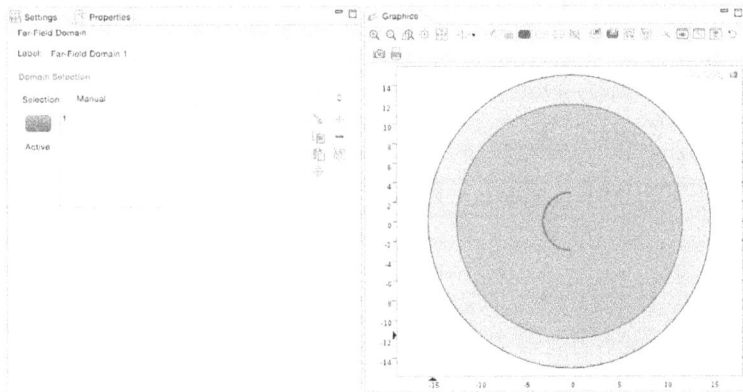

FIGURE 10.17a Far-field domain settings selection.

Figure 10.17a shows the Far-Field Domain Settings Selection.

Click > Far-Field Domain 1 twistie.

Click > Far-Field Calculation 1.

Replace the Default Boundary Selections, by doing the following operation,

Select: Mirror Boundaries from the Settings Far-Field Calculation > Boundary Selection > Selection Pop-up menu.

See Figure 10.17b.

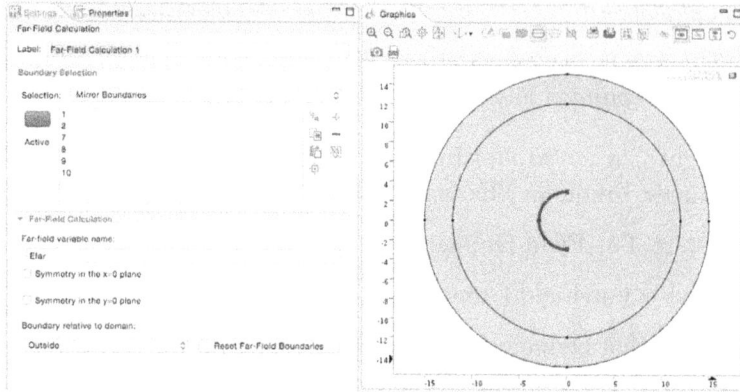

FIGURE 10.17b Far-field calculation settings selection.

Figure 10.17b shows the Far-Field Calculation Settings Selection.

Right-Click > Model Builder > Component 1 (*comp1*) > Electromagnetic Waves, Frequency Domain (emw).

Select: Perfect Electric Conductor.

Click > Perfect Electric Conductor 2.

Using the Clipboard,

Enter: 3 4 11 14 in Boundary Selection text field.

See Figure 10.18a.

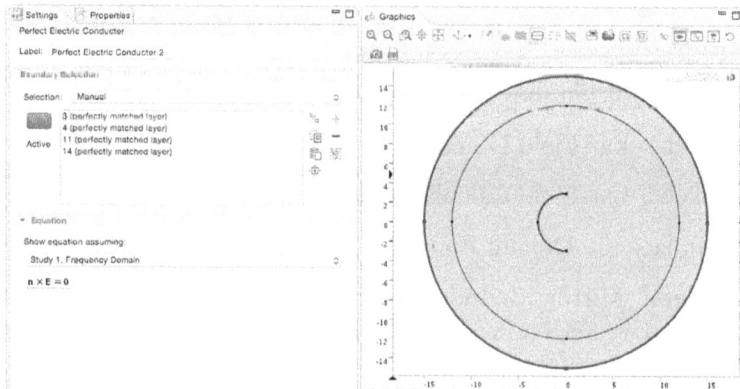

FIGURE 10.18a Perfect electric conductor settings selection.

Figure 10.18a shows the Perfect Electric Conductor Settings Selection.

Definitions

General Extrusion 1

_____
NOTE
_____

The General Extrusion {10.13} is a coupling operator that maps an expression defined on a source domain to an expression that can be evaluated on any domain where the destination map expressions are valid.

Right-Click > Model Builder > Component 1 (*comp1*) > Definitions.

Select: Nonlocal Couplings > General Extrusion.

Click > General Extrusion 1.

Enter: back (this label is a new operator name) in Settings General Extrusion > Operator Name text field.

Enter: 1, using the Clipboard to Select Domain 1 for the source Selection text field.

Enter: dm_x in Settings General Extrusion window > Destination Map tab > x-expression text field.

Enter > dm_y in Settings General Extrusion window > Destination Map tab > y-expression text field.

See Figure 10.18b.

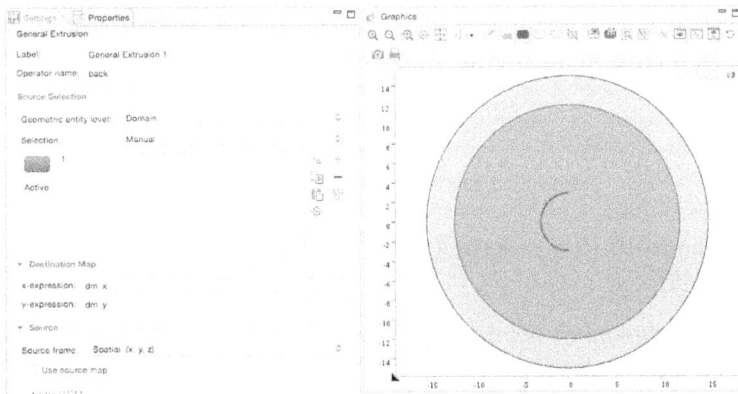

FIGURE 10.18b General extrusion settings selection.

Figure 10.18b shows the General Extrusion Settings Selection.

Mesh 1

Right-Click > Model Builder > Component 1 (*comp1*) > Mesh 1.

Select: Free Triangular from the Pop-up menu.

Click > Model Builder > Component 1 (*comp1*) > Mesh 1 > Size.

Click > Settings Size window > Element Size Parameters tab twistie.

Enter > me_s in Settings Size > Element Size Parameters tab > Maximum element size text field.

Click > Build All in Settings Size window toolbar.

See Figure 10.19.

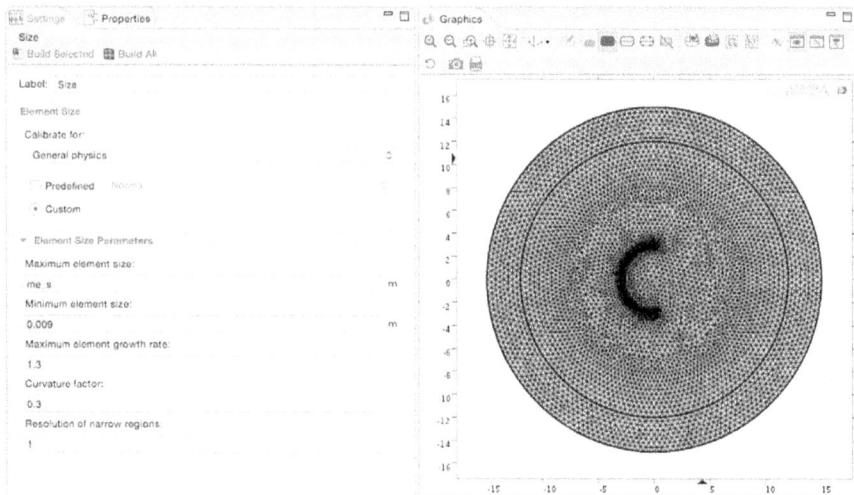

FIGURE 10.19 Graphics window and the size settings after mesh build.

Figure 10.19 shows the Graphics Window and the Size Settings after Mesh Build.

NOTE *After the mesh is built, the model should have 10254 domain elements and 478 boundary elements.*

Study 1

Frequencies

Click > Model Builder > Study 1 twistie.

Click > Model Builder –> Study 1 > Step 1: Frequency Domain.

Enter > f_a in Settings Frequency Domain window > Study Settings tab > Frequencies text field.

See Figure 10.20.

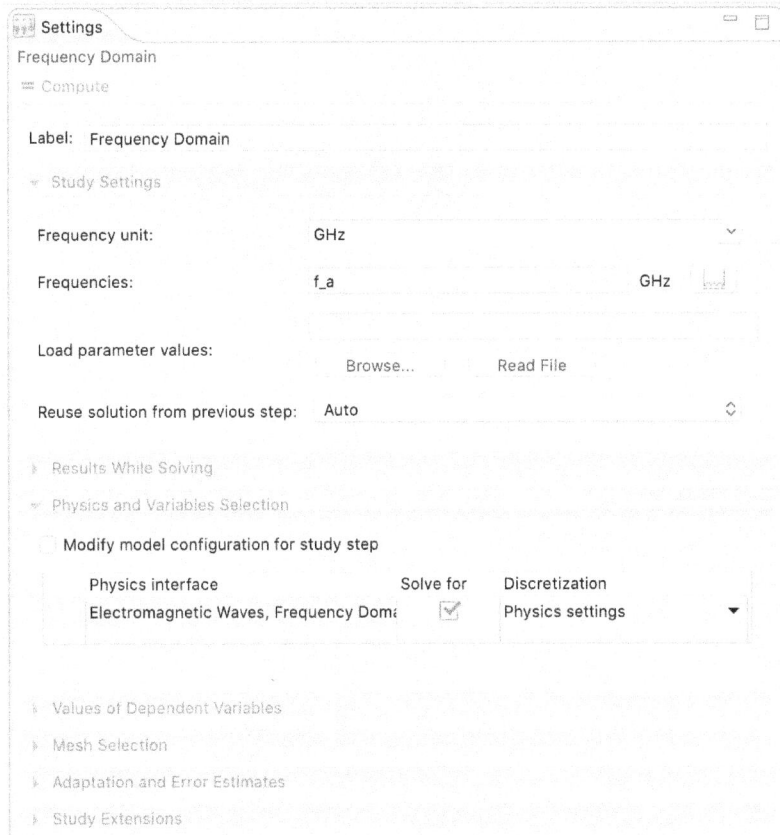

FIGURE 10.20 Settings – Frequency domain.

Figure 10.20 shows the Settings – Frequency Domain.

Right-Click > Model Builder > Study 1.

Select > Compute.

See Figure 10.21.

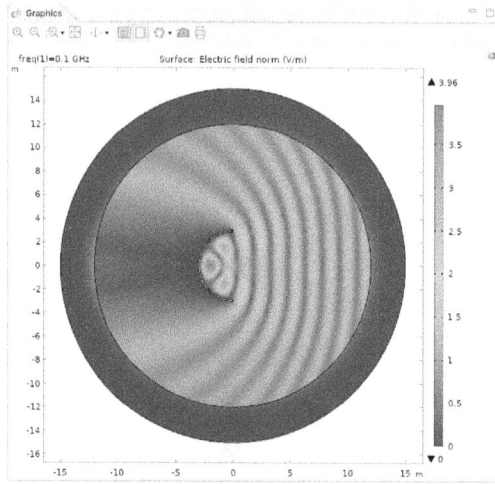

FIGURE 10.21 CMM computed solution.

Figure 10.21 shows the CMM Computed Solution.

Expanded Solution

Click > Zoom in twice.

See Figure 10.22.

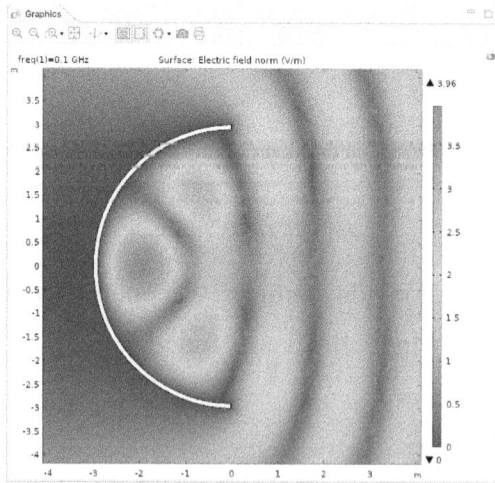

FIGURE 10.22 High electric field in CMM computed solution.

Figure 10.22 shows a High Electric Field in CMM Computed Solution.

NOTE *The modeler should note that a high electric field point forms immediately in front of the metallic mirror. The mirror contributes a gain of approximately four (4).*

2D Concave Metallic Mirror PML Model Summary and Conclusions

The 2D Concave Metallic Mirror PML Model is a powerful modeling tool that can be used to calculate the gain of various mirrors of different shapes and sizes. With this model, the modeler can easily vary all of the geometric parameters and optimize the design before the first prototype is physically built. These types of metallic mirrors are widely used in industry.

Energy Concentrator Mirror Systems

Metallic mirror energy concentration systems are widely used in renewable energy applications {10.14, 10.15, 10.16, 10.17}. This model demonstrates the building of a two-trough elliptical collector, using the perfectly matched layer method.

Building the 2D Energy Concentrator PML Model

NOTE *The elliptical mirror {10.18} is a design concept widely utilized in Optical Physics. In this application, the principles of Optics are applied to lower frequency electromagnetic waves to focus an impinging wavefront into the region of a sensor. The act of focusing the wavefront effectively increases the magnitude of the impinging signal, concentrating power in the region of interest, thus making the focused signal more easily usable.*

The focusing concept is demonstrated here to concentrate large-area, diffuse energy, from a renewable energy source (solar), into a smaller area, higher energy density source for more convenient application.

Startup 5.x, Click > Model Wizard button.

Select Space Dimension > 2D.

Click > Twistie for the Radio Frequency interface.

Click > Electromagnetic Waves, Frequency Domain (emw).

Click > Add button.

Click > Study (Right Pointing Arrow button).

Select > General Studies > Frequency Domain.

Click > Done (Checked Box button).

Click > Save As.

Enter MM2E5X_2D_PML_EC_1.mph.

Click > Save.

See Figure 10.23.

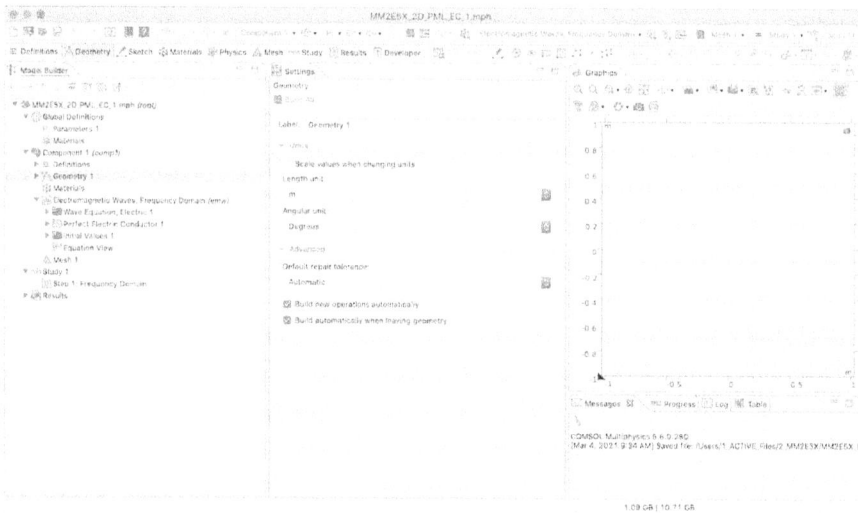

FIGURE 10.23 Desktop display for the MM2E5X_2D_PML_EC_1.mph model.

Figure 10.23 shows the Display for the MM2E5X_2D_PML_EC_1.mph model.

Global Definitions

Parameters

In Model Builder,

Click > Model Builder > Global Definitions > Parameters 1.

Enter all the Parameters, as shown, in Table 10.5.

The speed of light in free space (vacuum) is actually 299,792,458 m/s (exactly). However, for use in this model, the approximate value of 3.0e8 m/s will be used, as a good first approximation.

The wave number is proportional to the reciprocal of wavelength.

TABLE 10.5 EC Parameters.

Name	Expression	Description
f_a	100[MHz]	Frequency
c_L	3.0e8[m/s]	Speed of light in free space
k_0	2*pi*f_a/c_L	Free space wave number
E_z	1[V/m]	Electric field amplitude
me_s	c_L/f_a/6	Maximum element size

See Figure 10.24.

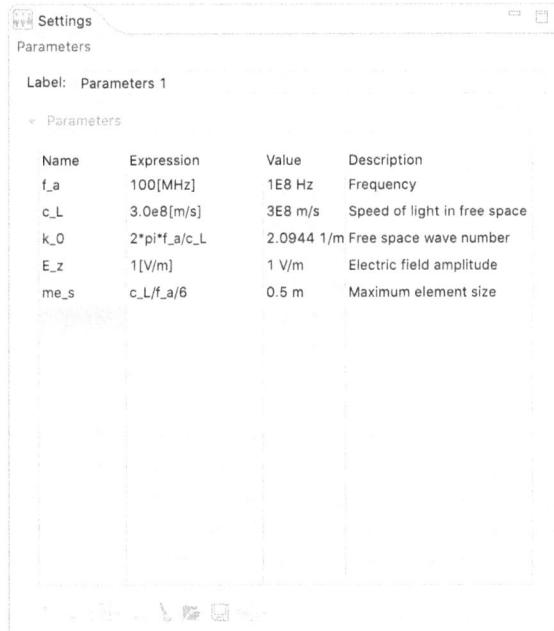

FIGURE 10.24 Settings parameters > Parameters tab text field.

Figure 10.24 shows the Settings Parameters > Parameters tab text field.

Variables

In Model Builder,

Right-Click > Model Builder – Global Definitions,

Select: Variables from the Pop-up menu.

Enter all the Variables, as shown, in Table 10.6.

TABLE 10.6 EC variables.

Name	Expression	Description
phideg	0	Initial angle
phi	phideg*pi/180	Angle of incidence, radians
E_b	exp(j*k_0*(x*cos(phi)+y*sin(phi)))*E_z	Electric field
dm_x	x*cos(phi)	Destination map x expression
dm_y	x*sin(phi)	Destination map y expression
Efar	1	Reciprocal backscatter

See Figure 10.25.

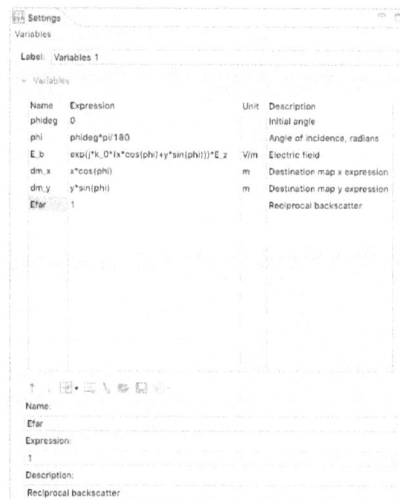

FIGURE 10.25 Settings variables > Variables text field.

Figure 10.25 shows the Settings Variables > Variables text field.

Geometry

PML Domain

In Model Builder,

Right-Click > Model Builder > Component 1 (*comp1*) > Geometry 1.

Select: Circle from the Pop-up menu.

Enter the coordinates shown in Table 10.7.

Click as instructed.

Repeat the sequence until completed.

TABLE 10.7 PML Domain.

Circle	Radius	Click
1	15[m]	Build Selected
2	12[m]	Build All Objects

See Figure 10.26.

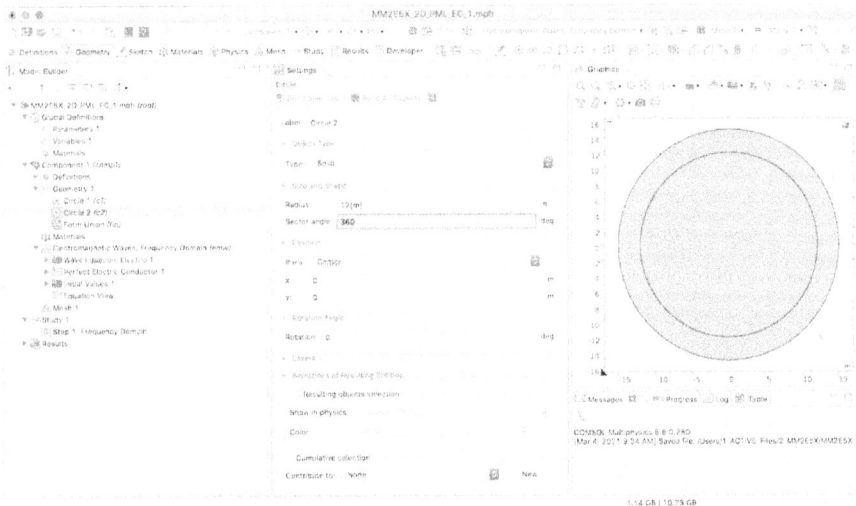

FIGURE 10.26 PML in the graphics window.

Figure 10.26 shows the PML in the Graphics window.

EC Domain

In Model Builder,

Right-Click > Model Builder > Component 1 (*comp1*) > Geometry 1.

Select: Ellipse from the Pop-up menu.

Enter the coordinates shown in Table 10.8.

Click as instructed.

Repeat the sequence until completed.

TABLE 10.8 EC Domains.

Ellipse	a-semiaxis	b-semiaxis	x	y	Click
1	3.0[m]	2.0[m]	0	−2	Build Selected
2	3.0[m]	2.0[m[	0	2	Build Selected
3	2.9[m]	1.9[m]	0	−2	Build Selected
4	2.9[m]	1.9[m]	0	2	Build Selected

See Figure 10.27.

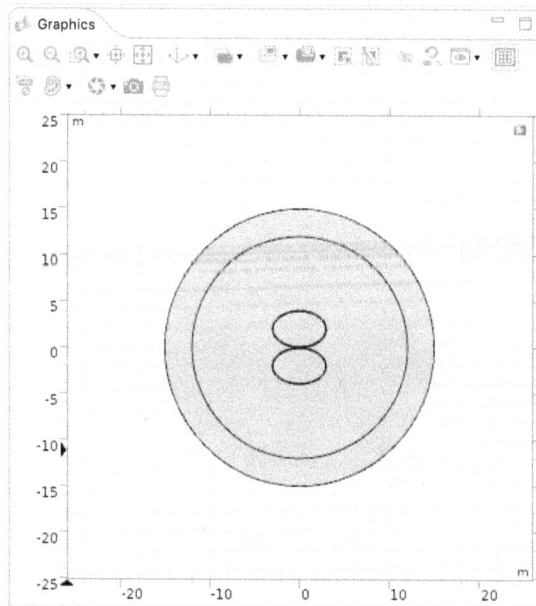

FIGURE 10.27 EC ellipses in the graphics window.

Figure 10.27 shows the EC Ellipses in the Graphics Window.

NOTE *In order to build the concentrators using the EC ellipses, the modeler needs to perform two Boolean Difference operations.*

EC Formation

Difference 1

      Right-Click > Model Builder > Component 1 (*comp1*) > Geometry 1.

      Select: Booleans and Partitions > Difference from the Pop-up menu.

      Activate the Objects to add text field (as needed).

      Use the Clipboard to Enter: e1 in the Objects to add: text field.

      Activate the Objects to subtract text field.

      Use the Clipboard to Enter: e3 in the Objects to subtract: text field.

      See Figure 10.28.

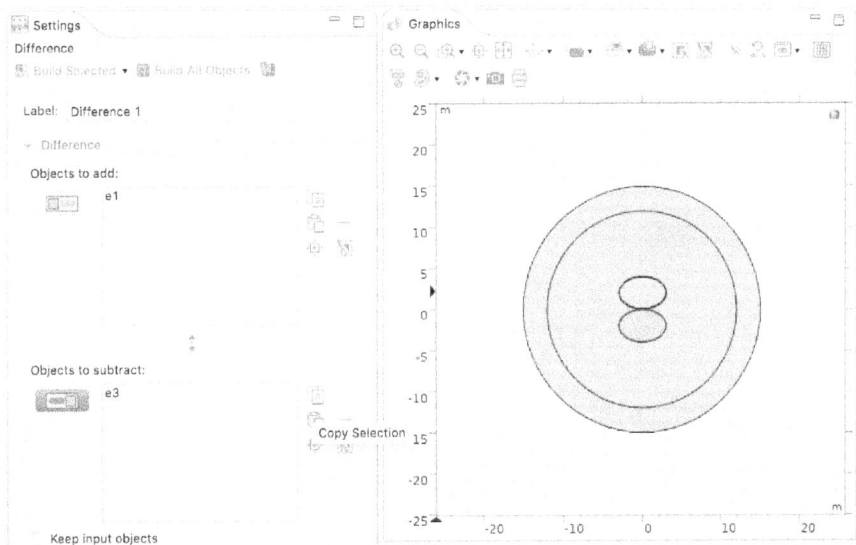

FIGURE 10.28 EC ellipses 1 & 3 in difference edit windows.

Figure 10.28 shows the EC Ellipses 1 & 3 in Difference edit windows.

Click > Build Selected

Difference 2

Right-Click > Model Builder > Component 1 (*comp1*) > Geometry 1.

Select: Booleans and Partitions > Difference from the Pop-up menu.

Activate the Objects to add window (as needed).

Use the Clipboard to Enter: e2 in the Objects to add: text field.

Activate the Objects to subtract text field.

Use the Clipboard to Enter: e4 in the Objects to subtract: text field.

See Figure 10.29.

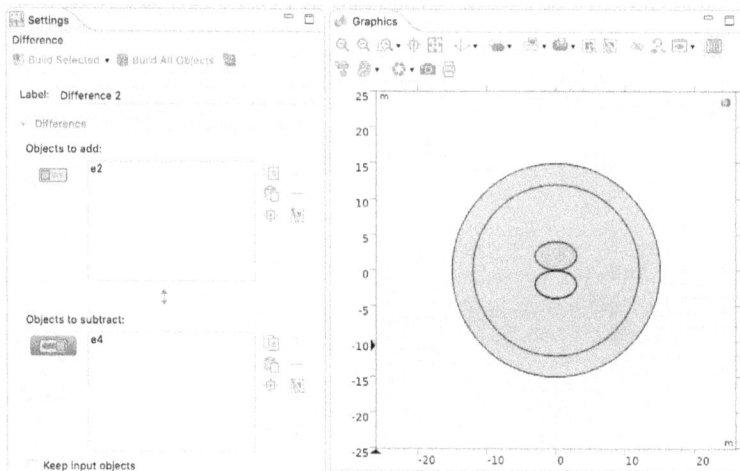

FIGURE 10.29 EC ellipses 2 & 4 in difference edit windows.

Figure 10.29 shows the EC Ellipses 2 & 4 in Difference edit windows.

Click > Build Selected

Rectangle

NOTE *A Rectangle is needed to remove the Right Half of the difference objects formed in the last steps.*

Right-Click > Model Builder > Component 1 (*comp1*) > Geometry 1.

Select: Rectangle from the Pop-up menu.

In the Settings Rectangle window,

Enter: Width = 3[m], Height = 8[m], Base = Corner, x = 0[m], and y = −4[m] in the appropriate text field.

Click > Build Selected

Difference 3

Right-Click > Model Builder > Component 1 (*comp1*) > Geometry 1.

Select: Booleans and Partitions > Difference from the Pop-up menu.

Use the Clipboard to Enter: dif1 dif2 in the Objects to add: text field.

Activate the Objects to subtract text field.

Use the Clipboard to Enter: r1 in the Objects to subtract: text field.

See Figure 10.30.

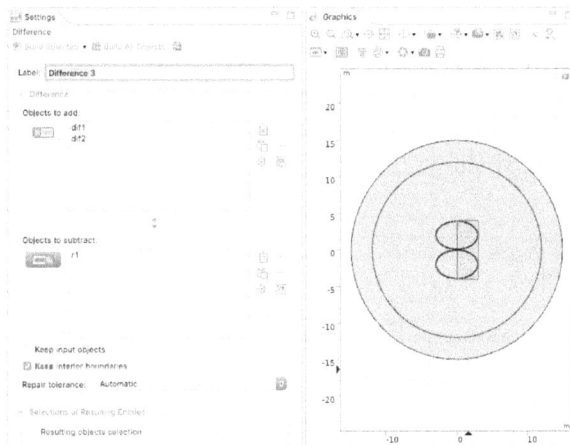

FIGURE 10.30 Settings – Difference > Difference prepared for the build selected operation.

Figure 10.30 shows Settings – Difference > Difference Prepared for the Build Selected operation.

Click > Build Selected in the Settings Toolbar.

Click > Zoom in twice (2).

See Figure 10.31.

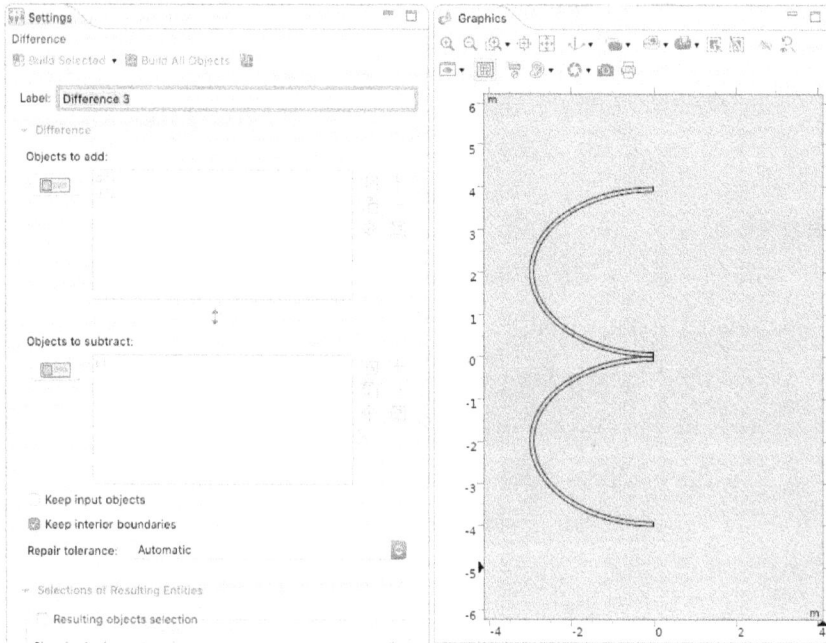

FIGURE 10.31 EC in the graphics window.

Figure 10.31 shows the EC in the Graphics window.

Selection 1

Right-Click > Model Builder > Component 1 (*comp1*) > Definitions.

Select: Selections > Explicit from the Pop-up menu.

Click > Model Builder > Component 1 (*comp1*) > Definitions > Explicit 1.

Right-Click > Model Builder > Component 1 (*comp1*) > Definitions > Explicit 1.

Select: Rename from the Pop-up menu.

Enter: EC Boundaries in the Rename Selection edit window.

Click > OK.

Select and Unselect > All domains Checkbox.

Select and Remove (minus sign) > Domains 1 and 4, the Input Entities edit window.

Click > Settings Explicit > Output Entities tab pop-up menu,

Select > Adjacent Boundaries from the pop-up menu.

See Figure 10.32.

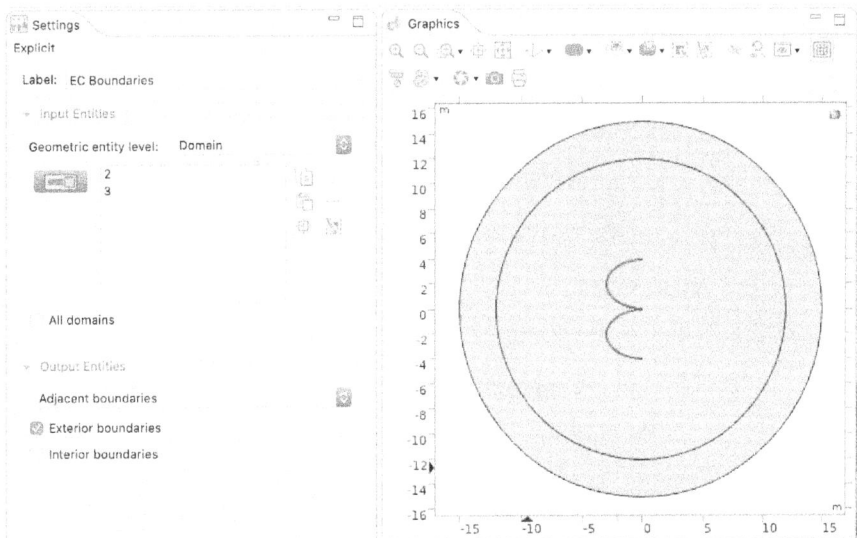

FIGURE 10.32 EC boundaries selection.

Figure 10.32 shows the EC Boundaries Selection.

Materials

Material 1

Right-Click > Model Builder > Component 1 (*comp1*) > Materials.

Select: Add Material from Library from the pop-up menu.

Click > Add Material > Built-in twistie.

Click > Air.

Right-Click > Air.

Select: Add to Component 1.

Select: Domain 2 and 3 in Settings Material window > Geometric Entity Selection tab, text field,

Click > Remove (minus sign) from Selection in Settings Material window > Geometric Entity Selection tab, text field.

Click > Close on the Add Material window (hollow X).

Material 2

Right-Click > Model Builder > Component 1 (*comp1*) > Materials.

Select: Add Material from Library from the pop-up menu.

Click > Add Material window > Built-in twistie.

Click > Aluminum.

Right-Click > Aluminum,

Select: Add to Component 1.

Click > Model Builder > Component 1 (*comp1*) > Materials > Aluminum (mat2).

Click > Geometric entity level pop-up menu in Settings Material > Geometric Entity Selection tab > Geometric entity level,

Select: Boundary from the pop-up menu.

Click > Close on the Add Material window (hollow X).

Selection

Click > Selection in Settings Material > Geometric Entity Selection tab > Selection pop-up menu,

Select: EC Boundaries from the Pop-up menu.

NOTE *Be sure to use the scroll bar on the selection text field to verify that all of the boundaries have been selected.*

See Figure 10.33.

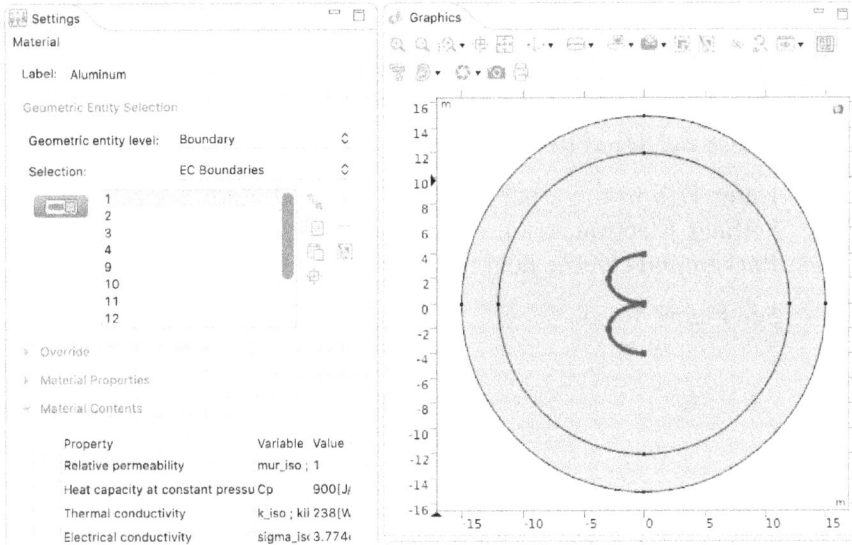

FIGURE 10.33 EC boundaries selection.

Figure 10.33 shows the EC Boundaries Selection.

Electromagnetic Waves (emw)

Domains

Click > Model Builder > Component 1 (*comp1*) > Electromagnetic Waves, Frequency Domain (emw).

Select: 2, 3 in Settings Electromagnetic Waves, Frequency Domain > Domain Selection tab > Selection text field.

Click> Remove from Selection button (Minus sign) in Settings Electromagnetic Waves, Frequency Domain.

NOTE   *This model uses the boundaries of the EC to represent the EC. Since that is the case, the field inside the EC is identically zero and requires no solution inside. This approach saves both time and effort.*

Settings

Click > Formulation pull-down menu in Settings Electromagnetic Waves, Frequency Domain > Formulation tab,

Select: Scattered field from the pull-down menu.

Enter E_b in the z position of the Background electric field: text field in Settings Electromagnetic Waves, Frequency Domain > Formulation tab > Background electric field text field.

See Figure 10.34.

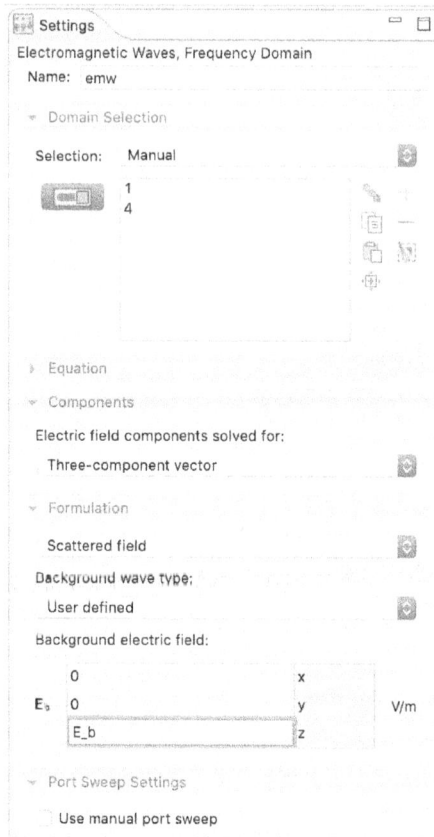

FIGURE 10.34 Electromagnetic waves settings selection.

Figure 10.34 shows the Electromagnetic Waves Settings Selection.

Perfectly Matched Layers 1

The Perfectly Matched Layer simulates an infinitely distant boundary layer. That eliminates possible interference of a back scattered wave with the waves in the region of interest at the center of the model.

Right-Click > Model Builder > Component 1 (*comp1*) > Definitions.

Select: Perfectly Matched Layer.

Click > Perfectly Matched Layer 1 (*pml1*).

Use the Clipboard to Enter 4 in the domain Selection text field.

Coordinates

Click > Type pull-down menu in Settings Perfectly Matched Layer > Geometry tab,

Select: Cylindrical from the pull-down menu.

See Figure 10.35.

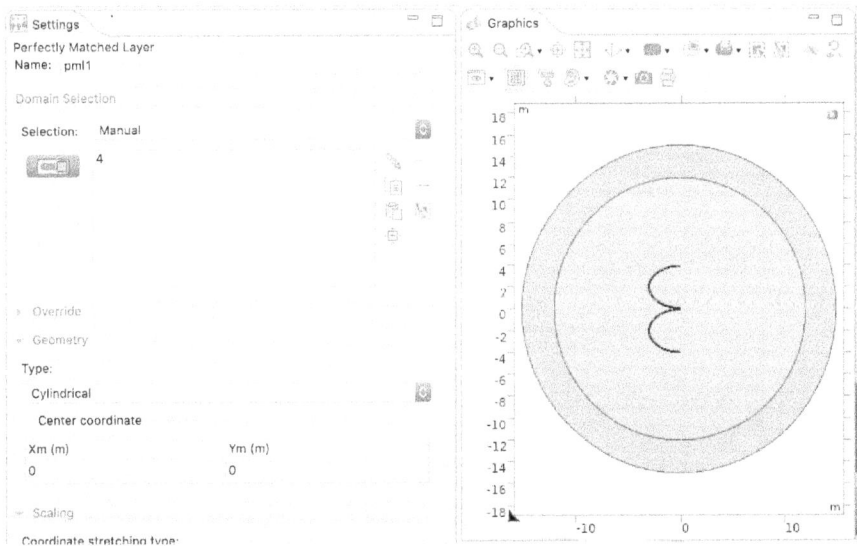

FIGURE 10.35 Perfectly matched layer settings selection.

Figure 10.35 shows the Perfectly Matched Layer Settings Selection.

Impedance Boundary Condition 1

NOTE *The Impedance Boundary Condition assumes that the skin depth in the material is significantly less than the material thickness. In this case it is on the order of microns.*

Right-Click > Model Builder > Component 1 (*comp1*) > Electromagnetic Waves, Frequency Domain (emw).

Select: Impedance Boundary Condition.

Click > Impedance Boundary Condition 1.

Click > Selection pull-down menu bar in Settings Impedance Boundary Condition window > Boundary Selection tab,

Select: EC Boundaries from the pull-down menu.

NOTE *Be sure to use the scroll bar on the selection text field to verify that all of the boundaries have been selected.*

See Figure 10.36.

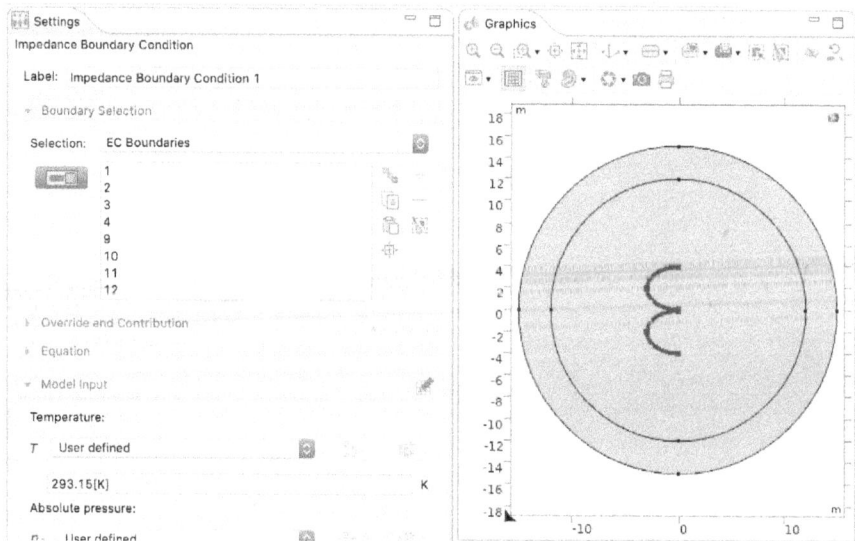

FIGURE 10.36 Impedance boundary condition settings selection.

Figure 10.36 shows the Impedance Boundary Condition Settings Selection.

Far-Field Calculation 1

NOTE *The requirement for the Far-Field Calculation is that all reflecting surfaces are surrounded, which they are.*

Right-Click > Model Builder > Component 1 (*comp1*) > Electromagnetic Waves, Frequency Domain (emw).

Select: Far-Field Domain.

Click > Far-Field Domain 1.

Use the Clipboard to Enter 1 in the Domain Selection text field (as needed).

See Figure 10.37.

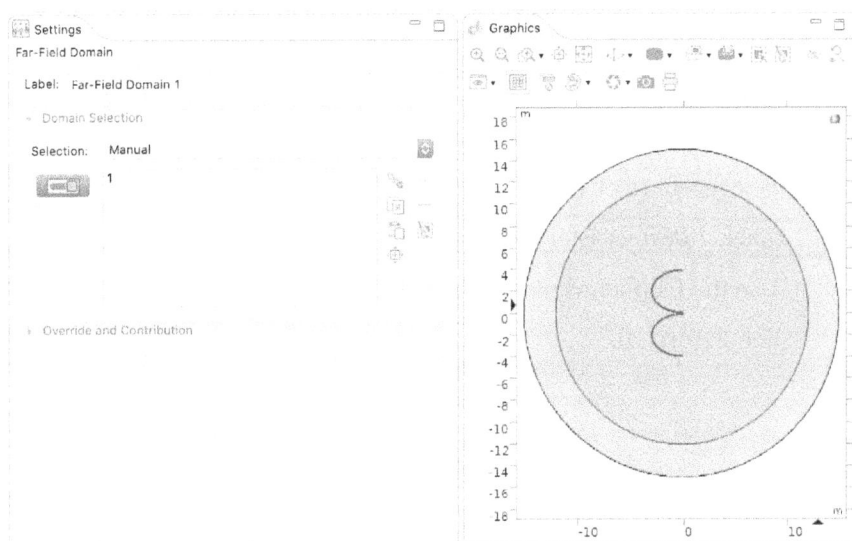

FIGURE 10.37 Far-field calculation settings selection.

Figure 10.37 shows the Far-Field Calculation Settings Selection.

Click > Far-Field Domain 1 twistie.

Click > Far-Field Calculation 1.

Select > EC Boundaries from the Settings Far-Field Calculation > Boundary Selection tab > Selection pop-up menu.

See Figure 10.38.

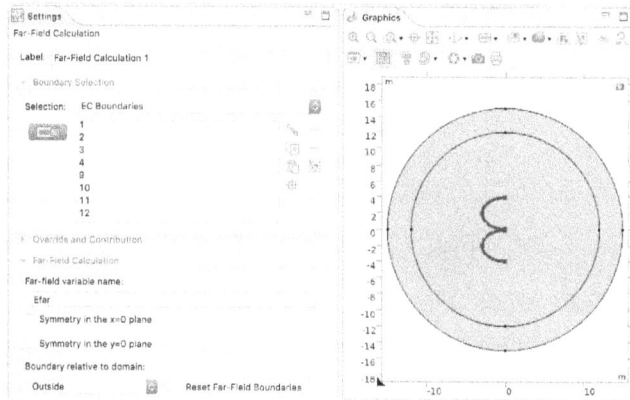

FIGURE 10.38 Far-field calculation settings selection.

Figure 10.38 shows the Far-Field Calculation Settings Selection.

Right-Click > Model Builder > Component 1 (*comp1*) > Electromagnetic Waves, Frequency Domain (emw).

Select > Perfect Electric Conductor.

Click > Perfect Electric Conductor 2.

Use the Clipboard to Enter 5 6 17 20 in the Boundary Selection text field.

See Figure 10.39.

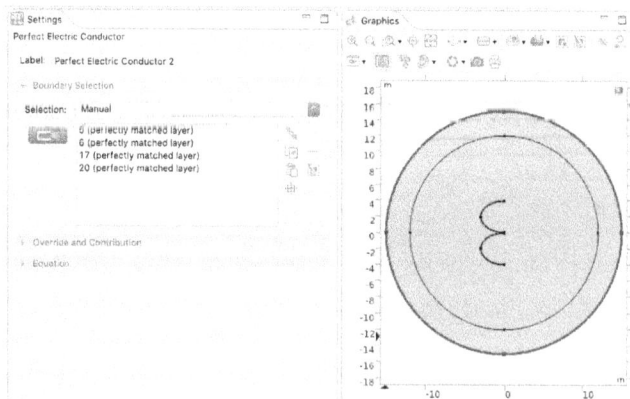

FIGURE 10.39 Perfect electric conductor settings selection.

Figure 10.39 shows the Perfect Electric Conductor Settings Selection.

Definitions

General Extrusion 1

NOTE *The General Extrusion is a coupling operator that maps an expression defined on a source domain to an expression that can be evaluated on any domain where the destination map expressions are valid.*

Right-Click > Model Builder > Component 1 (*comp1*) > Definitions.

Select: Nonlocal Couplings > General Extrusion.

Click > General Extrusion 1.

Enter: back in Settings General Extrusion > Operator Name:,

Use the Clipboard to Enter 1 in the Domain Selection text field.

Enter > dm_x in Settings General Extrusion > Destination Map tab > x-expression text field.

Enter > dm_y in Settings General Extrusion > Destination Map tab > y-expression text field.

See Figure 10.40.

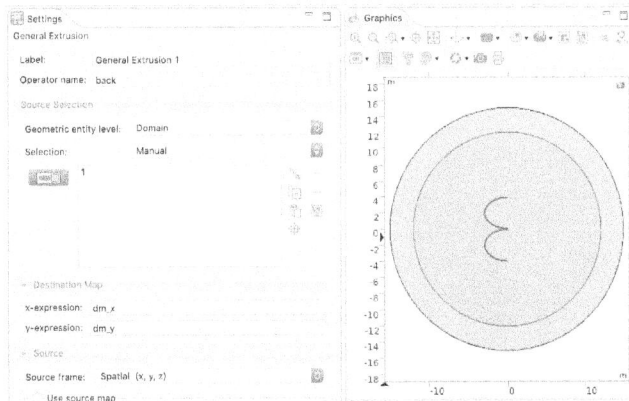

FIGURE 10.40 General extrusion settings selection.

Figure 10.40 shows the General Extrusion Settings Selection.

Mesh 1

Right-Click > Model Builder > Component 1 (*comp1*) > Mesh 1.

Select: Free Triangular from the Pop-up menu.

Click > Model Builder > Component 1 (*comp1*) > Mesh 1 > Size.

Click > Settings Size > Element Size Parameters Twistie.

Enter > me_s in Settings Size > Element Size Parameters tab > Maximum element size.

Click > Build All in the Settings Size toolbar.

See Figure 10.41.

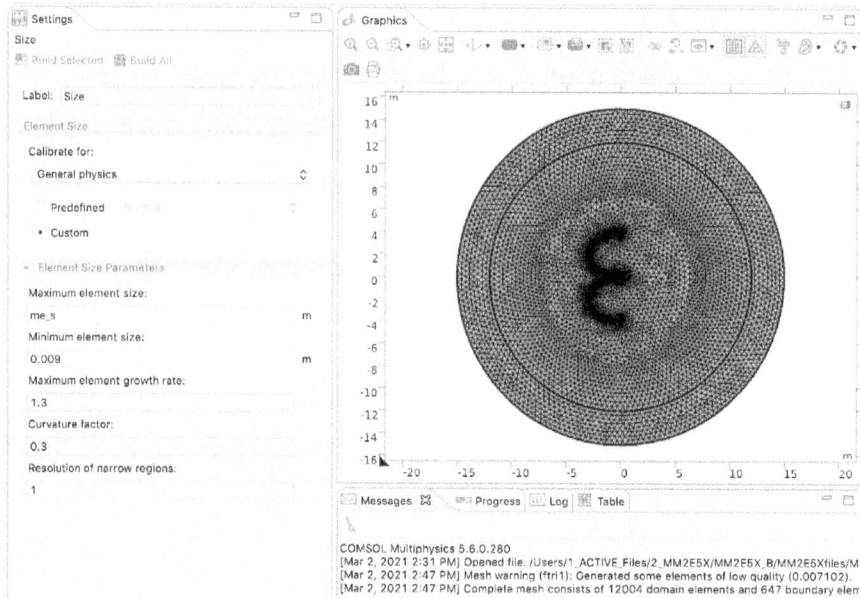

FIGURE 10.41 Graphics window after the mesh has been built.

Figure 10.41 shows the Graphics Window after the Mesh has been built.

NOTE *After the mesh is built, the model should have 12004 domain elements and 647 boundary elements.*

Study 1

Frequencies

Click > Model Builder > Study 1 twistie.

Click > Model Builder > Study 1 > Step 1: Frequency Domain.

Enter: f_a in Settings Frequency Domain > Study Settings tab > Frequencies text field.

See Figure 10.42.

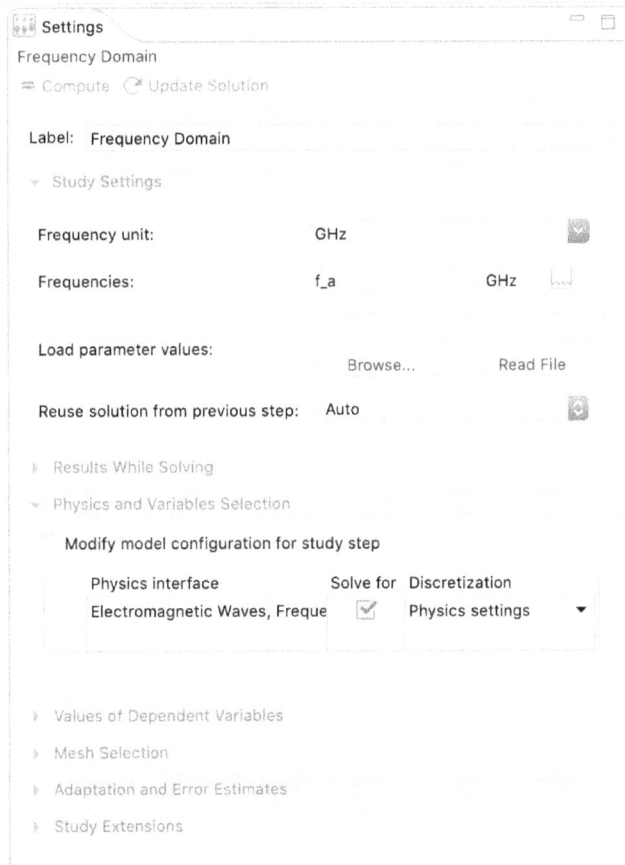

FIGURE 10.42 Settings frequency domain.

Figure 10.42 shows the Settings Frequency Domain.

Right-Click > Model Builder > Study 1.

Select: Compute.

See Figure 10.43.

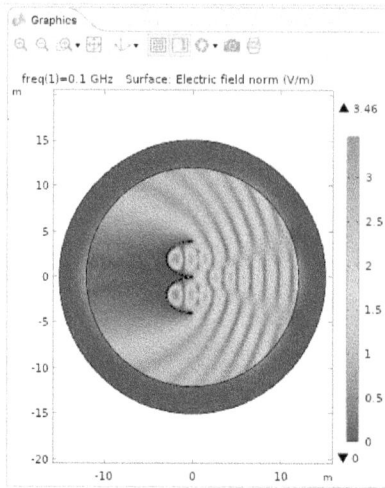

FIGURE 10.43 EC computed solution.

Figure 10.43 shows the EC Computed Solution.

Expanded Solution

Click > Zoom in twice.

See Figure 10.44.

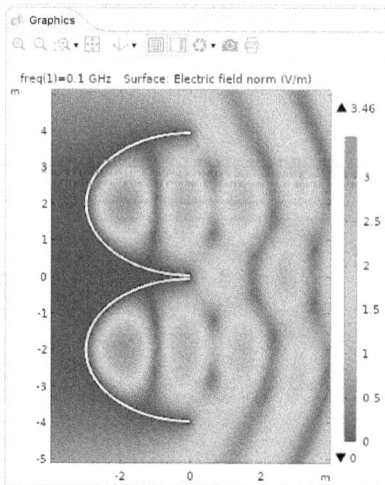

FIGURE 10.44 High electric field in EC computed solution.

Figure 10.44 shows a High Electric Field in EC Computed Solution.

The modeler should note that a high electric field point forms immediately in front of the EC. The EC contributes a gain of approximately three and one half (3.5).

2D Energy Concentrator PML Model Summary and Conclusions

The 2D Energy Concentrator PML Model is a powerful modeling tool that can be used to calculate the gain of various concentrators of different shapes and sizes. With this model, the modeler can easily vary all of the geometric parameters and optimize the design before the first prototype is physically built. These types of concentrators are widely used in industry.

FIRST PRINCIPLES AS APPLIED TO PML MODEL DEFINITION

First Principles Analysis derives from the fundamental laws of nature. In the case of models using this Classical Physics Analysis approach, the laws of conservation in physics require that what goes in (as mass, energy, charge, etc.) must come out (as mass, energy, charge, etc.) or must accumulate within the boundaries of the model.

The careful modeler must be knowledgeable of the implicit assumptions and default specifications that are normally incorporated into the COMSOL Multiphysics software model when a model is built using the default settings.

Consider, for example, the two PML models developed in this chapter. In these models, it is implicitly assumed there are no thermally related changes (mechanical, electrical, etc.), except as specified. It is also assumed the materials are homogeneous and isotropic, except as specifically indicated and there are no thin insulating contact barriers at the thermal junctions. None of these assumptions are typically true in the general case. However, by making such assumptions, it is possible to easily build a 2D First Approximation Model.

NOTE *A First Approximation Model is one that captures all the essential features of the problem that needs to be solved, without dwelling excessively on all of the small details. A good First Approximation Model will yield an answer that enables the modeler to determine if he needs to invest the time and the resources required to build a more highly detailed model.*

Also, the modeler needs to remember to name model parameters carefully as pointed out in Chapter 1.

REFERENCES

1. COMSOL Reference Manual, p. 254

2. *https://en.wikipedia.org/wiki/Maxwell%27s_equations*

3. *https://en.wikipedia.org/wiki/Boundary_value_problem*

4. *https://en.wikipedia.org/wiki/Perfectly_matched_layer*

5. *https://en.wikipedia.org/wiki/Finite-difference_time-domain_method*

6. *https://en.wikipedia.org/wiki/Finite_element_method*

7. *https://math.mit.edu/~stevenj/18.369/pml.pdf*

8. RF Module Users Guide, p. 13

9. RF Module Users Guide, p. 109

10. *https://en.wikipedia.org/wiki/Gain_(electronics)*

11. *https://en.wikipedia.org/wiki/Speed_of_light*

12. *https://en.wikipedia.org/wiki/Wavenumber*

13. COMSOL Reference Manual, p. 355

14. *https://en.wikipedia.org/wiki/Solar_thermal_collector*

15. *https://en.wikipedia.org/wiki/Solar_power_plants_in_the_Mojave_Desert*

16. *https://en.wikipedia.org/wiki/Solar_furnace*

17. *https://en.wikipedia.org/wiki/Concentrated_solar_power*

18. *https://en.wikipedia.org/wiki/Anidolic_lighting*

SUGGESTED MODELING EXERCISES

1. Build, mesh, and solve the 2D Concave Metallic Mirror PML Model as presented earlier in this chapter.

2. Build, mesh, and solve the 2D Energy Concentrator PML Model as presented earlier in this chapter.

3. Change the values of the materials parameters and then build, mesh, and solve the 2D Concave Metallic Mirror PML Model as an example problem.

4. Change the values of the materials parameters and then build, mesh, and solve the 2D Energy Concentrator PML Model as an example problem.

5. Change the value of the frequency parameter and then build, mesh, and solve the 2D Concave Metallic Mirror PML Model as an example problem.

6. Change the shape of the geometry and then solve the 2D Energy Concentrator PML Model as an example problem.

CHAPTER 11

BIOHEAT MODELS USING COMSOL MULTIPHYSICS 5.x

In This Chapter

- Guidelines for Bioheat Modeling in 5.x
  - Bioheat Modeling Considerations
- Bioheat Models
  - 2D Axisymmetric Tumor Laser Irradiation Model
  - 2D Axisymmetric Microwave Cancer Therapy Model
- First Principles as Applied to Bioheat Model Definition
- References
- Suggested Modeling Exercises

GUIDELINES FOR BIOHEAT MODELING IN 5.x

NOTE

In this chapter, two 2D Axisymmetric Bioheat models are presented. Bioheat models, in general, have proven to be very valuable in the study and application of energy locally to terminate cancer tumors. The science, engineering, and medical communities have employed this modeling technique successfully both in the past and currently. Such models serve as first-cut evaluations of potential systemic physical behavior under the influence of complex external stimuli without hazarding the life of a patient.

Bioheat model responses and other such ancillary information can be gathered and screened early in a project for a first-cut evaluation of the physical behavior of a planned prototype. Bioheat models are typically more conceptually and physically complex than the models that were presented in earlier chapters of this text. The calculated model (simulation) information can be used in the prototype fabrication stage as guidance in the selection of prototype geometry and materials.

Since the models in this chapter are more conceptually complex and are potentially more difficult to solve than the models presented thus far, it is important that the modeler have available the tools necessary to most easily utilize the powerful capabilities of the 5.x software. In order to do that, if you have not done this previously, the modeler should go to the main 5.x toolbar, Click > Options – Preferences – Show More Options. When the Preferences – Show More Options edit window is shown, Select > Equation view checkbox and Equation Section checkbox. Click > OK{11.1}.

Bioheat Modeling Considerations

Bioheat Equation Theory

For the new modeler or those readers unfamiliar with this topic, Bioheat Modeling is employed in the development of models to analyze heat transfer in materials (tissues, fluids, etc.) and other systems related to or derived from previously or currently living organisms. The solution of such Bioheat Equation models is most obviously important when those models are designed to explore techniques for potentially critical therapeutic applications (e.g. destroying cancer cells, killing tumors, etc.) in living entities (people, dogs, cats, cows, sheep, etc.).

Harry H. Pennes published his landmark paper "Analysis of tissue and arterial blood temperatures in the resting human forearm" {11.2} in August of 1948. He proposed in that paper that heat flow is proportional to the difference in temperature between the arterial blood and the local tissue. Pennes' work is considered foundational in this area of study and has since been cited extensively {11.3}.

In the COMSOL Multiphysics 5.x software, the Bioheat Equation (Pennes Equation) is found as a separate Interface within the Heat Transfer Interface. In the Bioheat Transfer (ht) Interface, the Bioheat Equation is formulated as follows:

$$\delta_{ts}\rho C\frac{\partial T}{\partial t} + \nabla \bullet (-\vec{k}\nabla T) = \rho_b C_b \omega_b (T_b - T) + Q_{met} + Q_{ext} \tag{11.1}$$

Where δ_{ts} = Time scaling coefficient (default value = 1) [dimensionless].

ρ = Tissue density [kg/m^3].

C = Tissue heat capacity [J/(kg×K)].

T = Temperature [K].

$\vec{k}$ = Tissue thermal conductivity tensor [W/(m×K)].

ρ_b = Blood density [kg/m^3].

C_b = Blood heat capacity [J/(kg×K)].

ω_b = Blood perfusion rate [m^3/(m^3×s)].

T_b = Temperature, Arterial blood [K].

Q_{met} = Metabolic heat source [W/m^3].

Q_{ext} = External environmental heat source [W/m^3].

The rate at which a fluid (e.g. blood) flows through a type of tissue (e.g. muscle, heart, liver, etc.) is the perfusion rate. It is very important, of course, to know the correct perfusion value for the tissue/fluid-type in question.

NOTE *The above equation is shown as formulated for blood flow. However, it can be also equally well employed for other fluids or fluid compositions under the appropriate circumstances (e.g., artificial blood, different animal life fluids, etc.). When the modeler employs variations of this formulation of the Bioheat Equation, he needs to carefully verify the underlying assumptions employed in his particular model.*

The Bioheat Equation is similar to the conduction heat equation. In the case of steady-state heat flow, the first term on the left vanishes. That is:

$$\delta_{ts}\rho C \frac{\partial T}{\partial t} = 0 \tag{11.2}$$

In the Bioheat Equation, what would normally be written as single heat source term on the left side of the heat conduction equation (Q) is now separated into three terms.

The perfusion term:

$$\rho_b C_b \omega_b (T_b - T) \tag{11.3}$$

The metabolic term:

$$Q_{met} \tag{11.4}$$

The external source term:

$$Q_{ext} \tag{11.5}$$

The division of what would typically be written as a single heat source term in the Bioheat Equation into three terms is done to facilitate for the user the conceptual linkage and to aid in the formulation of the PDE when creating models for this type of problem (biological).

NOTE

The Pennes Bioheat Equation constitutes a good First-Order Approximation to those physical processes (thermal conduction) involved in the solution of the heat transfer problems in biological specimens. The Pennes Bioheat Equation formulation is usually adequate for the modeling of most biological problems of this nature. More terms can, of course, be added, when desired, at the risk of increasing the complexity, the associated model size, and the computational time.

Since the Bioheat Equation, as configured, already delivers the needed level of accuracy for a typical decision, only slightly expanded knowledge will be gained by the addition of Second-Order Effects to the equation, considering the intrinsic fundamental limits of most biological system model problems.

Tumor Laser Irradiation Theory

The optical coefficient of absorption for laser photons (irradiation) of tumors is approximately the same as the optical coefficient of absorption for the surrounding tissue. This laser irradiation technique is implemented by raising the relative absorption coefficient locally by artificial means. The change in the local absorption coefficient is accomplished by injection of a designed highly optically absorbing material {11.4} into the tumor. Implementation of this type of procedure is typically considered as a minimally invasive procedure.

The absorbed laser beam photonic energy becomes a heat source for the Bioheat Equation in the region of the tumor as follows:

$$Q_{laser} = I_0\, a e^{az - \frac{r^2}{2\sigma^2}} \tag{11.6}$$

Where I_0 = Irradiation intensity [W/m^2].

   a = Absorbance [1/m].

   σ = Irradiated region width parameter [m].

Bioheat Transfer Models

NOTE *The Pennes Bioheat Equation is a valuable approach for calculating the potential efficacy of modeled treatment techniques. The fundamental principle needs to be that tumor cells are observed to die at elevated temperatures. The literature cites temperatures that range from 42°C (315.15 K) {11.5} to 60°C (333.15 K) {11.6}. If the modeled method raises the local temperature of the tumor cells, without excessively raising the temperature of the normal cells, then the modeled method will most probably be successful.*

The 2D Axisymmetric Tumor Laser Irradiation Model takes advantage of the transparency of human tissue in certain infrared (IR) wavelengths {11.7}. Figure 11.1 shows the structure of the modeling domain. Since the model is created as a 2D Axisymmetric Model, only the right half of the structure will (needs to) be used in the calculations.

See Figure 11.1.

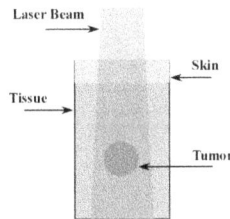

FIGURE 11.1 Tumor laser irradiation physical model.

Figure 11.1 shows the Tumor Laser Irradiation physical model.

Building the 2D Axisymmetric Tumor Laser Irradiation Model

Startup 5.x, Click > Model Wizard button.

Select: 2D Axisymmetric.

Click > Twistie for the Heat Transfer interface.

Click > Bioheat Transfer (ht).

Click > Add button.

Click > Study (Right Pointing Arrow button).

Select: General Studies > Time Dependent.

Click > Done (Checked Box button).

Click > Save As.

Enter MM2E5X_2DAxi_TLI_1.mph.

Click > Save.

See Figure 11.2.

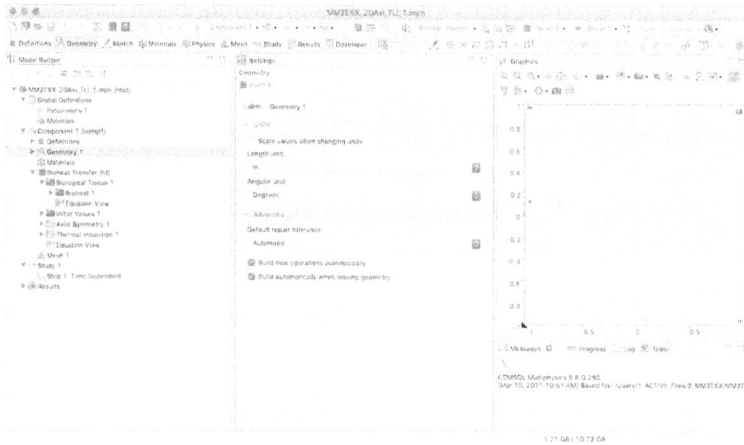

FIGURE 11.2 Desktop display for the MM2E5X_2DAxi_TLI_1.mph model.

Figure 11.2 shows the Display for the MM2E5X_2DAxi_TLI_1.mph model.

Global Definitions

Parameters

In Model Builder,

Click > Model Builder > Global Definitions.

Click > Parameters 1 from the Pop-up menu.

Enter all the Parameters, as shown, in Table 11.1.

TABLE 11.1 TLI Parameters.

Name	Expression	Description
rho_blood	1000[kg/m^3]	Density blood
C_blood	4200[J/(kg*K)]	Heat capacity blood
T_blood	37[degC]	Temperature blood

k_skin	0.2[W/(m*K)]	Thermal conductivity skin
rho_skin	1200[kg/m^3]	Density skin
C_skin	3600[J/(kg*K)]	Heat capacity skin
wb_skin	3e-3[1/s]	Blood perfusion rate skin
k_tissue	0.5[W/(m*K)]	Thermal conductivity tissue
rho_tissue	1050[kg/m^3]	Density tissue
C_tissue	3600[J/(kg*K)]	Heat capacity tissue
wb_tissue	6e-3[1/s]	Blood perfusion rate tissue
k_tumor	0.5[W/(m*K)]	Thermal conductivity tumor
rho_tumor	1050[kg/m^3]	Density tumor
C_tumor	3600[J/(kg*K)]	Heat capacity tumor
wb_tumor	6e-3[1/s]	Blood perfusion rate tumor
Q_met	400[W/m^3]	Metabolic heat generation
T0	37[degC]	Temperature reference blood
h_conv	10[W/(m^2*K)]	Heat transfer coefficient skin
T_inf	10[degC]	Temperature domain boundary
I0	1.4[W/mm^2]	Laser irradiation power
sigma	5[mm]	Laser beam width coefficient

See Figures 11.3 through 11.5.

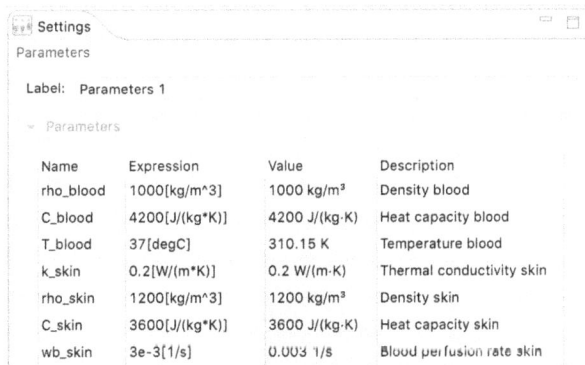

FIGURE 11.3 Settings – Parameters – Parameters edit window (Part 1).

Figure 11.3 shows the Settings – Parameters – Parameters edit window (part 1).

k_tissue	0.5[W/(m*K)]	0.5 W/(m·K)	Thermal conductivity tissue
rho_tissue	1050[kg/m^3]	1050 kg/m³	Density tissue
C_tissue	3600[J/(kg*K)]	3600 J/(kg·K)	Heat capacity tissue
wb_tissue	6e-3[1/s]	0.006 1/s	Blood perfusion rate tissue
k_tumor	0.5[W/(m*K)]	0.5 W/(m·K)	Thermal conductivity tumor
rho_tumor	1050[kg/m^3]	1050 kg/m³	Density tumor
C_tumor	3600[J/(kg*K)]	3600 J/(kg·K)	Heat capacity tumor

FIGURE 11.4 Settings – Parameters – Parameters edit window (Part 2).

Figure 11.4 shows the Settings – Parameters – Parameters edit window (part 2).

wb_tumor	6e-3[1/s]	0.006 1/s	Blood perfusion rate tumor
Q_met	400[W/m^3]	400 W/m³	Metabolic heat generation
T0	37[degC]	310.15 K	Temperature reference blood
h_conv	10[W/(m^2*K)]	10 W/(m²·K)	Heat transfer coefficient skin
T_inf	10[degC]	283.15 K	Temperature domain boundary
I0	1.4[W/mm^2]	1.4E6 W/m²	Laser irradiation power
sigma	5[mm]	0.005 m	Laser beam width coefficient

FIGURE 11.5 Settings – Parameters – Parameters edit window (Part 3).

Figure 11.5 shows the Settings – Parameters – Parameters edit window (part 3).

Geometry

> **NOTE** *The model geometry needs to be created before the local variables are entered into the model, so that the local variables can be assigned to specific domains.*

TLI Domains

In Model Builder,

Right-Click > Component 1 (*comp1*) > Geometry 1.

Select: Rectangle from the Pop-up menu.

Enter the coordinates shown in Table 11.2.

Click as instructed.

Repeat the sequence until completed.

TABLE 11.2 TLI Domains.

	Rectangles					
	Width	**Height**	**Base**	**r**	**z**	**Click**
1	0.1[m]	0.09[m]	Corner	−0.05[m]	−0.1[m]	Build Selected
2	0.1[m]	0.01[m]	Corner	−0.05[m]	−0.01[m]	Build Selected
	Circle					
	Radius	**Base**		**r**	**z**	**Click**
1	0.005[m]	Center		0[m]	−0.05[m]	Build All Objects

See Figure 11.6.

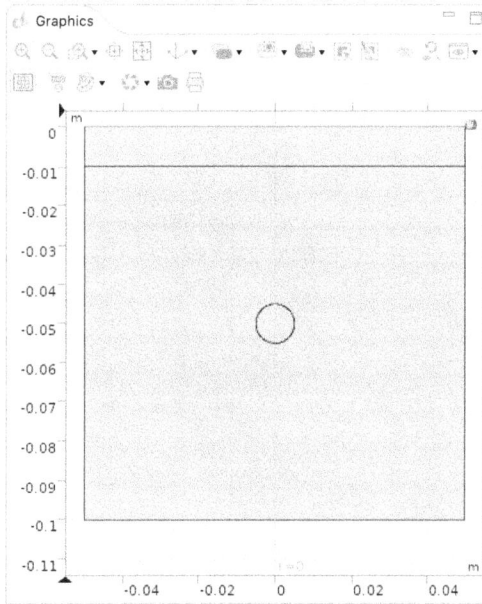

FIGURE 11.6 TLI Domains in the graphics edit window.

Figure 11.6 shows the TLI Domains in the Graphics edit window.

Boolean Operations

In Model Builder,

Right-Click > Model Builder > Component 1 (*comp1*) > Geometry 1.

Select: Rectangle from the Pop-up menu.

Enter the coordinates shown in Table 11.3.

Click as instructed.

TABLE 11.3 TLI Domains.

	Rectangle					
	Width	**Height**	**Base**	**r**	**z**	**Click**
1	0.06[m]	0.12[m]	Corner	−0.06[m]	−0.11[m]	Build Selected

See Figure 11.7.

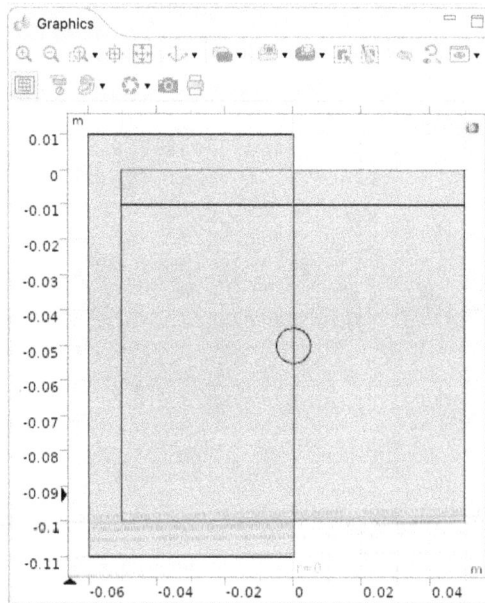

FIGURE 11.7 TLI domains in graphics edit window.

Figure 11.7 shows the TLI Domains in Graphics edit window.

NOTE *In order to build the final TLI domain using the TLI domains, the modeler needs to perform one Boolean Difference operation.*

Final TLI Domains Formation

Right-Click > Model Builder > Component 1 (*comp1*) > Geometry 1.

Select: Boolean and Partitions > Difference from the Pop-up menu.

Click > Select All in the Graphics Toolbar.

Click > r3 in the Objects to add: text field.

Click > Remove (Minus sign) (Rectangles r3 from the Objects to add window).

Click > Activate Selection in Settings – Difference – Difference – Objects to subtract.

Use the Clipboard to insert r3 in the Objects to subtract:

See Figure 11.8.

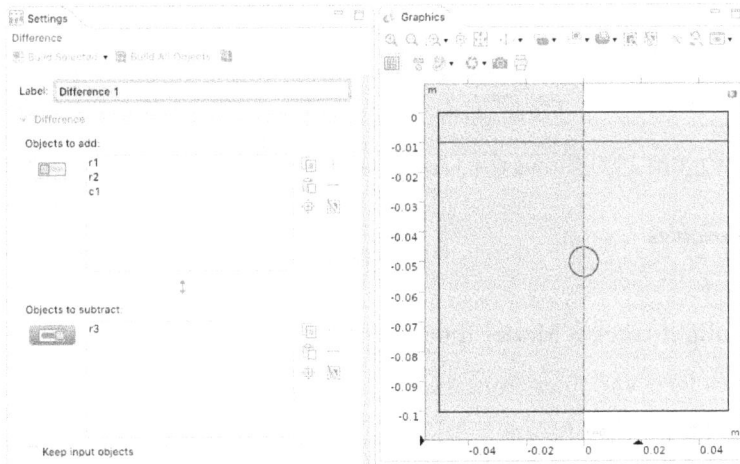

FIGURE 11.8 Settings – Difference – Difference prepared for the build selected operation.

Figure 11.8 shows Settings – Difference – Difference Prepared for the Build Selected operation.

Click > Build Selected in the Settings Toolbar.

See Figure 11.9.

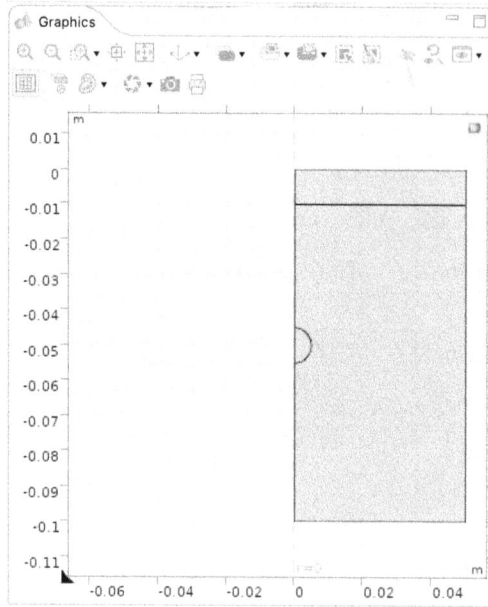

FIGURE 11.9 Graphics window with the final TLI domains built.

Figure 11.9 shows the Graphics Window with the Final TLI Domains Built.

Local Variables

Variables 1

Right-Click > Model Builder > Component 1 (*comp1*) > Definitions.

Select: Variables from the Pop-up menu.

Click > Variables 1.

Click > Geometric entity level in Settings Variables > Geometric Entity Selection tab > Geometric entity level: pop-up menu bar,

Select: Domain.

Click > Selection bar,

Select: All domains.

Click and Remove (minus sign) > Domain 2 > Geometric Entity Selection text field.

See Figure 11.10.

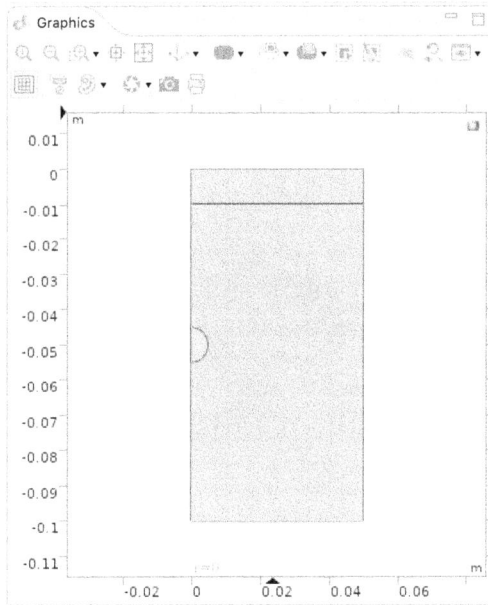

FIGURE 11.10 TLI domains 1 and 3 selected in the graphics window.

Figure 11.10 shows TLI Domains 1 and 3 selected in the Graphics Window.

Enter > Name = a, Expression = 0.1[1/m], Description = Absorbance, in the Settings Variables tab > Variables text field.

See Figure 11.11.

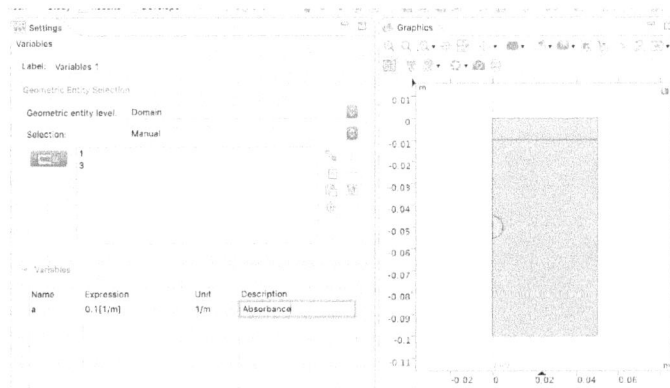

FIGURE 11.11 Settings – Variables – Variables text field.

Figure 11.11 shows the Settings – Variables – Variables text field.

Variables 2

Right-Click > Model Builder > Component 1 (*comp1*) > Definitions.

Select: Variables from the Pop-up menu.

Click > Variables 2.

Click > Geometric entity level in Settings Variables > Geometric Entity Selection tab > Geometric entity level: pop-up menu bar,

Select: Domain.

Use the Clipboard to insert 2 in the Selection text field.

See Figure 11.12.

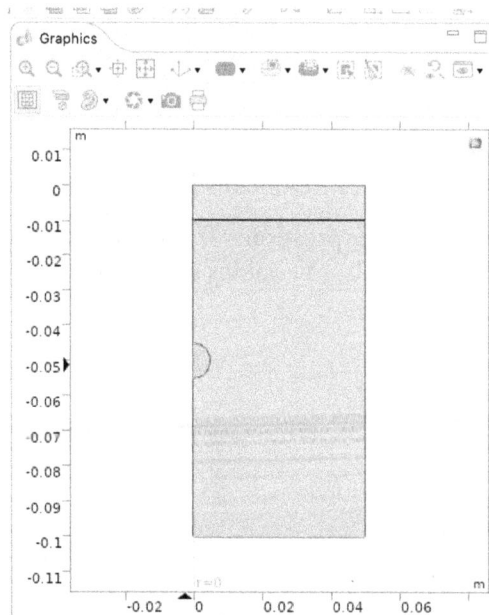

FIGURE 11.12 TLI domain 2 selected in graphics window.

Figure 11.12 shows TLI Domain 2 Selected in Graphics Window.

Enter > Name = a, Expression = 4[1/m], Description = Absorbance in Settings – Variables > Variables text field.

See Figure 11.13.

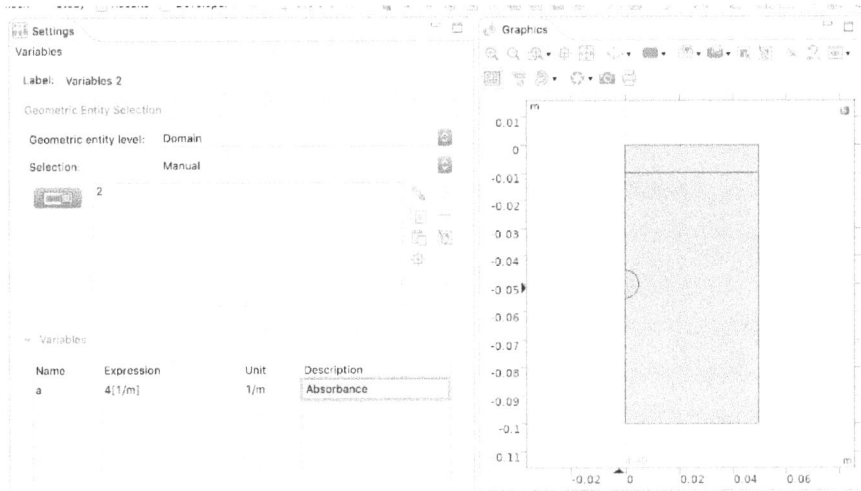

FIGURE 11.13 Settings variables > Variables text field.

Figure 11.13 shows the Settings Variables > Variables text field.

Variables 3

Right-Click > Model Builder > Component 1 (*comp1*) > Definitions.

Select: Variables from the Pop-up menu.

Click > Variables 3.

Enter: Name = Q_laser,

Expression = I0*a*exp(a*z-r^2/(2*sigma^2)),

Description = Laser energy distribution in the Settings Variables > Variables > Variables text field.

See Figure 11.14.

FIGURE 11.14 Settings variables > Variables text field.

Figure 11.14 shows the Settings Variables > Variables text field.

Bioheat Transfer (ht)

Click > Model Builder > Component 1 (*comp1*) > Bioheat Transfer (ht) twistie.

Domains

<u>NOTE</u> *The default condition is to apply all of these settings to the entire model. Please note that as the properties of other domains (tissues) are defined, the default settings will be overridden.*

Biological Tissue 1

Click > Biological Tissue 1.

Click > Thermal conductivity pop-up menu bar in Settings Biological Tissue > Heat Conduction, Solid.

Select: User defined.

Enter > k_skin in Settings Biological Tissue > Heat Conduction, Solid > Thermal conductivity (k) text field.

Click > Density pop-up menu bar in Settings Biological Tissue > Thermodynamics, Solid,

Select: User defined.

Enter > rho_skin in Settings Biological Tissue > Thermodynamics, Solid > Density > ρ text field.

Click > Heat capacity at constant pressure: pop-up menu bar in Settings Biological Tissue > Thermodynamics, Solid.

Select: User defined.

Enter > C_skin in Settings Biological Tissue > Thermodynamics, Solid > Heat capacity at constant pressure (C_p) text field.

See Figure 11.15.

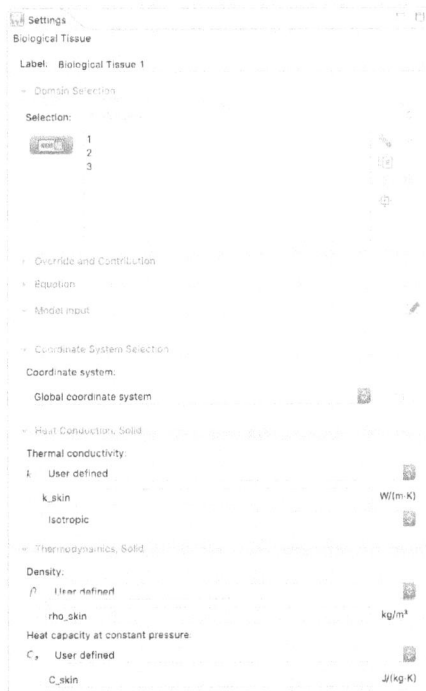

FIGURE 11.15 Settings biological tissue 1 text fields.

Figure 11.15 shows the Settings Biological Tissue 1 text fields.

Click > Model Builder > Component 1 (*comp1*) > Bioheat Transfer (ht) > Biological Tissue 1 twistie.

Click > Bioheat 1.

Enter: rho_blood in the Settings Bioheat > Bioheat > Density, blood (ρ_b) text field.

Enter: C_blood in the Settings Bioheat > Bioheat > Specific heat, blood ($C_{p,b}$) text field.

Enter: wb_skin in the Settings Bioheat > Bioheat > Blood perfusion rate (ω_b) text field.

Enter: T_blood in the Settings Bioheat > Bioheat > Arterial blood temperature (T_b) text field.

Enter: Q_met in the Settings – Bioheat > Bioheat > Metabolic heat source (Q_{met}) text field.

See Figure 11.16.

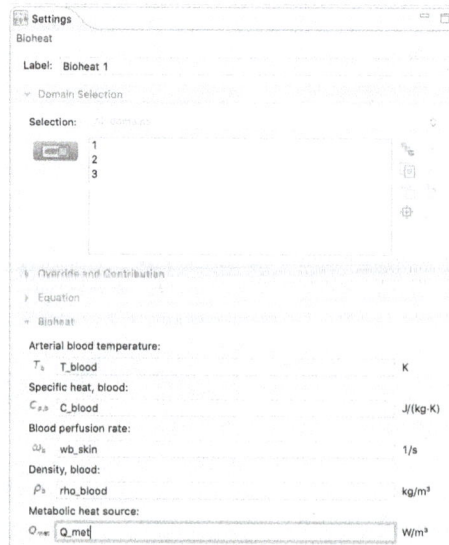

FIGURE 11.16 Settings – Bioheat 1 text fields.

Figure 11.16 shows the Settings – Bioheat 1 text fields.

Biological Tissue 2

Right-Click > Model Builder > Component 1 (*comp1*) > Bioheat Transfer (ht).

Select: Specific Media > Biological Tissue.

Click > Biological Tissue 2.

Use the Clipboard to insert 1 into the Domain Selection tab > Selection text field.

Click > Thermal conductivity pop-up menu bar in Settings Biological Tissue > Heat Conduction, Solid,

Select: User defined.

Enter: k_tissue in Settings Biological Tissue > Heat Conduction, Solid tab > Thermal conductivity (k) text field.

Click > Density pop-up menu bar in Settings Biological Tissue > Thermodynamics, Solid tab,

Select: User defined.

Enter > rho_tissue in Settings Biological Tissue > Thermodynamics, Solid > Density ρ text field.

Click > Heat capacity at constant pressure: pop-up menu bar in Settings Biological Tissue > Thermodynamics, Solid tab,

Select: User defined.

Enter: C_tissue in Settings Biological Tissue > Thermodynamics > Heat capacity at constant pressure (C_p) edit window.

See Figure 11.17.

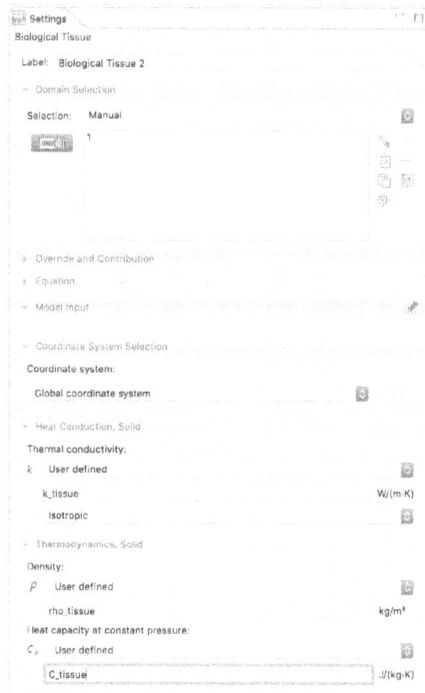

FIGURE 11.17 Settings biological tissue 2 text field.

Figure 11.17 shows the Settings Biological Tissue 2 text field.

Click > Model Builder > Component 1 (*comp1*) > Bioheat Transfer (ht) > Biological Tissue 2 twistie.

Click > Bioheat 1.

Enter > rho_blood in the Settings Bioheat > Bioheat tab > Density, blood (ρ_b) text field.

Enter > C_blood in the Settings Bioheat > Bioheat tab > Specific heat, blood ($C_{p,b}$) text field.

Enter > wb_tissue in the Settings Bioheat > Bioheat tab > Blood perfusion rate (ω_b) text field.

Enter > T_blood in the Settings Bioheat > Bioheat tab > Arterial blood temperature (T_b) text field.

Enter > Q_met in the Settings Bioheat > Bioheat tab > Metabolic heat source (Q_{met}) text field.

See Figure 11.18.

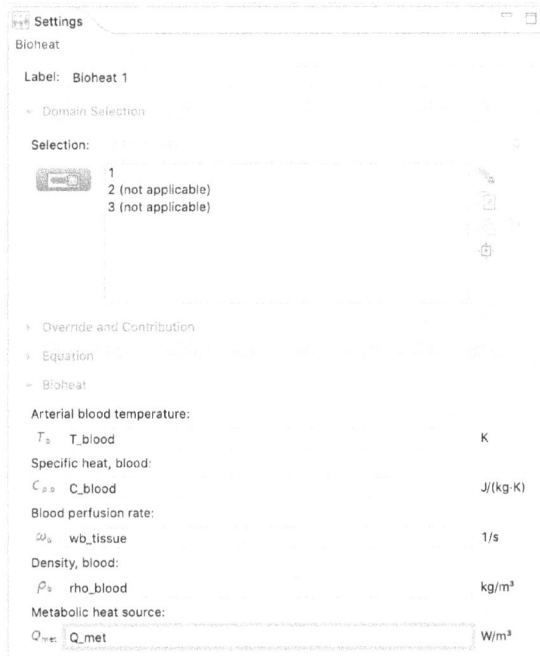

FIGURE 11.18 Settings bioheat 2 text field.

Figure 11.18 shows the Settings Bioheat 2 text field.

Biological Tissue 3

Right-Click > Model Builder > Component 1 (*comp1*) > Bioheat Transfer (ht).

Select: Specific Media > Biological Tissue.

Click > Biological Tissue 3.

Use the Clipboard to insert 2 into the Domain Selection tab > Selection text field.

Click > Thermal conductivity pop-up menu bar in Settings Biological Tissue > Heat Conduction, Solid tab,

Select: User defined.

Enter: k_tumor in Settings Biological Tissue > Heat Conduction, Solid tab > Thermal conductivity (k) text field.

Click > Density pop-up menu bar in Settings Biological Tissue > Thermodynamics, Solid tab,

Select: User defined.

Enter: rho_tumor in Settings Biological Tissue > Thermodynamics, Solid tab > Density (ρ) text field.

Click > Heat capacity at constant pressure pop-up menu bar in Settings Biological Tissue > Thermodynamics, Solid tab,

Select: User defined.

Enter > C_tumor in Settings Biological Tissue > Thermodynamics, Solid tab > Heat capacity at constant pressure (C_p) text field.

See Figure 11.19.

FIGURE 11.19 Settings biological tissue 3 text field.

Figure 11.19 shows the Settings Biological Tissue 3 text field.

Click > Model Builder > Component 1 (*comp1*) > Bioheat Transfer (ht) > Biological Tissue 3 twistie (as needed).

Click > Bioheat 1.

Enter > rho_blood in the Settings Bioheat > Bioheat tab > Density, blood (ρ_b) text field.

Enter > C_blood in the Settings Bioheat > Bioheat tab > Specific heat, blood (C_b) text field.

Enter > wb_tumor in the Settings Bioheat > Bioheat tab > > Blood perfusion rate (ω_b) text field.

Enter > T_blood in the Settings Bioheat > Bioheat tab > Arterial blood temperature (T_b) text field.

Enter > Q_met in the Settings Bioheat > Bioheat tab > Metabolic heat source (Q_{met}) text field.

See Figure 11.20.

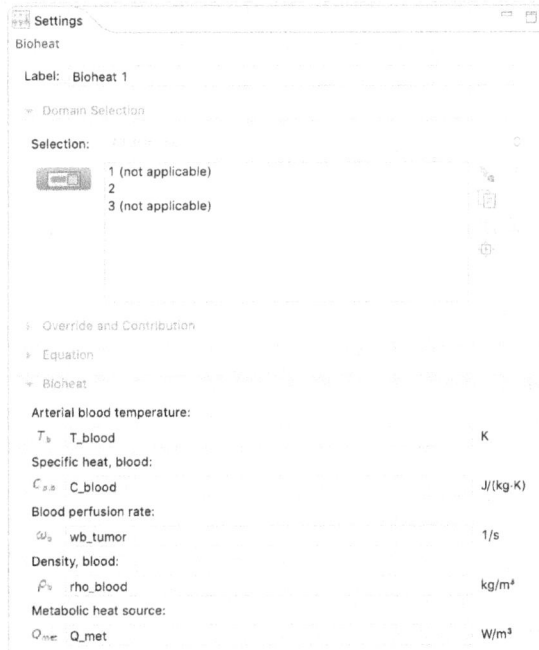

FIGURE 11.20 Settings bioheat 3 text field.

Figure 11.20 shows the Settings - Bioheat 3 text field.

Initial Values

Click > Model Builder > Component 1 (*comp1*) > Bioheat Transfer (ht) > Initial Values 1.

Enter: T0 in Settings Initial Values > Initial Values tab > Temperature.

See Figure 11.21.

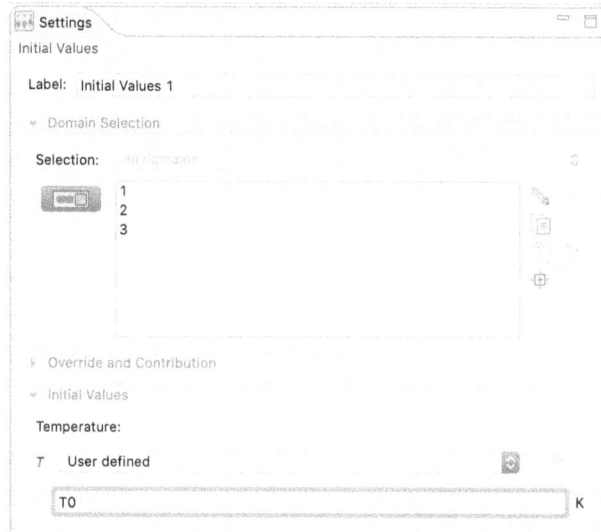

FIGURE 11.21 Settings initial values text field.

Figure 11.21 shows the Settings Initial Values text field.

Heat Source

Right-Click > Model Builder > Component 1 (comp1) > Bioheat Transfer (ht).

Select > Heat Source from the Pop-up menu.

Click > Heat Source 1.

Click > Settings Heat Source > Domain Selection tab, > Selection pop-up menu bar,

Select: All Domains.

Enter > Q_laser in the Settings Heat Source > Heat Source tab > General source text field.

See Figure 11.22.

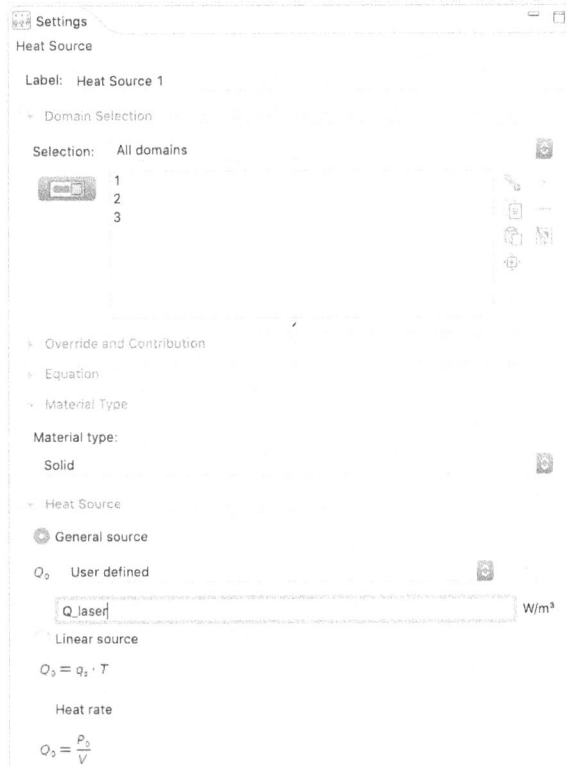

FIGURE 11.22 Settings heat source text field.

Figure 11.22 shows the Settings Heat Source text field.

Heat Flux

Right-Click > Model Builder > Component 1 (comp1) > Bioheat Transfer (ht).

Select: Heat Flux from the Pop-up menu.

Click > Heat Flux 1.

Use the Clipboard to insert 7 into the Boundary Selection tab > Selection text field.

See Figure 11.23.

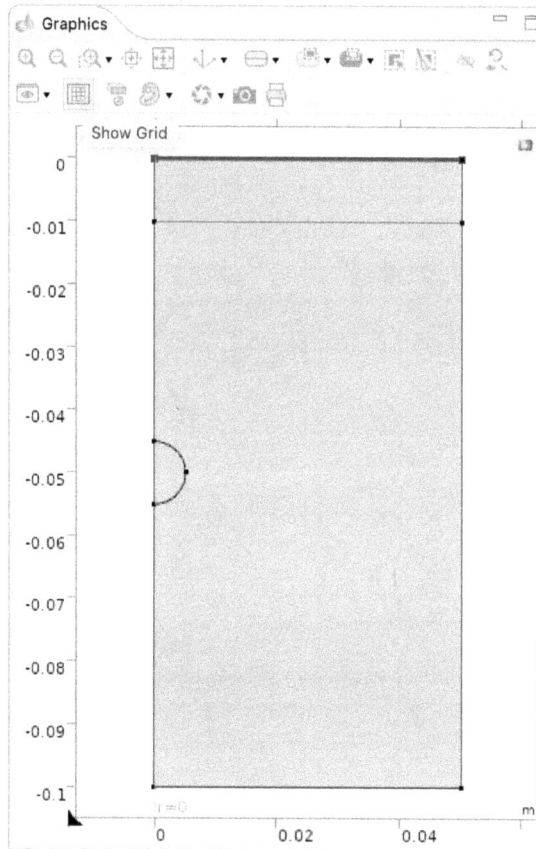

Figure 11.23 shows the Boundary 7 in the Graphics window.

Click > Convective heat flux button in Settings Heat Flux > Heat Flux tab,

Enter: h_conv in the Settings Heat Flux > Heat Flux tab > Heat transfer coefficient text field.

Enter: T_inf in the Settings Heat Flux > Heat Flux tab > External temperature (T_{ext}) text field.

See Figure 11.24.

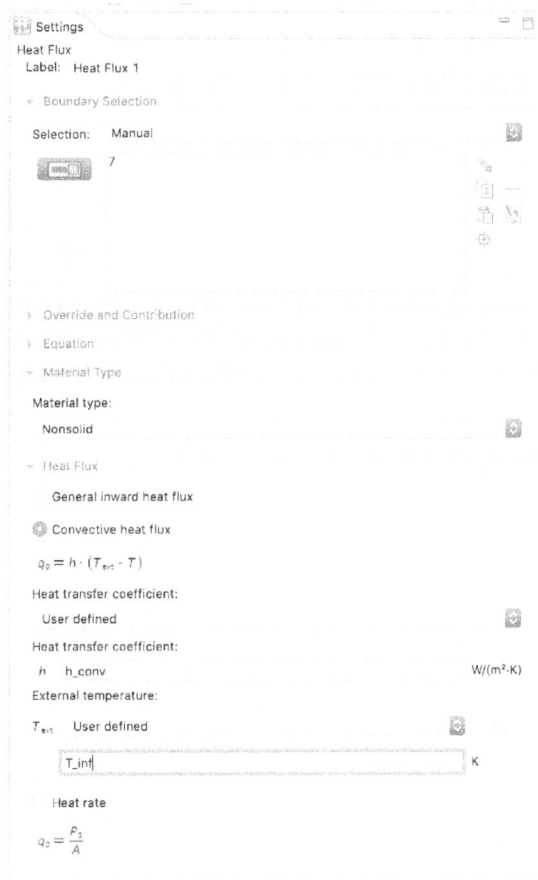

FIGURE 11.24 Settings heat flux text field.

Figure 11.24 shows the Settings Heat Flux text field.

Mesh 1

Right-Click > Model Builder > Component 1 (*comp1*) > Mesh 1.

Select: Free Triangular from the Pop-up menu.

Click > Build All in the Settings Free Triangular window toolbar.

See Figure 11.25.

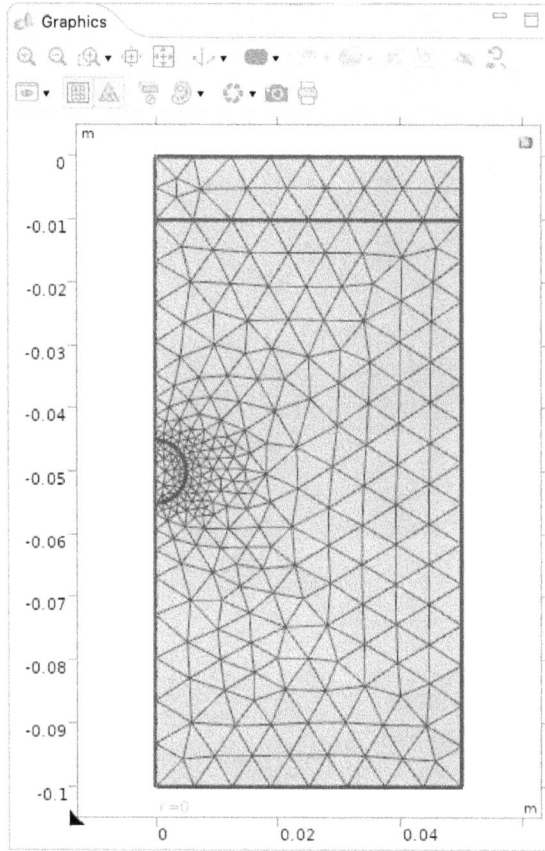

FIGURE 11.25 Graphics window after the mesh build.

Figure 11.25 shows the Graphics Window after the Mesh Build.

NOTE *After the mesh is built, the model should have 536 domain elements and 76 boundary elements.*

Study 1

Time Dependent Solver

Click > Model Builder > Study 1 twistie.

Click > Model Builder > Study 1 > Step 1: Time Dependent.

Click > Range button in Settings Time Dependent > Study Settings tab,

Enter: Start = 0, Step = 10, Stop = 600 in the pop-up Range text field.

Click > Replace button.

See Figure 11.26.

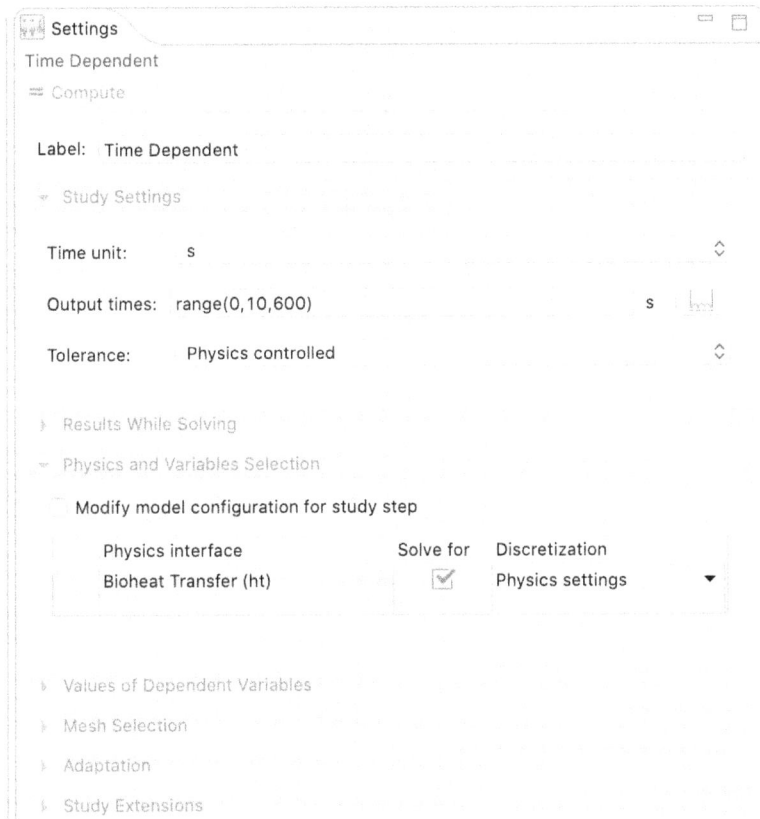

FIGURE 11.26 Settings time dependent > Study settings.

Figure 11.26 shows the Settings Time Dependent > Study Settings.

Right-Click > Model Builder > Study 1.

Select : Compute.

TLI Solution

See Figure 11.27.

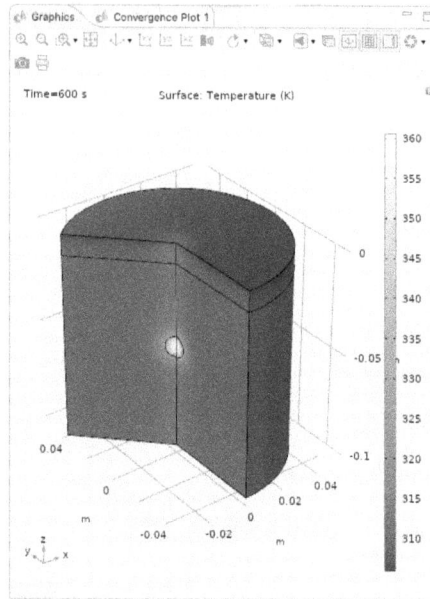

FIGURE 11.27 TLI computed solution, default parameter plot.

Figure 11.27 shows the TLI Computed Solution, Default Parameter Plot.

Degrees C Plotted Solution

Click > Model Builder > Results twistie (as needed).

Right-Click > Model Builder > Results > Temperature, 3D (ht),

Select: Duplicate from the Pop-up menu.

Right-Click > Model Builder > Results > Temperature, 3D (ht) 1,

Select: Rename from the Pop-up menu.

Enter: Temperature, 3D (ht) C

Click > OK

Click > Model Builder > Results > Temperature, 3D (ht) C twistie

Click > Surface.

Click > Settings Surface window > Expression tab > Unit pop-up menu,

Select: degC from the Settings Surface > Expression tab > Unit pop-up menu.

Click > Plot in the Settings Surface window toolbar.

See Figure 11.28.

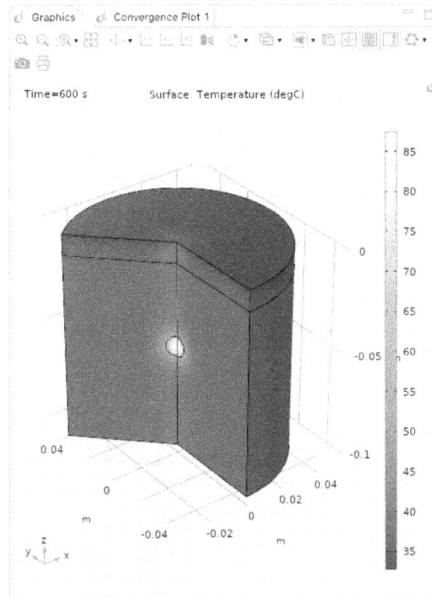

FIGURE 11.28 TLI computed solution degrees C plot.

Figure 11.28 shows the TLI Computed Solution Degrees C Plot.

NOTE *The modeler should note that the temperature of the tumor exceeds the desired 60C needed to destroy the tumor.*

2D Axisymmetric Tumor Laser Irradiation Model Summary and Conclusions

The 2D Axisymmetric Tumor Laser Irradiation Model is a powerful modeling tool that can be used to calculate the gain in temperature under laser irradiation for different injected absorbance materials. With this model, the modeler can easily vary all of the parameters and optimize the design before the first prototype is physically built or used. These types of models are widely used in medicine and industry.

Microwave Cancer Therapy Theory

Hyperthermic oncology (high-temperature cancer and/or tumor treatment) {11.8} is the use of elevated temperatures to kill cancer and other tumor cells. As discussed and prototyped in the TLI model, in treatment it is necessary to locally raise the temperature of the cancer/tumor cells, while doing minimal damage to the normal (healthy) cells surrounding the tumor. In the TLI model, the energy was supplied as photothermal energy using laser irradiation. In this model, the externally applied energy is supplied through the use of a specialized microwave antenna and the application of Ohm's and Joule's Laws {11.9, 11.10}. This type of procedure is typically designated as a minimally invasive procedure {11.11}.

Figure 11.29 shows the microwave antenna in cross-section.

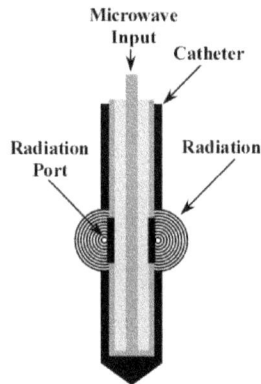

FIGURE 11.29 MCT antenna cross-section.

Figure 11.29 shows the MCT Antenna Cross-section.

2D Axisymmetric Microwave Cancer Therapy Model

The following Multiphysics model solution is derived from a model that was originally developed by COMSOL as a Heat Transfer Interface Tutorial Model for the demonstration of the solution of a Bioheat Equation model. That model was developed for distribution with the Heat Transfer Module software as part of the COMSOL Heat Transfer Interface Model Library.

This model takes advantage of the conductivity of human tissue. Figure 11.30 shows the microwave antenna in cross-section radiating power, imbedded in the modeling domain (liver tissue).

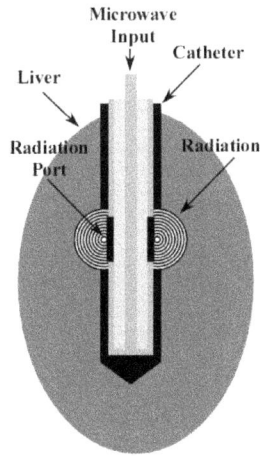

FIGURE 11.30 MCT antenna imbedded in liver tissue.

Figure 11.30 shows the MCT Antenna Imbedded in Liver Tissue.

Since the model is created as a 2D Axisymmetric Model, only the right half of the structure is used in the calculations.

Bioheat Transfer Models

NOTE

As referenced in the TLI model, the Pennes Bioheat Equation is a valuable approach for calculating the potential efficacy of modeled treatment techniques. The fundamental principle needs to be that tumor cells are observed to die at elevated temperatures. The literature cites temperatures that range from 42°C (315.15 K) to 60°C (333.15 K). If the modeled method raises the local temperature of the tumor cells, without excessively raising the temperature of the normal cells, then the modeled method will most probably be successful.

Building the 2D Axisymmetric Microwave Cancer Therapy Model

Startup 5.x, Click > Model Wizard button.

Select: 2D Axisymmetric.

Click > Twistie for the Radio Frequency interface.

Select > Electromagnetic Waves, Frequency Domain (emw).

Click > Add button.

Click > Twistie for the Heat Transfer interface.

Select > Bioheat Transfer (ht).

Click > Add button.

See Figure 11.31.

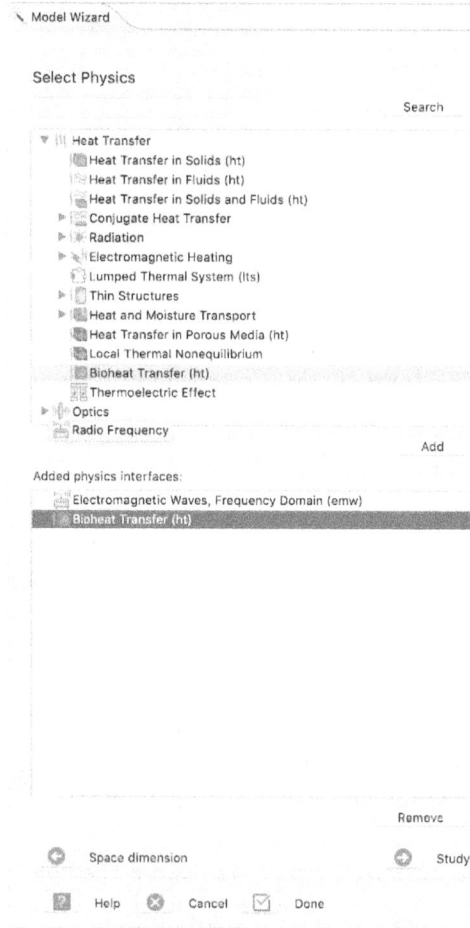

FIGURE 11.31 Model wizard > Add physics selection windows.

Figure 11.31 Model Wizard > Add Physics Selection Windows.

Click > Study (Right Pointing Arrow button).

Click > Twistie for the Preset Studies for Some Physics Interfaces.

Select: Frequency Domain in Preset Studies for Some Physics interfaces.

Click > Done (Checked Box button).

Click > Save As.

Enter MM2E5X_2DAxi_MCT_1.mph.

Click > Save.

See Figure 11.32.

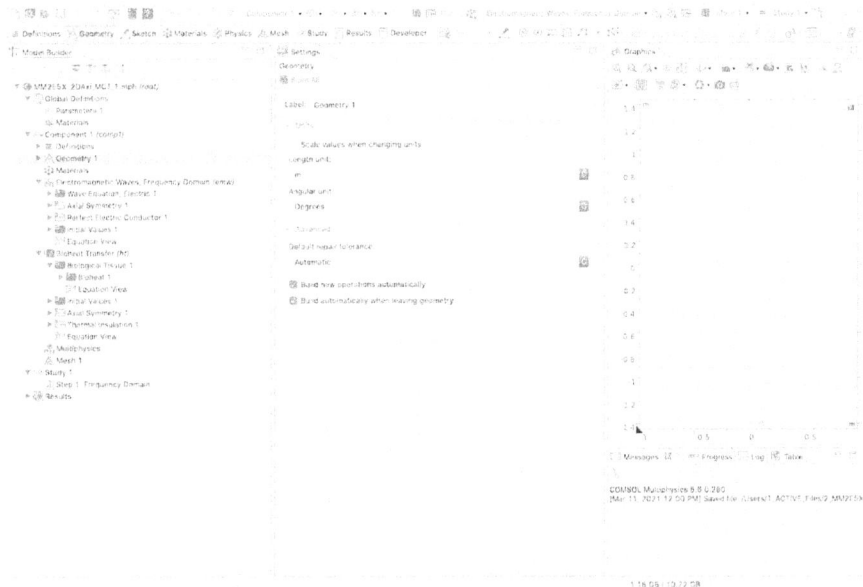

FIGURE 11.32 Desktop display for the MM2E5X_2DAxi_MCT_1.mph model.

Figure 11.32 shows the Display for the MM2E5X_2DAxi_MCT_1.mph model.

Global Definitions

Parameters

In Model Builder,

Click > Model Builder > Global Definitions > Parameters 1.

Enter all the Parameters, as shown, in Table 11.4.

TABLE 11.4 MCT Parameters.

Name	Expression	Description
k_liver	0.56[W/(m*K)]	Thermal conductivity liver
rho_blood	1e3[kg/m^3]	Density blood
Cp_blood	3639[J/(kg*K)]	Heat capacity blood
omega_blood	3.6e-3[1/s]	Blood perfusion rate
T_blood	37[degC]	Temperature blood
P_in	10[W]	Microwave power input
f	2.45[GHz]	Microwave frequency
eps_diel	2.6[1]	Dielectric relative permittivity
eps_cat	2.6[1]	Catheter relative permittivity
eps_liver	43.03[1]	Liver relative permittivity
sigma_liver	1.69[S/m]	Conductivity liver
rho_liver	1038[kg/m^3]	Liver density
Cp_liver	4187[J/(kg*K)]	Liver heat capacity
sigma_cat	1e-6[S/m]	Catheter conductivity
mur_cat	1[1]	Catheter relative permeability
k_ptfe	0.25[W/(m*K)]	Thermal conductivity PTFE
rho_ptfe	2200[kg/m^3]	Density PTFE
Cp_ptfe	1300[J/(kg*K)]	Heat capacity PTFE
sigma_diel	1e-6[S/m]	Microwave center feed
mur_diel	1[1]	Relative permeability dielectric
mur_liver	1[1]	Relative permeability liver

See Figures 11.33 through 11.35.

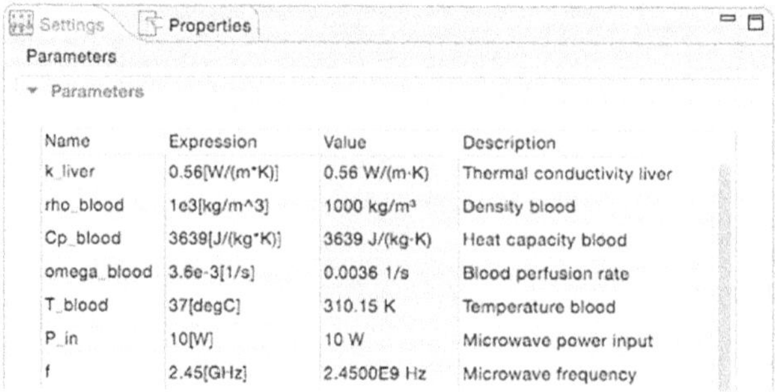

Name	Expression	Value	Description
k_liver	0.56[W/(m*K)]	0.56 W/(m·K)	Thermal conductivity liver
rho_blood	1e3[kg/m^3]	1000 kg/m³	Density blood
Cp_blood	3639[J/(kg*K)]	3639 J/(kg·K)	Heat capacity blood
omega_blood	3.6e-3[1/s]	0.0036 1/s	Blood perfusion rate
T_blood	37[degC]	310.15 K	Temperature blood
P_in	10[W]	10 W	Microwave power input
f	2.45[GHz]	2.4500E9 Hz	Microwave frequency

FIGURE 11.33 Settings parameters > Parameters text field (Part 1).

Figure 11.33 shows the Settings Parameters > Parameters text field (part 1).

Name	Expression	Value	Description
eps_diel	2.6[1]	2.6	Dielectric relative permittivity
eps_cat	2.6[1]	2.6	Catheter relative permittivity
eps_liver	43.03[1]	43.03	Liver relative permittivity
sigma_liver	1.69[S/m]	1.69 S/m	Conductivity liver
rho_liver	1038[kg/m^3]	1038 kg/m³	Liver density
Cp_liver	4187[J/(kg*K)]	4187 J/(kg-K)	Liver heat capacity
sigma_cat	1e-6[S/m]	1.0000E-6 S/m	Catheter conductivity

FIGURE 11.34 Settings parameters > Parameters text field (Part 2).

Figure 11.34 shows the Settings Parameters > Parameters text field (part 2).

mur_cat	1[1]	1	Catheter relative permeability
k_ptfe	0.25[W/(m*K)]	0.25 W/(m·K)	Thermal conductivity of PTFE
rho_ptfe	2200[kg/m^3]	2200 kg/m³	Density PTFE
Cp_ptfe	1300[J/(kg*K)]	1300 J/(kg·K)	Heat capacity PTFE
sigma_diel	1e-6[S/m]	1E-6 S/m	Microwave center feed
mur_diel	1[1]	1	Relative permeability dielectric
mur_liver	1[1]	1	Relative permeability liver

FIGURE 11.35 Settings parameters > Parameters text field (Part 3).

Figure 11.35 shows the Settings Parameters > Parameters text field (part 3).

Geometry

MCT Domains

In Model Builder,

Right-Click > Model Builder > Component 1 (comp1) > Geometry 1.

Select: Rectangle from the pop-up menu.

Enter the coordinates shown in Table 11.5.

Click as instructed.

Repeat the sequence until completed.

TABLE 11.5 MCT Domains.

Rectangles						
	Width	**Height**	**Base**	**r**	**z**	**Click**
1	0.595e-3[m]	0.01[m]	Corner	0[m]	0[m]	Build Selected
2	29.405e-3[m]	0.08[m]	Corner	0.595e-3[m]	0[m]	Build Selected

Boolean Operations – Union

Right-Click > Model Builder > Component 1 (comp1) > Geometry 1.

Select: Booleans and Partitions > Union from the Pop-up menu.

Click > Select All in the Graphics Toolbar.

Uncheck the Keep interior boundaries checkbox.

See Figure 11.36.

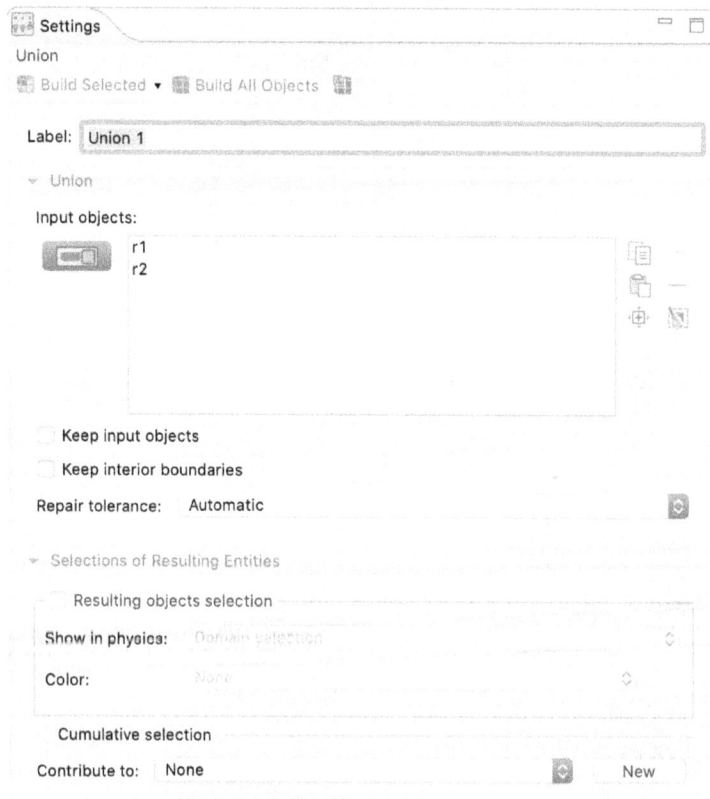

FIGURE 11.36 MCT settings union > Union text field.

Figure 11.36 shows the MCT Settings Union > Union text field.

Click > Build Selected in the Settings Union toolbar.

See Figure 11.37.

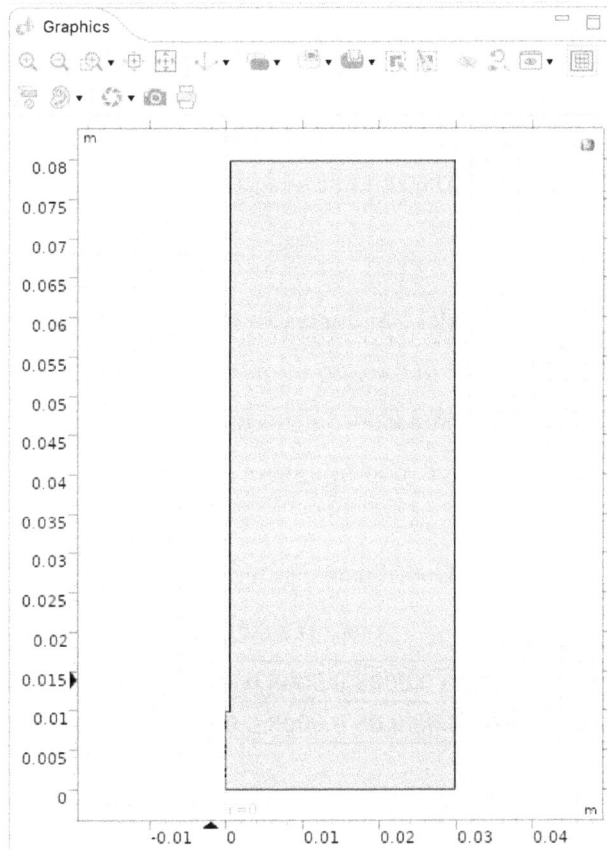

FIGURE 11.37 MCT result of union operation on r1 and r2.

Figure 11.37 shows the MCT Result of Union Operation on r1 and r2.

More MCT Domains

In Model Builder,

Right-Click > Model Builder > Component 1 (comp1) > Geometry 1.

Select > Rectangle from the Pop-up menu.

Enter the coordinates shown in Table 11.6.

Click as instructed.

Repeat the sequence until completed.

TABLE 11.6 MCT Domains.

	Rectangles					
	Width	**Height**	**Base**	**r**	**z**	**Click**
3	0.125e-3[m]	1.0e-3[m]	Corner	0.47e-3[m]	0.0155[m]	Build Selected
4	0.335e-3[m]	0.0699[m]	Corner	0.135e-3[m]	0.0101[m]	Build Selected

Polygon

Right-Click > Model Builder > Component 1 (comp1) > Geometry 1.

Select: Polygon from the pop-up menu.

In Settings Polygon > Coordinates tab,

Click > Data Source: pop-up menu,

Select: Vectors.

Enter the Vector Coordinates, as indicated in Table 11.7.

TABLE 11.7 MCT Vector Coordinates.

r	0.0 0.03 0.03 0.03 0.03 0.0008 0.0008 0.0008 0.0008 0.0 0.0 0.0	**(meters)**
z	0.0 0.0 0.0 0.08 0.08 0.08 0.08 0.0098 0.0098 0.0092 0.0092 0.0	**(meters)**

See Figure 11.37a.

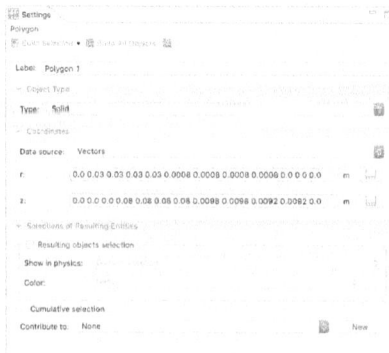

FIGURE 11.37a MCT settings polygon vector coordinates.

Figure 11.37a shows the MCT Settings Polygon Vector Coordinates.

Click > Build All Objects in Settings Polygon toolbar.

Boolean Operations > Union

Right-Click > Model Builder > Component 1 (comp1) > Geometry 1.

Select: Booleans and Partitions > Union from the pop-up menu.

Click > Select All in the Graphics Toolbar.

Click and Remove (minus sign) > r3 and r4.

Check the Keep interior boundaries checkbox (as needed).

See Figure 11.38.

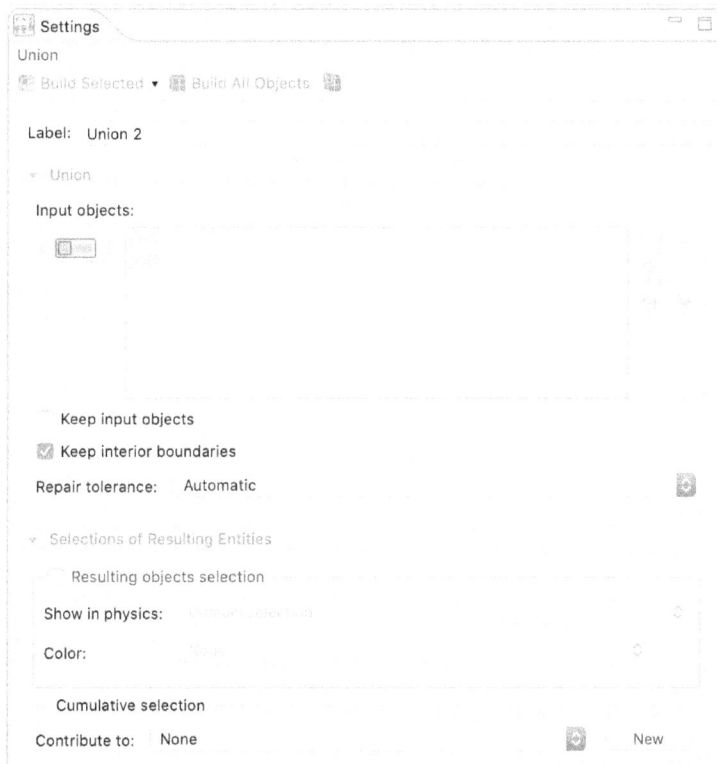

FIGURE 11.38 MCT settings union > Union text field.

Figure 11.38 shows the MCT Settings Union > Union text field.

Click > Build Selected, in the Settings Union toolbar.

See Figure 11.39.

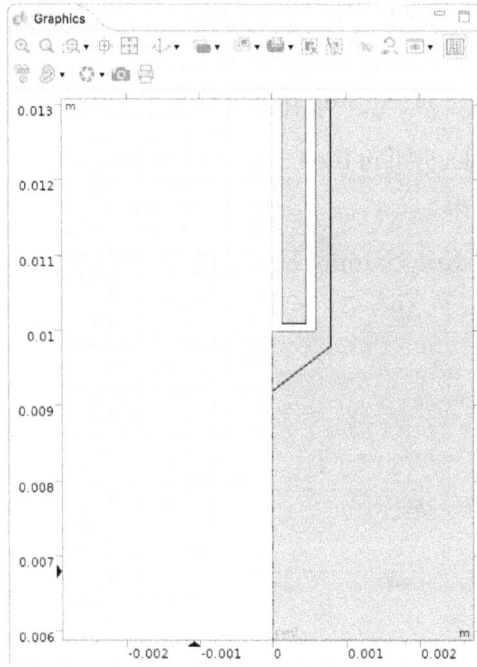

FIGURE 11.39 MCT result of union operation on uni1 and pol1.

Figure 11.39 shows the MCT Result of Union Operation on uni1 and pol1.

Materials

Material 1

Right-Click > Model Builder > Component 1(*comp1*) > Materials.

Select: Blank Material from the pop-up menu.

Click > Material 1.

Right-Click > Material 1.

Select > Rename from the Pop-up menu.

Enter > Liver Tissue in the Rename Material edit window.

Click > OK.

Click > Selection pop-up menu bar, in Settings Material > Geometric Entity Selection tab,

Select: Manual from the pop-up menu.

Click and Remove > Domains 2, 3, 4 in Settings Material > Geometric Entity Selection tab, Selection text field.

See Figure 11.40.

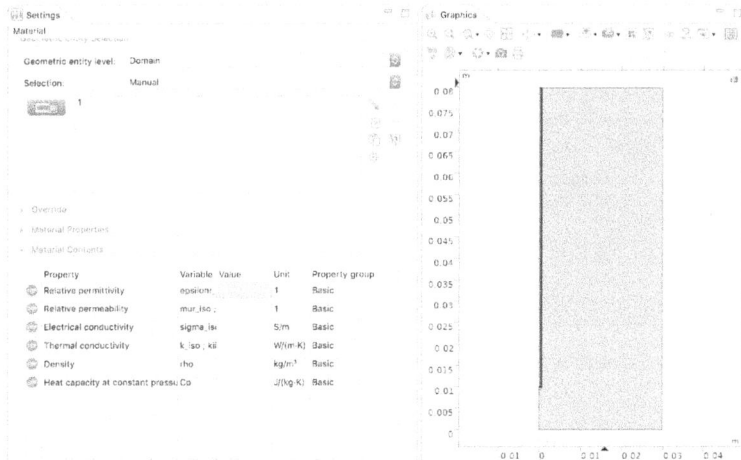

FIGURE 11.40 MCT material 1 domain selection.

Figure 11.40 shows the MCT Material 1 Domain Selection.

In Settings Material > Material Contents tab, text fields are available for the entry of physical property parameters.

Enter the Liver Tissue Value Names as shown in Table 11.8.

TABLE 11.8 Liver Tissue.

Property	Name	Value
Relative permittivity	epsilonr	eps_liver
Relative permeability	mur	mur_liver
Electric conductivity	sigma	sigma_liver
Thermal conductivity	k	k_liver
Density	rho	rho_liver
Heat capacity at constant pressure	Cp	Cp_liver

See Figure 11.41.

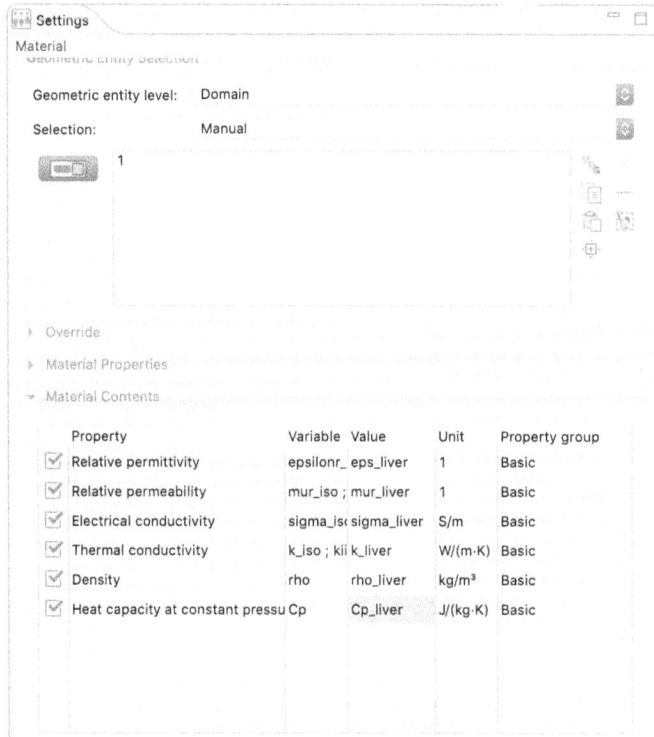

FIGURE 11.41 MCT material 1 liver tissue properties.

Figure 11.41 shows the MCT Material 1 Liver Tissue Properties.

Material 2

Right-Click > Model Builder > Component 1 (*comp1*) > Materials.

Select: Blank Material from the pop-up menu.

Click > Material 2.

Right-Click > Material 2.

Select: Rename from the pop-up menu.

Enter > Catheter in the Rename Material edit window.

Click > OK.

Using the Clipboard, insert 2 in the Settings Material > Geometric Entity Selection tab > Selection text field.

In Settings Material > Material Contents tab, text fields are available for the entry of the Catheter physical property parameters.

Enter: the Catheter Value Names as shown in Table 11.8.

TABLE 11.8 Catheter.

Property	Name	Value
Relative permittivity	epsilonr	eps_cat
Relative permeability	mur	mur_cat
Electric conductivity	sigma	sigma_cat
Thermal conductivity	k	k_ptfe
Density	rho	rho_ptfe
Heat capacity at constant pressure	Cp	Cp_ptfe

See Figure 11.42.

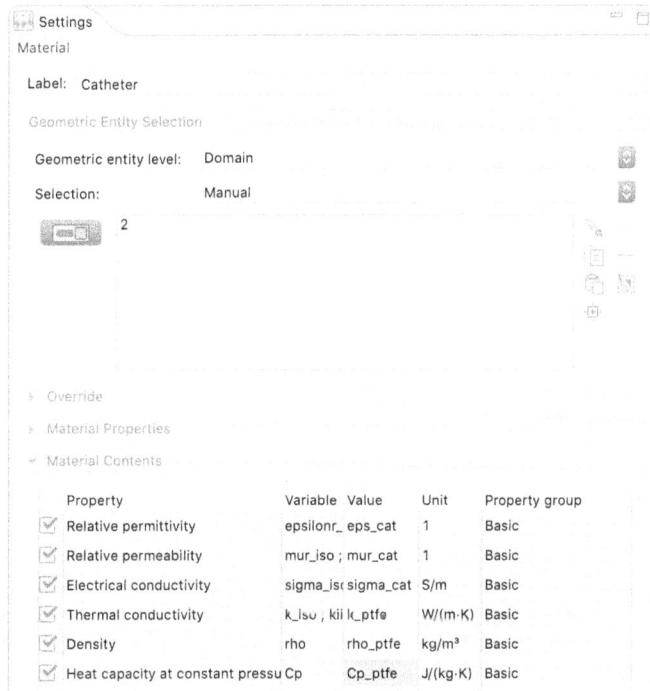

FIGURE 11.42 MCT material 2 catheter properties.

Figure 11.42 shows the MCT Material 2 Catheter Properties.

Material 3

Right-Click > Model Builder > Component 1 (*comp1*) > Materials.

Select > Blank Material from the pop-up menu.

Click > Material 3.

Right-Click > Material 3.

Select: Rename from the pop-up menu.

Enter > Dielectric in the Rename Material text field.

Click > OK.

Using the Clipboard, insert 3 in the Settings Material > Geometric Entity Selection tab > Selection text field.

In Settings Material > Material Contents tab, text fields are available for the entry of the Dielectric physical property parameters.

Enter: the Dielectric Value Names as shown in Table 11.9.

TABLE 11.9 Dielectric.

Property	Name	Value
Relative permittivity	epsilonr	eps_diel
Relative permeability	mur	mur_diel
Electric conductivity	sigma	sigma_diel
Thermal conductivity	k	k_ptfe
Density	rho	rho_ptfe
Heat capacity at constant pressure	Cp	Cp_ptfe

See Figure 11.43.

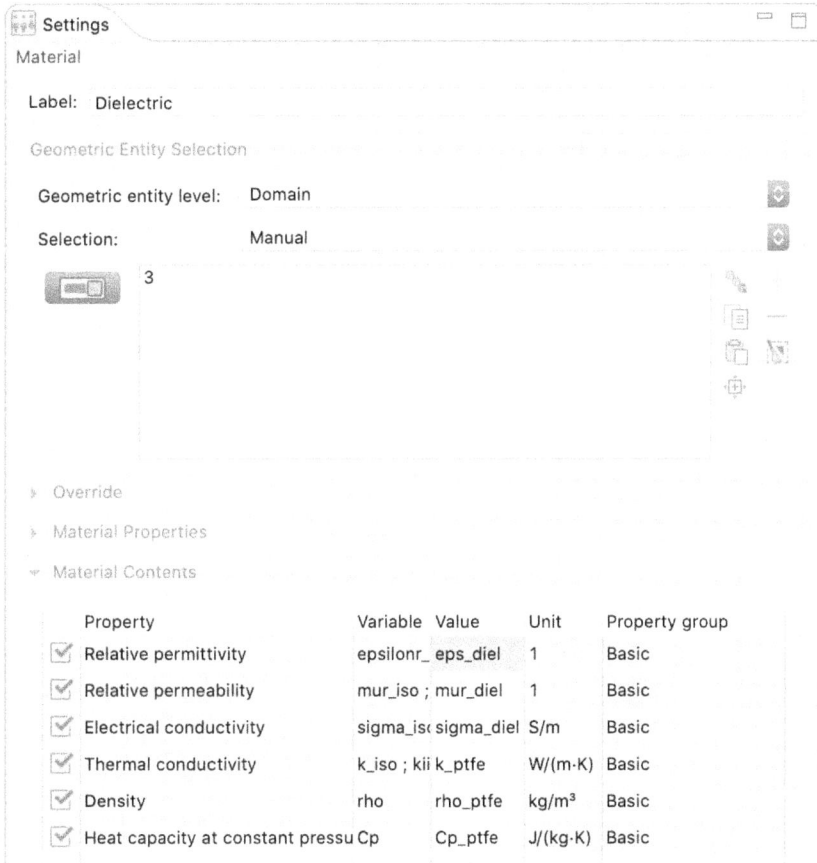

FIGURE 11.43 MCT material 3 dielectric properties.

Figure 11.43 shows the MCT Material 3 Dielectric Properties.

Material 4

Right-Click > Model Builder > Component 1 (*comp1*) > Materials.

Select: Add Material from Library from the pop-up menu.

Click > Add Material > Built-in twistie.

Right-Click > Add Material > Built-in > Air.

Select > Add to Component.

Click > Model Builder > Component 1 (*comp1*) > Materials > Air.

Using the Clipboard, insert 4 in the Settings Material > Geometric Entity Selection tab > Selection text field.

Click > Close Add Material (hollow X).

See Figure 11.44.

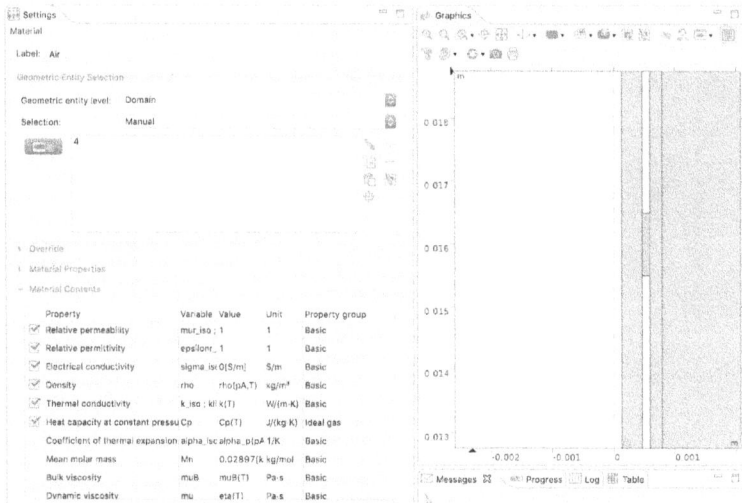

FIGURE 11.44 MCT material 4 air properties.

Figure 11.44 shows the MCT Material 4 Air Properties.

Electromagnetic Waves, Frequency Domain (emw)

NOTE

The electromagnetic wave step introduces the power used to heat the tumor at a single frequency.

Click > Model Builder > Component 1 (comp1) > Electromagnetic Waves, Frequency Domain (emw).

Click > Settings Electromagnetic Waves, Frequency Domain > Equation tab twistie.

Click > Equation form: pop-up menu bar in Settings Electromagnetic Waves, Frequency Domain > Equation tab,

Select: Frequency Domain from the pop-up menu.

Click > Frequency: pop-up menu bar in Settings Electromagnetic Waves, Frequency Domain > Equation tab,

Select: User defined from the pop-up menu.

Enter > f in the Settings Electromagnetic Waves, Frequency Domain > Equation tab > Frequency text field.

See Figure 11.45.

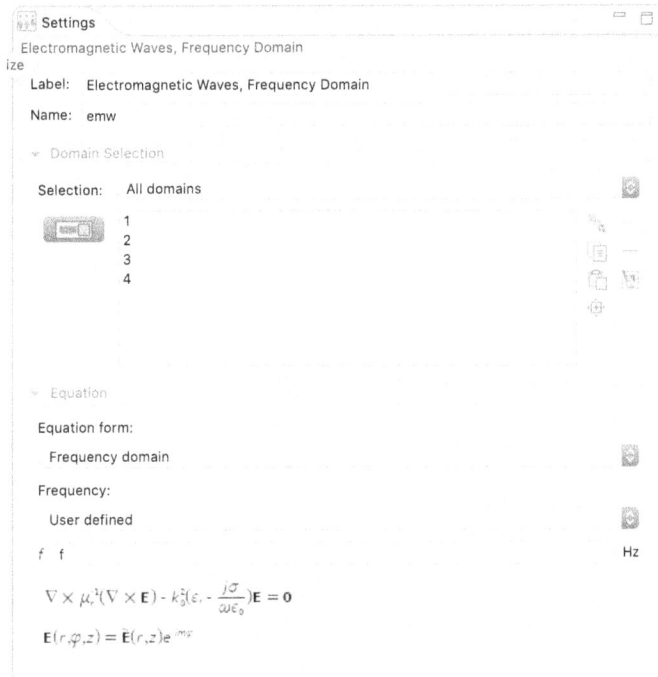

FIGURE 11.45 MCT electromagnetic waves settings.

Figure 11.45 shows the MCT Electromagnetic Waves Settings.

Port 1

Right-Click > Model Builder > Component 1 (comp1) > Electromagnetic Waves, Frequency Domain (emw),

Select: Port from the pop-up menu.

Using the Clipboard, insert 8 in the Settings Port > Boundary Selection tab > Selection text field (Click > Zoom Extents to see the selection).

Click > Type of Port: pop-up menu bar in Settings Port > Port Properties tab,

Select: Coaxial.

Click > Wave excitation at this port: pop-up menu bar in Settings Port >Port Properties tab,

Select: On (as needed).

Enter: P_in in the Settings Port > Port Properties tab > Port input power: (P_{in}) text field.

See Figure 11.46.

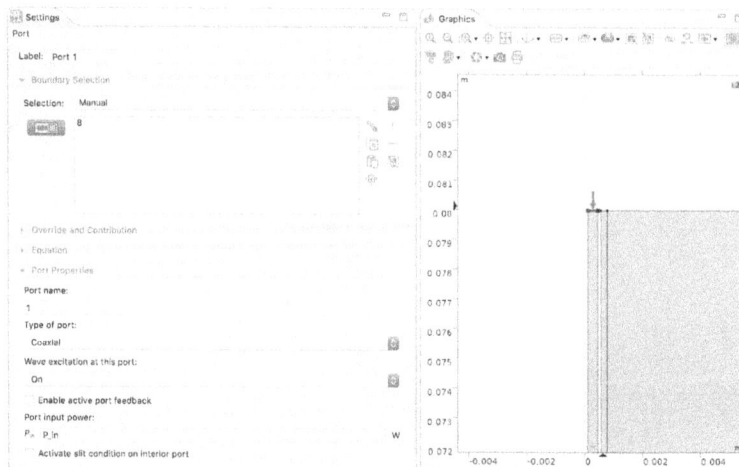

FIGURE 11.46 MCT electromagnetic waves port settings.

Figure 11.46 shows the MCT Electromagnetic Waves Port Settings.

Scattering Boundary Condition 1

Right-Click > Model Builder > Component 1 (comp1) > Electromagnetic Waves, Frequency Domain (emw).

Select: Scattering Boundary Condition from the pop-up menu.

NOTE *The easiest method that the modeler can use to implement the next step is to use the Clipboard and insert the correct boundaries.*

Use the Clipboard and insert 14 18 20 21 in the Boundary Selection tab > Selection text field.

See Figure 11.47.

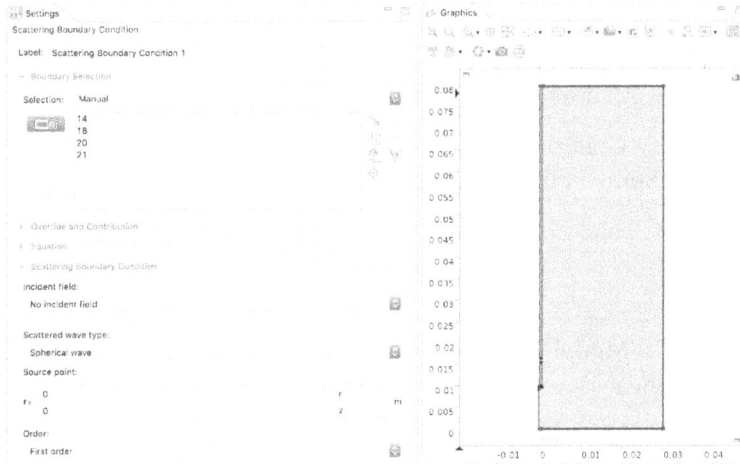

FIGURE 11.47 MCT electromagnetic waves scattering boundary conditions settings.

Figure 11.47 shows the MCT Electromagnetic Waves Scattering Boundary Conditions Settings.

Heat Transfer (ht)

Click > Model Builder > Component 1 (comp1) > Bioheat Transfer (ht) twistie.

Domains

NOTE *The default condition is to apply all of these settings to the entire model. Please note that as the properties of other domains (tissues) are defined, the default settings will be overridden.*

Click > Model Builder > Component 1 (comp1) > Bioheat Transfer (ht).

Click > Select All in the Graphics Toolbar (as needed).

Click and Remove (minus sign) > All Domains except 1 in the Domain Selection tab > Selection text field.

Heat Source 1

Right-Click > Model Builder > Component 1 (comp1) > Bioheat Transfer (ht).

Select: Heat Source from the pop-up menu.

Click > Heat Source 1.

Click > Selection pop-up menu bar in Settings Heat Source > Domain Selection tab >,

Select: All domains from the Pop-up menu.

Click > General source pop-up menu bar (Q_0) in Settings Heat Source > Heat Source tab, under the General source radio button.

Select: Total power dissipation density (emw/wee1) from the pop-up menu.

Bioheat 1

Click > Model Builder > Component 1 (*comp1*) > Bioheat Transfer (ht) > Biological Tissue 1 twistie.

Click > Model Builder > Component 1 (*comp1*) > Bioheat Transfer (ht) > Biological > Tissue 1, > Bioheat 1.

Enter: ρ_b = rho_blood, $_{Cp,b}$ = Cp_blood and ω_b = omega_blood in the Settings Bioheat > Bioheat tab > text fields.

See Figure 11.48.

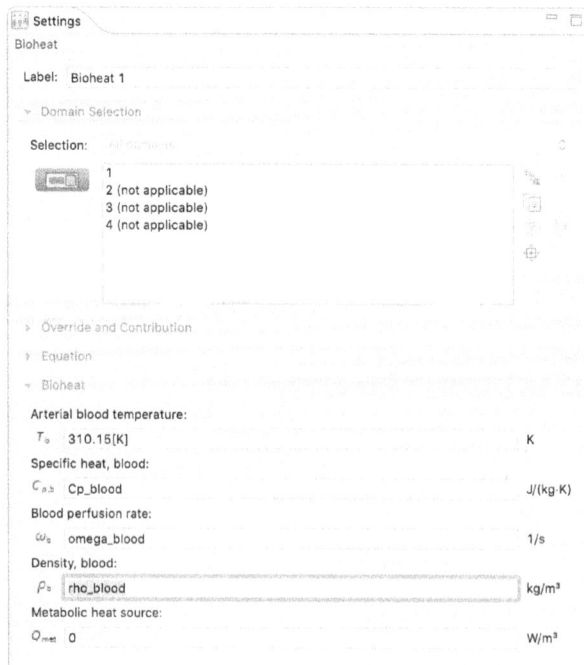

FIGURE 11.48 Settings bioheat > Bioheat text fields.

Figure 11.48 shows the Settings Bioheat > Bioheat text fields.

Mesh 1

Right-Click > Model Builder > Component 1 (*comp1*) > Mesh 1.

Select: Free Triangular from the pop-up menu.

Click > Model Builder > Component 1 (*comp1*) > Mesh 1 > Size.

Click > Settings Size > Element Size Parameters twistie.

Enter: 3[mm] in Settings Size > Element Size Parameters tab > Maximum element size.

See Figure 11.49.

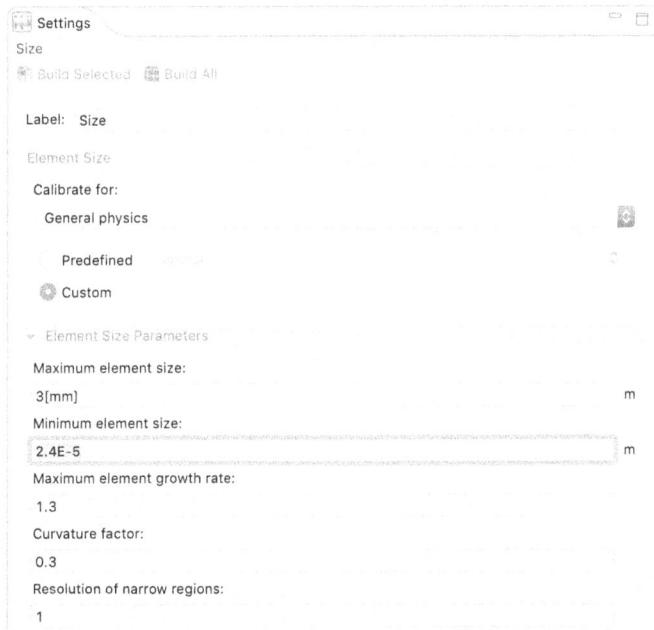

FIGURE 11.49 Size settings.

Figure 11.49 shows the Size Settings.

Right-Click > Model Builder > Component 1 (*comp1*) > Mesh 1 > Free Triangular 1.

Select > Size from the pop-up menu.

Click > Size 1.

Click > Geometric entity level in Settings Size > Geometric Entity Selection tab.

Select > Domain from the Pop-up menu.

Use the Clipboard and insert 3 in the Settings Size > Geometric Entity Selection tab > Selection text field.

Click > Custom radio button in Settings Size > Element Size tab.

Click > Maximum element size check box in Settings Size > Element Size Parameters,

Enter: 0.15[mm].

See Figure 11.50.

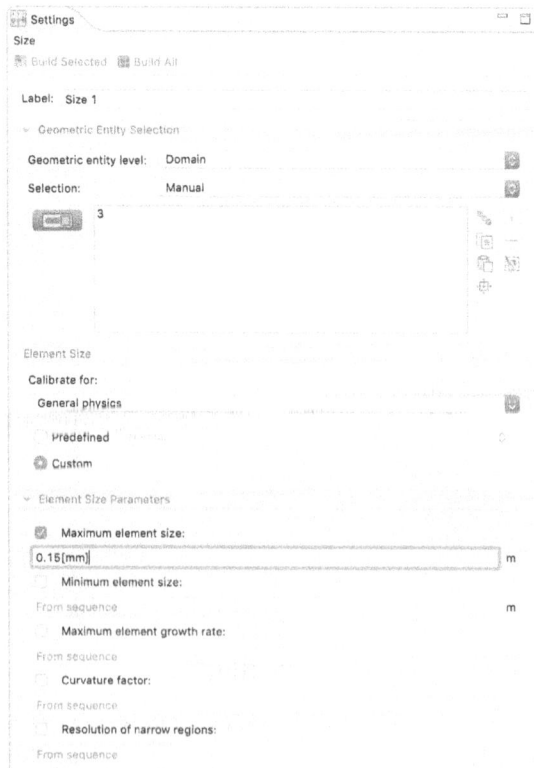

FIGURE 11.50 Size 1 settings and domain 3.

Figure 11.50 shows the Size 1 Settings and Domain 3.

Click > Build All.

See Figure 11.51.

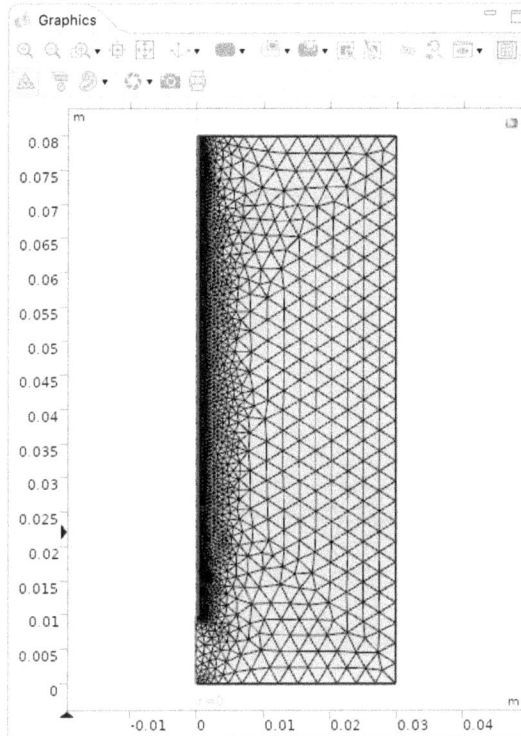

FIGURE 11.51 MCT meshed model.

Figure 11.51 shows the MCT Meshed Model.

NOTE   *After the mesh is built, the model should have 7086 domain elements and 1523 boundary elements.*

Study 1

Frequency Domain Solver

Click > Model Builder > Study 1 twistie.

Click > Model Builder > Study 1 > Step 1: Frequency Domain.

Enter: f in Settings Frequency Domain > Study Settings tab > Frequencies text field.

Heat Transfer Stationary Solver

Right-Click > Model Builder > Study 1.

Select > Study Steps > Stationary from the pop-up menu.

Right-Click > Model Builder > Study 1.

Select: Compute.

Results

Electric Field (emw)

See Figure 11.52.

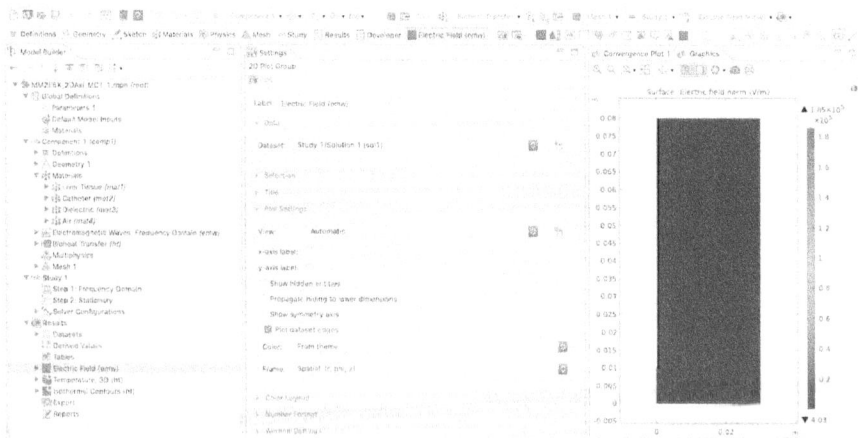

FIGURE 11.52 MCT model solution using default plot parameters.

Figure 11.52 shows the MCT Model Solution using Default Plot Parameters.

NOTE *The solution shown in Figure 11.52 is the sequential solution, first for the microwave power dissipation and then for the resultant heating. This solution assumes that the material properties are not a function of temperature.*

Right-Click > Results > Electric Field (emw),

Select > Duplicate.

Right – Click > Results > Electric Field (emw) 1,

Select > Rename.

Enter > log10 (Electric Field (emw))

Click > OK.

Click > log10 (Electric Field (emw)) twisty.

Click > Surface.

Enter: log10 (emw.normE) in Settings Surface > Expression tab > Expression text field.

Click > Plot in the Settings Surface toolbar.

See Figure 11.53.

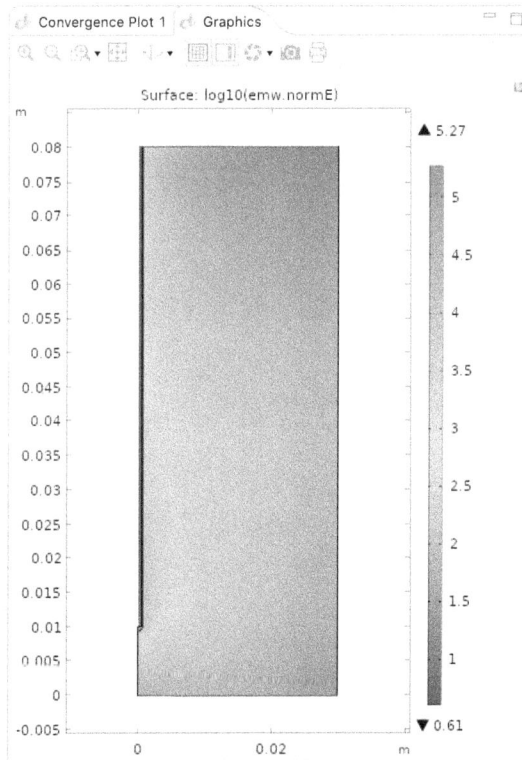

FIGURE 11.53 MCT model solution using log10(normal electric field).

Figure 11.53 shows the MCT Model Solution using log10(normal electric field).

Surface Temperature Plot

Click > Results > Temperature, 3D (ht) twistie.

Click > Surface.

Click > Unit pop-up menu bar in Settings Surface > Expression tab,

Select: degC.

Click > Plot.

See Figure 11.54.

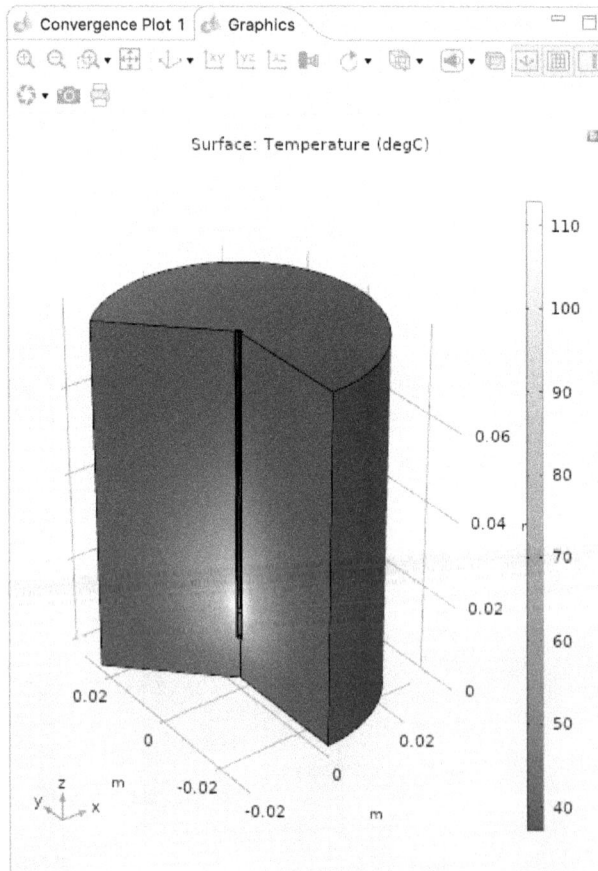

FIGURE 11.54 MCT model solution local temperatures.

Figure 11.54 shows the MCT Model Solution Local Temperatures.

NOTE *The modeler should note that the temperature of the tumor exceeds the desired minimum temperature of 60C needed to destroy the tumor.*

2D Axisymmetric Microwave Cancer Therapy Model Summary and Conclusions

The 2D Axisymmetric Microwave Cancer Therapy Model is a powerful modeling tool that can be used to calculate the gain in temperature under microwave dissipation for different geometry designs, materials, and power levels. With this model, the modeler can easily vary all of the parameters and optimize the design before the first prototype is physically built or used. These types of models are widely used in industry.

FIRST PRINCIPLES AS APPLIED TO BIOHEAT MODEL DEFINITION

First Principles Analysis derives from the fundamental laws of nature. In the case of models using this Classical Physics Analysis approach, the laws of conservation in physics require that what goes in (as mass, energy, charge, etc.) must come out (as mass, energy, charge, etc.) or must accumulate within the boundaries of the model.

The careful modeler must be knowledgeable of the implicit assumptions and default specifications that are normally incorporated into the COMSOL Multiphysics software model when a model is built using the default settings.

Consider, for example, the two Bioheat models developed in this chapter. In these models, it is implicitly assumed that there are no thermally related changes (mechanical, electrical, etc.), except as specified. It is also assumed the materials are homogeneous and isotropic, except as specifically indicated and there are no thin insulating contact barriers at the thermal junctions. None of these assumptions are typically true in the general case. However, by making such assumptions, it is possible to easily build a 2D Axisymmetric First Approximation Model.

NOTE *A First Approximation Model is one that captures all the essential features of the problem that needs to be solved, without dwelling excessively on all of the small details. A good First Approximation Model will yield an answer that enables the modeler to determine if he needs to invest the time and the resources required to build a more highly detailed model.*

Also, the modeler needs to remember to name model parameters carefully as pointed out in Chapter 1.

REFERENCES

1. COMSOL Reference Manual, p. 254

2. H.H. Pennes, J. Appl. Physiology, V. 1, No.2, pp. 93-122

3. *https://en.wikipedia.org/wiki/Biomedical_engineering*

4. L.R. Hirsch, et al., Engineering in Medicine and Biology, 2002. 24th Annual Conference, pp. 530-531

5. L.R. Hirsch, et al., Proceedings of the 25' Annual International Conference of the IEEE EMBS Cancun, Mexico. September 17–21, 2003

6. Saito, et al., Antennas, Propagation and EM Theory, 2000. Proceedings. ISAPE 2000. 5th International Symposium

7. D.P. O'Neal et al., j.canlet.2004.02.004

8. *https://www.cancer.gov/about-cancer/treatment/types/surgery/hyperthermia-fact-sheet*

9. *https://en.wikipedia.org/wiki/Ohm%27s_law*

10. *https://en.wikipedia.org/wiki/Joule_effect*

11. *https://en.wikipedia.org/wiki/Minimally_invasive_ procedure*

SUGGESTED MODELING EXERCISES

1. Build, mesh, and solve the 2D Axisymmetric Tumor Laser Irradiation Model as presented earlier in this chapter.

2. Build, mesh, and solve the 2D Axisymmetric Microwave Cancer Therapy Model as presented earlier in this chapter.

3. Change the values of the materials parameters and then build, mesh, and solve the 2D Axisymmetric Tumor Laser Irradiation Model as an example problem.

4. Change the values of the materials parameters and then build, mesh, and solve the 2D Axisymmetric Microwave Cancer Therapy Model as an example problem.

5. Change the value of the absorbance parameter and then build, mesh, and solve the 2D Axisymmetric Tumor Laser Irradiation Model as an example problem.

6. Change the design of the geometry and then solve the 2D Axisymmetric Microwave Cancer Therapy Model as an example problem.

A Brief Introduction to LiveLink™ for MATLAB® Using COMSOL Multiphysics 5.x

In This Appendix

- Guidelines for LiveLink Exploration through Modeling in 5.x
- Getting Started using LiveLink with COMSOL Multiphysics 5.x
- First Principles as Applied to Multiphysics 5.x Model Definition
- References
- Suggested Modeling Exercises

GUIDELINES FOR LIVELINK EXPLORATION THROUGH MODELING IN 5.x

NOTE

In this appendix, a brief introduction to exploring COMSOL Multiphysics modeling through the LiveLink for MATLAB approach to modeling is presented. MATLAB script-based models, in general, have proven to be very valuable in the study and exploration of models for engineers and scientists with an extensive level of experience in problem solution through the application of script-based solutions. The combined connection of COMSOL Multiphysics and the MATLAB family of programming tools significantly expands the power and versatility of both systems. The science, engineering, and medical communities have employed this modeling technique successfully both in the past and currently. Such models can serve as first-cut evaluations of potential systemic physical behavior under the influence of complex external stimuli.

Getting Started using LiveLink for MATLAB with COMSOL Multiphysics 5.x on a Windows® 10 platform

Getting Started

LiveLink for MATLAB is employed to allow the development of script-based models, to analyze any of the physical functions in materials that COMSOL Multiphysics models can analyze. The solution of such script-based models is most obviously important when those models are designed to explore any of a broad range of techniques for problem solution. Conversely, a COMSOL Modeler can take advantage of the many powerful module functions available through the MATLAB script-based analysis system.

More detailed guidance on the use of LiveLink for MATLAB is available in the literature supplied with the COMSOL Multiphysics {A.1, A.2, A.3} software. The purpose of this appendix is to introduce the reader to the availability of the LiveLink Module and to briefly explore that module in actual use.

The fundamental requirement that needs to be satisfied in order to minimize the difficulty in the establishment of the MATLAB/COMSOL LiveLink connection is that each respective software needs to be properly installed in the correct sequence. For both the Macintosh and the Windows 10 platforms, it works best when the most recent version of MATLAB is installed on the platform before the COMSOL 5.x is installed. With MATLAB installed first, the COMSOL Multiphysics 5.x installation goes very smoothly.

After Installation

Once both MATLAB and COMSOL Multiphysics 5.x are installed, Double-Clicking the COMSOL with MATLAB.app icon activates both COMSOL Multiphysics and MATLAB. Figure A.1 shows the contents of the MATLAB Command Window with both applications and LiveLink for MATLAB running.

See Figure A.1.

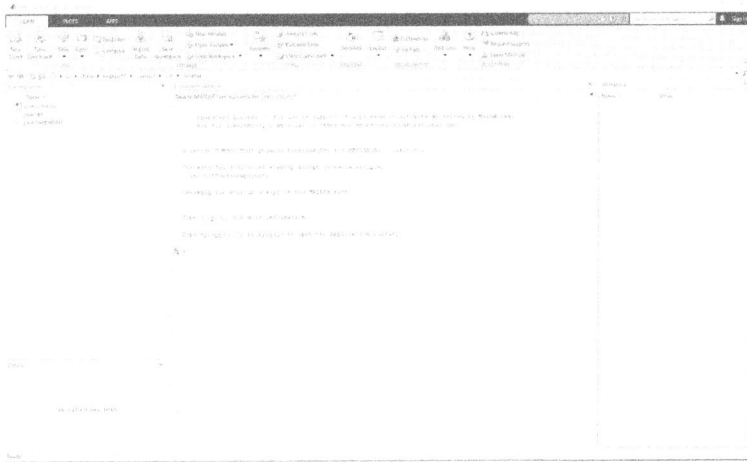

FIGURE A.1 MATLAB command window with livelink activated.

Figure A.1 shows the MATLAB Command Window with LiveLink activated.

With both applications running and talking to each other, the reader has several path options that can be followed next, based upon the reader's relative skill level with either or both COMSOL Multiphysics 5.x and the MATLAB_R2021a.app. Those readers with little MATLAB experience may want to take the "NEW to MATLAB?" path, through the active "Getting Started" link shown at the top of the Command Window, as in Figure A.2.

See Figure A.2.

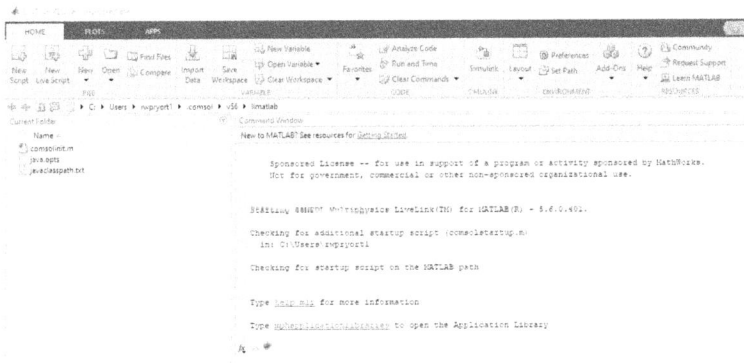

FIGURE A.2 MATLAB command window with getting started link.

Figure A.2 shows the MATLAB Command Window with Getting Started link.

When the reader clicks on the Getting Started link, a new window is opened with access to a broad range of resources, as shown in Figure A.3.

See Figure A.3.

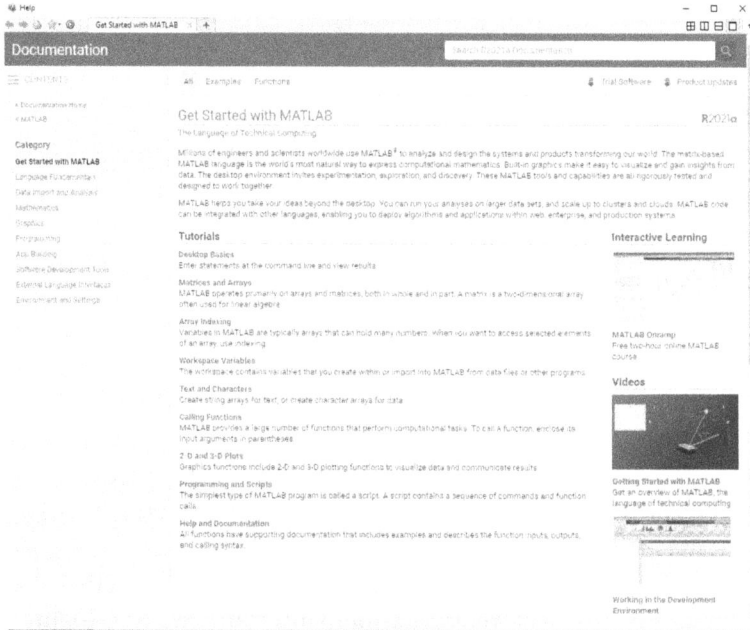

FIGURE A.3 MATLAB resource window from following the getting started link.

Figure A.3 shows the MATLAB Resource Window from following the Getting Started link.

NOTE

The purpose of this Appendix is to introduce the COMSOL LiveLink for MATLAB Module and expose the reader to the expanded modeling capabilities available through the use of the scripting-based capability of the MATLAB system functionality. The author recommends and leaves to the reader the further exploration of the Getting Started resources in MATLAB.

Further Exploration (from COMSOL to MATLAB)

The next step in using COMSOL Multiphysics with MATLAB is to explore the COMSOL Multiphysics – LiveLink – MATLAB interface. It should be noticed that there is an activated link mphapplicationlibraries in the MATLAB Command Window, which when clicked, brings up a new COMSOL window with access to the application library.

See Figure A.4.

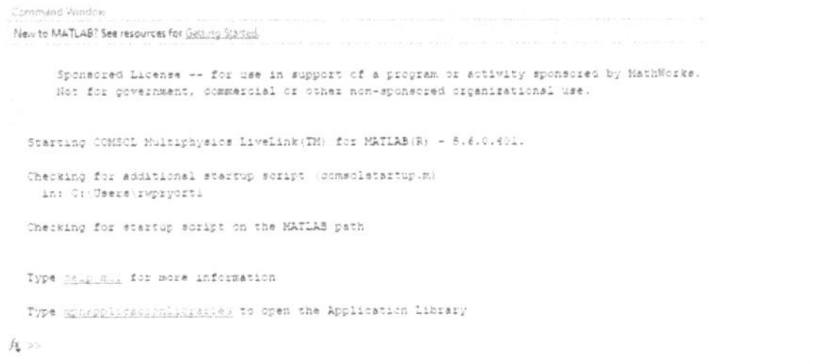

FIGURE A.4 COMSOL application library window link.

Figure A.4 shows the COMSOL Application Library Window Link.

Click > LiveLink for MATLAB crossbox.

Click > Tutorials crossbox.

Once the crossbox has been Clicked, a full display of the Tutorial Models becomes available, as shown in Figure A.5.

NOTE *These Tutorial Models are a collection of COMSOL Models, some of which have been saved in COMSOL format(.mph files) and others in MATLAB scripting format (.m files).*

See Figure A.5.

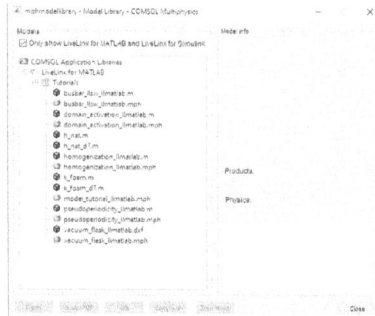

FIGURE A.5 COMSOL tutorials library window.

Figure A.5 shows the COMSOL Tutorials Library Window.

The first Tutorial Model shown in the COMSOL Model Library window is one version of the busbar (physical electrical connecting link) model. This busbar library model is named busbar_llsw_llmatlab.m (the llsw stands for LiveLink for SolidWorks).

Click on the busbar_llsw_llmatlab.m link and the Edit button is activated. Click on the Edit button and the script for busbar_llsw_llmatlab.m is displayed in the Editor Window of MATLAB, as seen in Figure A.6.

See Figure A.6.

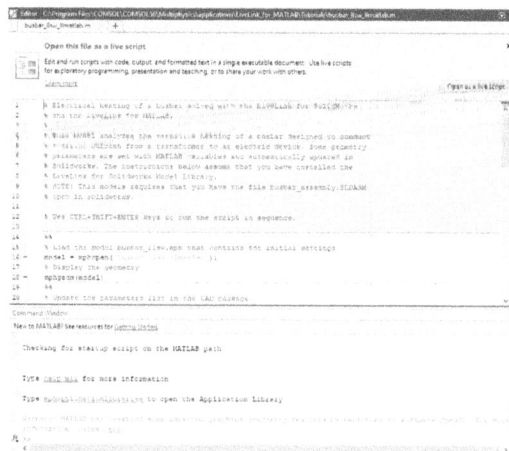

FIGURE A.6 MATLAB editor window.

Figure A.6 shows the MATLAB Editor Window.

As the reader can clearly see in the description of the program, the busbar_ llsw_llmatlab.m model utilizes both LiveLink for SolidWorks and LiveLink for MATLAB in the development of this model.

Close all MATLAB related windows by clicking on the small x in the upper-righthand corner.

COMSOL and MATLAB Only (from MATLAB to COMSOL)

For the reader who wishes to build a model using COMSOL Multiphysics and MATLAB only, the next step is to shut down the presently running applications and start a new model build.

First, Double-Click (Start-up) the COMSOL with MATLAB.app.

Second, start the modeling process by entering

model = ModelUtil.create('Model');

at the MATLAB Command Window prompt,

Enter > mphlaunch > "return" key.

NOTE | *When the reader enters the mphlaunch command, COMSOL Multiphysics desktop will become available.*

See Figure A.6a.

FIGURE A.6A COMSOL multiphysics desktop.

Figure A.6a Shows the COMSOL Multiphysics Desktop.

The COMSOL – LiveLink – MATLAB has now been established and model building can now start.

As a test, Enter > model.param.set('L', '9[cm]', 'Length of the busbar'); at the MATLAB Command Window prompt and hit the "return" key.

The contents of the script command ('L', '9[cm]', 'Length of the busbar') will be stored in the COMSOL Multiphysics Parameters 1 file and displayed in the Parameters 1 edit window.

See Figure A.7.

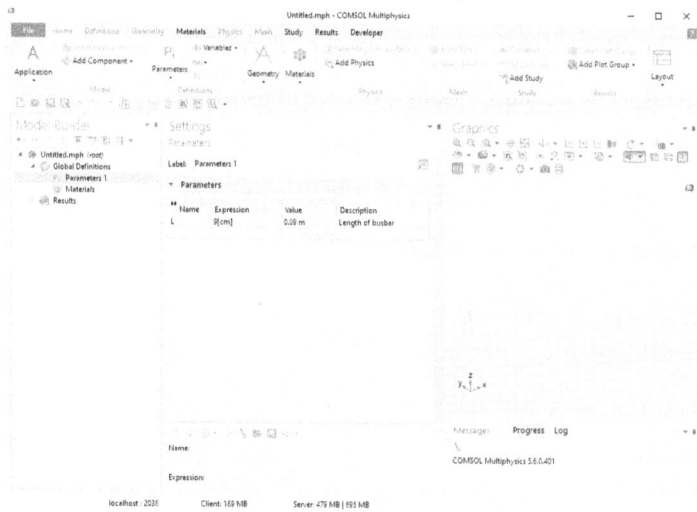

FIGURE A.7 COMSOL multiphysics parameters 1 settings window.

Figure A.7 shows the COMSOL Multiphysics Parameters 1 Settings Window.

On the COMSOL Multiphysics Desktop, in the Model Builder pane,

Click > Model Builder > Global Definitions > Parameters 1,

See Figure A.8.

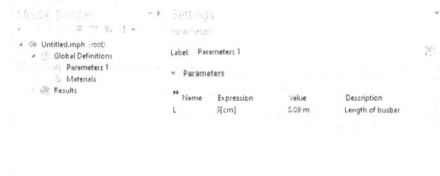

FIGURE A.8 COMSOL parameters 1 edit window.

Figure A.8 shows the COMSOL Parameters 1 Edit Window.

As the reader can see, each new instruction from Table A1 entered into the MATLAB Command Window enters one of the parametric values needed to establish the parametric criteria essential to build a busbar in the COMSOL Multiphysics Desktop.

Table A1

model.param.set('rad_1', '6[mm]', 'Radius of the fillet');

model.param.set('tbb', '5[mm]', 'Thickness of the busbar');

model.param.set('wbb', '5[cm]', 'Width of the busbar');

model.param.set('mh', '6[mm]', 'Maximum element size');

model.param.set('htc', '5[W/m^2/K]', 'Heat transfer coefficient');

model.param.set('Vtot', '20[mV]', 'Applied electric potential');

See Figure A.9.

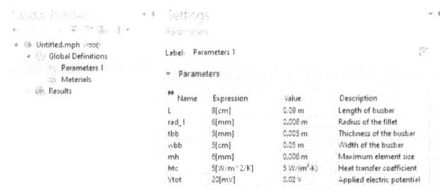

FIGURE A.9 COMSOL parameters 1 edit window completed.

Figure A.9 shows the COMSOL Parameters 1 Edit Window Completed.

Now that the reader has established a sound link to the COMSOL Multiphysics Desktop, the remainder of the code needs to be entered, see reference {A.1}.

When properly scripted, the results of the modeling will be exactly the same as if they had been created through the COMSOL Multiphysics Desktop.

FIRST PRINCIPLES AS APPLIED TO SCRIPTING AND GUI MODEL DEFINITION

First Principles Analysis derives from the fundamental laws of nature. In the case of models using this Classical Physics Analysis approach, the laws of conservation in physics require that what goes in (as mass, energy, charge, etc.) must come out (as mass, energy, charge, etc.) or must accumulate within the boundaries of the model.

The careful modeler must be knowledgeable of the implicit assumptions and default specifications that are normally incorporated into the COMSOL Multiphysics software model when a model is built using the default settings. In the case of a free-standing script-based MATLAB model, it is the responsibility of the modeler to supply the correct default settings (e.g., NO! division by zero, etc.)

In any properly designed models, it is implicitly assumed that there are no thermally related changes (mechanical, electrical, etc.), except as specified. It is also assumed the materials are homogeneous and isotropic, except as specifically indicated and there are no thin insulating contact barriers at the thermal junctions. None of these assumptions are typically true in the general case. However, by making such assumptions, it is possible to easily build a GUI or Scripted First Approximation Model.

NOTE *A First Approximation Model is one that captures all the essential features of the problem that needs to be solved, without dwelling excessively on all of the small details. A good First Approximation Model will yield an answer that enables the modeler to determine if he needs to invest the time and the resources required to build a more highly detailed model.*

Also, the modeler needs to remember to name model parameters carefully as pointed out in Chapter 1.

REFERENCES

A.1 IntroductionToLiveLinkForMATLAB, p. 5

A.2 LiveLinkForMATLABUsersGuide, p.3

A.3 COMSOL Programming Reference Manual, p. 3

SUGGESTED MODELING EXERCISES

1. Explore the BUSBAR model detailed in the COMSOL *Introduction to LiveLink for MATLAB* document.

2. Explore the tutorial models in the COMSOL Model Library.

INDEX